Methods in Molecular Biology

Series Editor
John M. Walker
School of Life and Medical Sciences
University of Hertfordshire
Hatfield, Hertfordshire, UK

For further volumes:
http://www.springer.com/series/7651

For over 35 years, biological scientists have come to rely on the research protocols and methodologies in the critically acclaimed Methods in Molecular Biology series. The series was the first to introduce the step-by-step protocols approach that has become the standard in all biomedical protocol publishing. Each protocol is provided in readily-reproducible step-by step fashion, opening with an introductory overview, a list of the materials and reagents needed to complete the experiment, and followed by a detailed procedure that is supported with a helpful notes section offering tips and tricks of the trade as well as troubleshooting advice. These hallmark features were introduced by series editor Dr. John Walker and constitute the key ingredient in each and every volume of the Methods in Molecular Biology series. Tested and trusted, comprehensive and reliable, all protocols from the series are indexed in PubMed.

Manuscript Instructions https://drive.google.com/file/d/1K4Az5vpAFVzYadJ3jKx2-QSo7Dqa_SNA/view?usp=sharing

Computational Methods and Data Analysis for Metabolomics

Second Edition

Edited by

Michael Witting

Metabolomics and Proteomics Core, Helmholtz Zentrum München, Neuherberg, Germany;
Chair of Analytical Food Chemistry, TUM School of Life Sciences, Technical University of Munich,
Munich, Germany,

Humana Press

Editor
Michael Witting
Metabolomics and Proteomics Core
Helmholtz Zentrum München
Neuherberg, Germany

Chair of Analytical Food Chemistry
TUM School of Life Sciences
Technical University of Munich
Munich, Germany

ISSN 1064-3745 ISSN 1940-6029 (electronic)
Methods in Molecular Biology
ISBN 978-1-0716-5451-4 ISBN 978-1-0716-5452-1 (eBook)
https://doi.org/10.1007/978-1-0716-5452-1

This Humana imprint is published by the registered company Springer Science+Business Media, LLC, part of Springer Nature.
The registered company address is: 1 New York Plaza, New York, NY 10004, U.S.A.

Preface

Data analysis in metabolomics can be a complicated endeavor. Multiple software tools and pipelines exist, each offering distinct features and advantages. Based on the data collected and the data analysis task ahead, both features need to be combined. The first edition of this book, edited by Shuzhao Li, was very timely and offered multiple protocols for metabolomics data analysis. This volume now represents the second edition and offers revised and new chapters of multiple state-of-the-art tools for the analysis of liquid chromatography–mass spectrometry-based (LC-MS) metabolomics data. There is a specific focus on LC-MS-based metabolomics for this edition. This book is sectioned into three different parts following a typical metabolomics data analysis workflow.

This book only covers a small portion of the metabolomics software and tool universe, and many more great tools exist. With improved interoperability, we are getting closer to more reproducible data analysis. This is also the topic of the first chapter, focusing on data science in interactive notebooks using Jupyter.

Part I: LC-MS Data Processing

The first part in the metabolomics workflow is the extraction of detected metabolite features from the LC-MS raw data. LC-MS and LC-MS/MS are multidimensional containing information on a metabolite's retention time, mass-to-charge ratio (m/z), and its fragmentation in tandem MS experiments. Optionally, data might also contain an ion mobility dimension offering additional separation. This multidimensional data needs to be processed, which includes steps such as peak picking, isotope and adduct grouping, and alignment between multiple samples. Multiple software tools exist for this task. The first part of this book covers prominent tools such as xcms (Chap. 3), mzmine (Chap. 4), OpenMS (Chap. 5), asari (Chap. 6), MSDial (Chaps. 7 and 8), and a suite of tools dedicated to processing of ion mobility data (Chap. 9). The breath of tools ranges from fully flexible command line-based to integrated GUIs and are useful for beginners and experienced users.

Part II: Metabolite and Lipid Annotation

The annotation process is necessary to convert detected metabolite features into metabolites, which can be used for biological interpretation of the data. However, this task typically represents a major bottleneck in metabolomics workflows. Multiple spectral databases in various formats exist and are available for this task. Chapter 10 gives a brief overview on the commonly used ones. While multiple tools from Part I also allow to perform annotation, scientists might want to establish their own pipeline, independent of the processing tool. The MetaboAnnotation R package presented in Chap. 11 allows to generate such a workflow. Metabolites are normally not detected in isolation but are part of large metabolic network. This can be used for annotation of metabolites and analysis of their potential biological context. The MetDNA suite is presented in Chap. 12. Another important task is the confident and correct annotation of lipid without overannotation. The Lipid Data Analyzer tool presented in Chap. 13 exactly enables this task. Furthermore, oxidized lipids gain increasing attention as important entities. The LPPtiger tool described in Chap. 14 deals with the prediction and annotation of such modified (epi)lipids.

Part III: Data Analysis

Part III of this book covers multiple tools for statistical analysis of metabolomics data and data interpretation. In order to obtain meaningful results, proper normalization of metabolomics data must be performed, e.g., normalizing drifts along the analytical sequence or between batches. Chapters 15 and 16 present the packages notame and TIGER developed for this task. After normalization, data can be used for statistical analysis. One easy-to-use tool for this purpose is MetaboAnalyst, presented in Chap. 17. Besides statistical analysis, it also offers some processing and identification features as well as integrated metabolic pathway analysis. In Chap. 18, a set of R-based tools is presented for multi-omics (proteomics and metabolomics) data and network analysis. Last but not least, the obtained information needs to be interpreted. The Human Metabolome Database (HMDB) is one of the premier databases for human metabolomics. Besides storing information on metabolites, it also offers support for metabolomics data analysis, which is presented in Chap. 19.

Neuherberg, Germany *Michael Witting*

Contents

Contributors

HARSH BHOTIKA • *Pacific Northwest National Laboratory, Richland, WA, USA*
AIVETT BILBAO • *Pacific Northwest National Laboratory, Richland, WA, USA; US Department of Energy Agile BioFoundry, Emeryville, CA, USA*
BUJINLKHAM BUYANTOGTOKH • *Department of Biotechnology and Life Science, Tokyo University of Agriculture and Technology, Naka-cho, Koganei-shi, Tokyo, Japan*
JASMINE CHONG • *The Jackson Laboratory for Genomic Medicine, Farmington, CT, USA*
MARILYN DE GRAEVE • *Institute for Biomedicine, Eurac Research, Bolzano, Italy*
PAYAM EMAMI • *Department of Biochemistry and Biophysics, National Bioinformatics Infrastructure Sweden, Science for Life Laboratory, Stockholm University, Solna, Sweden*
MARIA FEDOROVA • *Center of Membrane Biochemistry and Lipid Research, University Hospital Carl Gustav Carus and Faculty of Medicine of TU Dresden, Dresden, Germany*
RETU HAIKONEN • *Institute of Public Health and Clinical Nutrition, Faculty of Health Sciences, University of Eastern Finland, Kuopio, Finland*
SIYU HAN • *Institute of Translational Genomics, Helmholtz Zentrum München, German Research Center for Environmental Health, Neuherberg, Germany; German Center for Diabetes Research (DZD), München-Neuherberg, Germany*
KATI HANHINEVA • *Food Sciences Unit, Faculty of Technology, University of Turku, Turku, Finland; Institute of Public Health and Clinical Nutrition, Faculty of Health Sciences, University of Eastern Finland, Kuopio, Finland; Afekta Technologies Ltd, Kuopio, Finland*
JÜRGEN HARTLER • *Institute of Pharmaceutical Sciences, University of Graz, Graz, Austria; Field of Excellence BioHealth, University of Graz, Graz, Austria*
STEFFEN HEUCKEROTH • *mzio GmbH, Bremen, Germany*
FIONA HUI • *Institute of Parasitology, McGill University, Ste-Anne-de-Bellevue, QC, Canada*
OLLI KÄRKKÄINEN • *Afekta Technologies Ltd, Kuopio, Finland; School of Pharmacy, Faculty of Health Sciences, University of Eastern Finland, Kuopio, Finland*
OLIVER KOHLBACHER • *Applied Bioinformatics, Department of Computer Science, University of Tübingen, Tübingen, Germany; Institute for Bioinformatics and Medical Informatics, University of Tübingen, Tübingen, Germany*
VILLE KOISTINEN • *Food Sciences Unit, Faculty of Technology, University of Turku, Turku, Finland; Institute of Public Health and Clinical Nutrition, Faculty of Health Sciences, University of Eastern Finland, Kuopio, Finland; Afekta Technologies Ltd, Kuopio, Finland*
EFTYCHIA E. KONTOU • *Enveda, Boulder, CO, USA*
ANSGAR KORF • *mzio GmbH, Bremen, Germany*
PRASHANTHI KOVUR • *Department of Biological Sciences, University of Alberta, Edmonton, AB, Canada*
LEO LAHTI • *Department of Computing, Faculty of Technology, University of Turku, Turku, Finland*
LEONIDA M. LAMP • *Institute of Pharmaceutical Sciences, University of Graz, Graz, Austria*

SHUZHAO LI • *University of Connecticut School of Medicine, Farmington, CT, USA; The Jackson Laboratory for Genomic Medicine, Farmington, CT, USA*
ATTE LIHTAMO • *Afekta Technologies Ltd, Kuopio, Finland*
YAMILÉ LÓPEZ-HERNÁNDEZ • *Department of Biological Sciences, University of Alberta, Edmonton, AB, Canada; SECIHTI-Metabolomics and Proteomics Laboratory, Academic Unit of Biological Sciences, Autonomous University of Zacatecas, Zacatecas, Mexico*
PHILIPPINE LOUAIL • *Institute for Biomedicine, Eurac Research, Bolzano, Italy; Chair for Bioinformatics, Friedrich Schiller University Jena, Jena, Germany*
SUSAN MCCLATCHY • *The Jackson Laboratory for Genomic Medicine, Farmington, CT, USA*
JOSHUA MITCHELL • *The Jackson Laboratory for Genomic Medicine, Farmington, CT, USA*
NATHALIE MUÑOZ • *Pacific Northwest National Laboratory, Richland, WA, USA; US Department of Energy Agile BioFoundry, Emeryville, CA, USA*
THOMAS NAAKE • *European Molecular Biology Laboratory, Heidelberg, Germany; European Molecular Biology Laboratory, Hamburg, Germany; Physikalisch-Technische Bundesanstalt, Braunschweig, Germany; Department of Bioinformatics and Biochemistry, Faculty of Life Sciences, Technische Universität Braunschweig, Braunschweig, Germany*
ZHIXU NI • *Institute of Data and Information, Tsinghua Shenzhen International Graduate School, Tsinghua University, Shenzhen, China; Center of Membrane Biochemistry and Lipid Research, University Hospital Carl Gustav Carus and Faculty of Medicine of TU Dresden, Dresden, Germany*
ZHIQIANG PANG • *Department of Microbiology and Immunology, McGill University, Montreal, QC, Canada*
TOMÁŠ PLUSKAL • *mzio GmbH, Bremen, Germany; Institute of Organic Chemistry and Biochemistry of the Czech Academy of Sciences, Prague, Czech Republic*
JOHANNES RAINER • *Institute for Biomedicine, Eurac Research, Bolzano, Italy*
VALENTINA RAMUNDI • *Metabolomics and Proteomics Core, Helmholtz Zentrum München, Neuherberg, Germany*
DYLAN H. ROSS • *Pacific Northwest National Laboratory, Richland, WA, USA*
ROBIN SCHMID • *mzio GmbH, Bremen, Germany*
CHRISTINA SCHMIDT • *Heidelberg University, Faculty of Medicine, and Heidelberg University Hospital, Institute for Computational Biomedicine, Heidelberg, Germany*
VILHELM SUKSI • *Department of Computing, Faculty of Technology, University of Turku, Turku, Finland*
GERHARD G. THALLINGER • *Institute of Biomedical Informatics, Graz University of Technology, Graz, Austria; Omics Center Graz, BioTechMed-Graz, Graz, Austria*
YOSHIMASA TODOROKI • *Department of Biotechnology and Life Science, Tokyo University of Agriculture and Technology, Naka-cho, Koganei-shi, Tokyo, Japan*
HIROSHI TSUGAWA • *Department of Biotechnology and Life Science, Tokyo University of Agriculture and Technology, Naka-cho, Koganei-shi, Tokyo, Japan; Graduate School of Pharmaceutical Sciences, Kyoto University, Kyoto, Japan*
AXEL WALTER • *Applied Bioinformatics, Department of Computer Science, University of Tübingen, Tübingen, Germany; Institute for Bioinformatics and Medical Informatics, University of Tübingen, Tübingen, Germany; Department of Peptide-based Immunotherapy, Institute of Immunology, University and University Hospital Tübingen, Tübingen, Germany; Quantitative Biology Center (QBiC), University of Tübingen, Tübingen, Germany*

RUI WANG-SATTLER • *Institute of Translational Genomics, Helmholtz Zentrum München, German Research Center for Environmental Health, Neuherberg, Germany; German Center for Diabetes Research (DZD), München-Neuherberg, Germany*
DAVID S. WISHART • *Department of Biological Sciences, University of Alberta, Edmonton, AB, Canada; Department of Computing Science, University of Alberta, Edmonton, AB, Canada; Department of Laboratory Medicine and Pathology, University of Alberta, Edmonton, AB, Canada; Department of Pharmacy and Pharmaceutical Sciences, University of Alberta, Edmonton, AB, Canada*
MICHAEL WITTING • *Metabolomics and Proteomics Core, Helmholtz Zentrum München, Neuherberg, Germany; Chair of Analytical Food Chemistry, TUM School of Life Sciences, Technical University of Munich, Munich, Germany*
JIANGUO XIA • *Department of Microbiology and Immunology, McGill University, Montreal, QC, Canada; Institute of Parasitology, McGill University, Ste-Anne-de-Bellevue, QC, Canada*
HAOSONG ZHANG • *Interdisciplinary Research Center on Biology and Chemistry, Shanghai Institute of Organic Chemistry, Chinese Academy of Sciences, Shanghai, P. R. China*
XUEYUN ZHENG • *Pacific Northwest National Laboratory, Richland, WA, USA*
ZHENG-JIANG ZHU • *Interdisciplinary Research Center on Biology and Chemistry, Shanghai Institute of Organic Chemistry, Chinese Academy of Sciences, Shanghai, P. R. China*

Chapter 1

Practicing Data Science in Interactive Notebooks

Jasmine Chong, Susan McClatchy, Joshua Mitchell and Shuzhao Li

Abstract

The Jupyter Notebook is a platform for interactive computing that displays code and results in the same browser, making it valuable for teaching, prototyping, data analysis, and collaboration. Its explicit and transparent structure greatly reproducibility while its backend server supports flexible deployment. In the past few years, Jupyter notebooks and similar tools have become increasingly popular. In this chapter, we will review key aspects of data analysis in a cloud environment and demonstrate common tasks for analyzing metabolomics data using template notebooks. This is an accompaniment to the basic bioinformatics tools and essential data science toolkit introduced in the first edition.

Keywords Jupyter, Python, R, Metabolomics, Notebooks, Data science, Data analysis

1 Introduction

The "-omics" in "metabolomics" requires data science. In the first edition, two chapters were included to introduce the basics: "A Bioinformatics Primer to Data Science, with Examples for Metabolomics" [1] and "The Essential Toolbox of Data Science: Python, R, Git, and Docker" [2]. These chapters covered topics on data management, scripting, version control, data wrangling, quality control, statistical analysis, and visualization. Readers are encouraged to refer to them to get started.

Since the first edition was published, the technological landscape has changed considerably over the past 5 years. Python has become indispensable for deep learning. Artificial intelligence now writes computer code—it is more important now than ever to understand the fundamentals, to ensure that code works as expected, and projects are authenticated. Jupyter notebooks and similar tools are now commonplace. They are very useful for interactive data analysis, collaboration, sharing, reproducibility,and software prototyping, and will be the focus of this chapter.

Michael Witting (ed.), *Computational Methods and Data Analysis for Metabolomics*, Methods in Molecular Biology, vol. 3063,
https://doi.org/10.1007/978-1-0716-5452-1_1,

1.1 What Are Jupyter Notebooks?

Computational notebooks combine code, code output, and narrative in a single document. Graphs and interactive plots can be embedded. Popular examples include MATLAB, RStudio, and Jupyter Notebook. Because the result is shown next to the code, they lower the barrier of abstract thinking in programming and are a great fit for data analysis. Thus, the notebooks deliver a convenient and interactive computing environment for developing and executing code, documenting analyses, visualizing data, and communicating results with others.

Jupyter Notebook has its roots in the IPython project (hence the .ipynb extension) but has evolved into a tool of choice for data scientists [3]. Today, Jupyter Notebook supports numerous computer languages and is utilized in millions of instances on GitHub. Its design includes a backend server and an editor in a web browser, which gives the flexibility of both local and cloud deployments. The latter allows use of the notebooks without local installation, while possibilities abound in connecting the notebooks to powerful and sophisticated applications.

1.2 Jupyter Notebooks Can Be Run Locally or on the Cloud

The Jupyter notebook editor runs in any major web browser, with output and results displayed below the code that generated them. Support for code editing includes automatic syntax highlighting, indentation, and tab completion. The accompanying narrative is written in Markdown, a markup language for formatting text. Mathematical notation can be included using LaTeX for formatting mathematical symbols. Notebooks are stored internally as JavaScript Object Notation (JSON) files with the .ipynb extension. JSON is a plain text format and, as such, can be shared and maintained under version control. Notebooks can be exported to HTML, LaTeX, PDF, slide shows, and other formats for easy communication and sharing.

The backend server of Jupyter can be run locally on a user's computer (browser address starts with http://localhost: or http://127.0.0.1:). In fact, Jupyter Notebook can be embedded in other applications such as Visual Studio Code and Anaconda, where the server may not even be visible to users. The backend server can also be hosted and managed remotely. Lead cloud providers offer various flavors of Jupyter notebooks. For example, Google Cloud Platform offers Colab ("Colaboratory"), which will be used in this chapter to work through example notebooks. A major consideration of using a cloud service is data security, as data are exchanged between users and the cloud resources. Users should perform their own due diligence on how data are protected and whether it satisfies project requirements.

1.3 Tools and Data Used in This Chapter

Users are recommended to install a local version of Jupyter Notebook, using one of several options: standalone Jupyter [4], Anaconda [5], or Visual Studio Code [6]. For simplicity, this chapter will illustrate examples using Google Colab on the cloud, which can be accessed and authenticated with a free Google account. Colab requires no installation other than having a modern web browser. All example notebooks and datasets for this chapter are accessible from this GitHub repository (https://github.com/shuzhao-li-lab/2025_Chapter_DataScience). These example notebooks have built-in clickable buttons to run on Colab.

2 Jupyter Notebook Basics

Notebook documents contain a complete computational record of a session, and consist of a sequence of cells that contain either code or text formatted in Markdown. A cell is a self-contained input field that can be executed according to the cell's type. Code cells produce executable code and display the resulting output, while Markdown cells generate formatted text and mathematical notation. Code cells are executed with a play button on the left cell margin, with code output appearing directly below the cell. Markdown cells are good for documenting the notebook by explaining bits of code, steps in a pipeline, or notebook logic in general.

One important characteristic of these notebooks is that when executing a sequence of cells, code cells do not automatically update based on changes to one of the code cells in the sequence. When changes are made to a code cell, the preceding cells must be executed again to update any values or variables they might share.

3 Using Google Colab to Run Jupyter Notebooks

While Jupyter notebooks can be run locally, doing so requires installing software, managing dependencies, and configuring your computer's environment. Google Colab offers a simpler alternative by hosting Jupyter notebooks in the cloud for free with a Google account. With a single click you can open, edit, and share notebooks with zero setup required (example in Fig. 1).

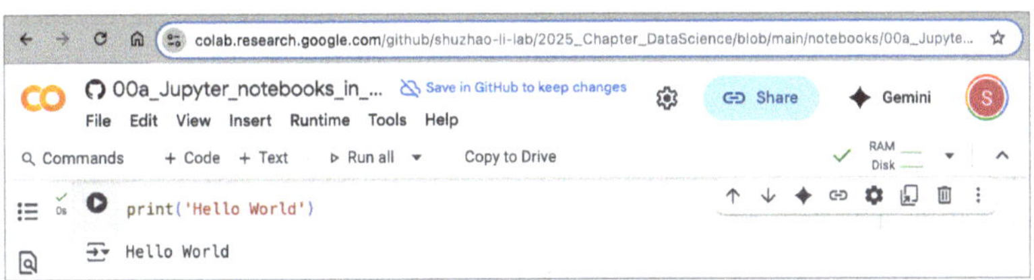

Fig. 1 A screenshot of running a Jupyter notebook in a web browser using Google Colab. This is the top of the first notebook in the chapter's GitHub repository

Google Colab provides access to free computing resources. A runtime is a virtual machine where your code is executed. A runtime environment includes the CPU (a general-purpose processor for running most code) or GPU (graphics processing unit, suitable for specialized tasks), the programming language (Python by default), and any variables, packages, or files created during the session. However, free versions of Google Colab have limits on resource availability, especially for GPU access. Disconnecting runtimes when not in use, avoiding unnecessary GPU usage, and focusing on interactive computing tasks instead of long-running computations are all good strategies to make the most of free Google Colab resources (*see* **Note 2**). Google Colab is best suited for teaching and learning; intensive computing tasks should not be run on the free versions.

3.1 Navigating Google Colab

You can find Google Colab at https://colab.research.google.com. You will be greeted with a Welcome to Colab notebook that has helpful introductory information.

3.1.1 How to Open a Jupyter Notebook

There are several options for opening notebooks in Google Colab. A new blank notebook can be opened in the user's Google Drive with *File → New notebook in Drive*. Existing notebooks in Drive can also be opened with *File → Open notebook → Google Drive*. Notebooks on the user's local computer can be uploaded with *File → Upload notebook*. Finally, notebooks can be opened from GitHub with *File → Open Notebook → GitHub*. Enter the GitHub URL directly in the search bar, and Colab will list all notebooks from that URL under *Path*.

For a beginner-friendly introduction to Google Colab, Jupyter Notebooks, and Python, navigate to the *notebooks* folder in the GitHub repository for this chapter (https://github.com/shuzhao-li-lab/2025_Chapter_DataScience). There are two introductory notebooks: one provides a quick overview for understanding notebooks in just 5 minutes, and the other provides a more comprehensive introduction to notebooks and Python. At the top of either of these notebooks, you will find a blue "Open in Colab" button. Click this button to launch the notebook in Google Colab, where you can save a copy to your Google Drive and make your own edits (*see* **Note 1**).

3.1.2 The Google Colab Interface

The main notebook menu, located at the upper left, includes the options: *File, Edit, View, Insert, Runtime, Tools,* and *Help* (Fig. 1). *File* includes options for creating, opening, saving, and downloading notebooks. *Edit* contains commands for selecting, cutting, copying, pasting, or deleting cells, clearing output, finding and replacing text or code, and reviewing notebook settings. *File →Revision history* tracks changes to notebook contents, such as edits to code and text cells. This can be used to view earlier versions, see how the notebook has changed, or recover deleted content. *View* provides access to the notebook table of contents, notebook information, a history of executed code, and any comments provided by notebook users or editors. *Insert* contains commands to insert code cells, text cells, or section headers. Code and text cells can also be inserted with either the + *Code* or + *Text* options beneath the main menu.

Runtime provides options to run all cells, all preceding cells, all selected cells, or all following cells. When code in one cell is changed, *Runtime* → *Run cell and below* ensures that all following cells are updated with these changes. As mentioned earlier, cells do not automatically update when changes are made to code in one cell. They must be re-executed to be updated with new values or variables. *Runtime* → *Interrupt execution* halts cell execution when code takes too long to execute and needs revision. *Runtime* → *Change runtime type* permits the user to switch between Python, R, or Julia, or to switch from CPU to GPU for greater processing power. There are also options to restart a session or manage multiple sessions. A session is a connection between the browser and the Google Colab runtime through Hypertext Transfer Protocol Secure (https), the main protocol for transferring data over the internet. Managing sessions by terminating those not in use can be helpful for freeing up computational resources (*see* **Note 2**).

The *Tools* menu provides access to a searchable command palette, keyboard shortcuts, Colab settings, and a tool to find differences between two notebooks. The *Help* menu provides frequently asked questions, release notes, terms of service, submission of bug reports and feedback, and a way to search code snippets. To the right of the main menu is a blue *Share* button for sharing a notebook by searching for other Google accounts or copying a shareable link. Choose appropriate permission levels (view, comment, or edit) when sharing.

4 Exploring LC–MS Metabolomics Data

Now that we are comfortable with the Google Colab environment, let us dive into the fun stuff. This series of notebooks is designed to help users become familiar with running Jupyter notebooks in Colab and to demonstrate good practices for formatting and structuring notebooks (*see* **Note 3**). It is not intended to teach Python or R programming—please refer to Chap. 15 of the first edition of this book [2] for tutorials on learning and using these languages. The focus here is to equip readers with the tools and basic skills needed to explore LC–MS data and perform statistical analysis in a notebook environment. We will begin with exploring a few common concepts in LC-MS (liquid chromatography–mass spectrometry) metabolomics data, including mass spectrum, mass track, total ion chromatograms, and extracted ion chromatograms. These visualizations help understand the data, assess data quality, and build workflows.

1. Navigate to the GitHub repository for this chapter (https://github.com/shuzhao-li-lab/2025_Chapter_DataScience), open the notebooks folder, and select 01_Python_Exploring_Raw_Metabolomics_Data.ipynb. At the top of the notebook, you will see a blue button labeled "Open in Colab." Click this button to directly open the notebook in Google Colab. If desired, you can save a copy to your Google Drive and make your own versions (*see* **Note 1**).
2. Notebook setup: Google Colab comes pre-installed with several core Python libraries commonly used in data science. For this notebook, we only need to install the asari-metabolomics library [7], which bundles useful tools for working with raw metabolomics data such as *pymzml* [8] and *mass2chem*. Click the *play* button on the left margin of code cell 1 to install and import the necessary libraries. Note that this should not take much time (< 5 min). The cell successfully executed if you see something like this line at the end of the output: Successfully installed asari-metabolomics-1.13.1.
3. Data retrieval: Following setup, we will download the raw data files (.mzML format, centroided) from a GitHub repository for this tutorial. The dataset consists of two reference plasma samples in replicates, using HILIC (Hydrophilic Interaction Liquid Chromatography) in positive ion mode. Click the *play* button on code cell 2. You should see a list of messages starting with "Downloading ..." appear sequentially under the cell. Note that these downloaded files, as well as the installed packages in the previous step, live in the cloud virtual machine, not the user's computer.
4. Execute code cell 3, which prints all the downloaded files in our working directory. Next, execute code cell 4, which places all the mzML files into a list. There should be a total of eight mzML files in the list.
5. Reading and visualizing spectra: A mass spectrum is a list of ions (in m/z, mass-to-charge ratio) with intensity values. A metabolomics experiment collects a combination of mass spectra, e.g., a series of scans of profiling spectra in LC-MS ordered by chromatography, and/or fragment spectra of selected ions in MS/MS (tandem mass spectrometry). We first plot a mass spectrum, using m/z as the x-axis and ion intensity on the y-axis. In this plot, each peak represents a detected ion, with its height corresponding to its relative abundance.

   a. Execute code cell 5, which will read in the first sample from our list of downloaded files and load all of its spectra into memory.
   b. Execute code cell 6, which generates a spectrum for scan number 100 (Fig. 2, full spectrum).

```
[ ] # setup plot background and size
    sns.set_style("white")
    plt.figure(figsize=(10, 5))

    # plot each peak as a vertical line
    plt.vlines(mz, [0], intensity, color="steelblue", alpha=0.7)

    # add labels and title
    plt.xlabel("m/z")
    plt.ylabel("Intensity")
    plt.title(f"Spectrum at RT={spectrum.scan_time[0]:.2f} min, Scan #{spectrum.ID}")

    # adjust the layout so everything fits nicely in the plot
    plt.tight_layout()
    plt.show()
```

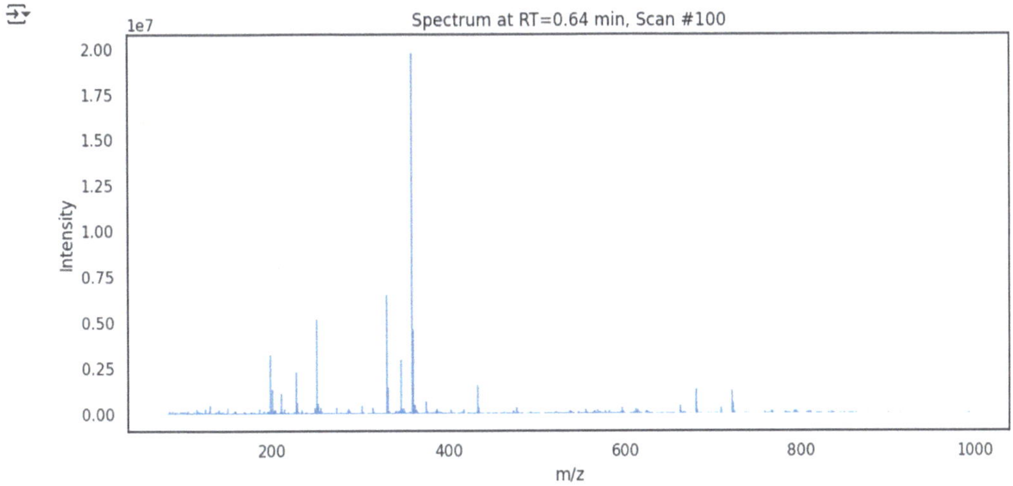

Fig. 2 Plotting a mass spectrum in a Jupyter notebook. The top panel shows the code cell; the graph is the output below.

c. Sometimes we want to focus on a specific region of a spectrum. In our case, let us zoom into the 200–400 m/z range, where many common small polar plasma metabolites can be found. Execute code cell 7, which plots a zoomed-in spectrum of this m/z range.

d. In the zoomed-in spectrum, we observe a single peak with the highest intensity at roughly 360 m/z. Execute code cell 8 to get the actual m/z value and intensity.

6. Plotting mass tracks: As LC–MS data are composed of a series of snapshots along elution time, a metabolite ion appears in multiple consecutive scans. Therefore, processing LC–MS data involves the reconstruction of the extracted ion chromatogram (EIC or XIC). This is extended into the concept of mass tracks in the Asarisoftware [7] to robustly identify recurrent feature patterns. A mass track differs from an EIC by the full range of values, including zeros, across retention time. This makes mass

alignment independent from elution peak detection. For this notebook, we will focus on exploring the structure of a mass track and visualizing it.

a. Execute code cell 9 to extract the mass track from our first sample. A mass track is represented as a Python dictionary, containing three keys: (1) *rt_numbers*—scan numbers/indices, (2) *rt_times*—corresponding retention times for each scan number, and (3) *tracks*—a list of EICs, one for each detected m/z value.

b. Execute code cell 10 to inspect the structure of a single track. Each track is a tuple, made of two elements. The first is a float, representing the m/z value of the track. The second is an array, which contains the intensity values for that m/z across the entire retention time.

c. A track is linked with *rt_numbers* and *rt_times*, as the length of the intensity array matches the number of scans and retention times. This means that each index in the intensity array corresponds directly to a specific scan and its retention time. Execute code cell 11 to confirm this relationship.

d. Finally, execute code cell 12 to visualize the track (Fig. 3—mass track). The plot illustrates how ion intensity at 106.0684 m/z varies over the course of the chromatographic run.

Note that we look at the raw data points here. When software tools "beautify" the data, they may lead to false perceptions and even research reproducibility issues. Now that we have a good understanding of LC-MS concepts, let us move on to sample QC.

7. TIC and BPC: Two important metrics for evaluating intensity profiles across all files are TIC and BPC. Both track scan intensities over the course of the run, but in slightly different ways. TIC sums the intensities of all detected ions in each scan, representing total signal intensity at each time point. In contrast, BPC only records the intensity of the most abundant ion (base peak) per scan. Both metrics are commonly used to identify samples with abnormally low or noisy signals. Execute code cell 13 to calculate TIC and BPC values for all samples and plot their respective chromatograms (Fig. 4—TIC and BPC).

    a. A good chromatogram should show consistent shapes and retention time ranges across samples. Signal intensities should be comparable, and flat or noisy curves indicate failed injections or low-abundance samples. TIC and BPC plots should look similar, but the BPC will have a lower y-axis scale and can appear spikier as it reflects only a single ion versus the sum of all ions.

```
# 12 - visualize track 222
this_mz = mass_tracks['tracks'][222][0]
intensity = mass_tracks['tracks'][222][1]
rt_times = mass_tracks['rt_times']

# setup DataFrame
df = pd.DataFrame({
    'Retention Time': rt_times,
    'Intensity': intensity
})

# set figure size
plt.figure(figsize=(10, 5))
sns.scatterplot(data=df, x='Retention Time', y='Intensity')

plt.title(f"Mass Track at m/z {this_mz:.4f}")
plt.xlabel("Retention Time (s)")
plt.ylabel("Intensity")
plt.show()
```

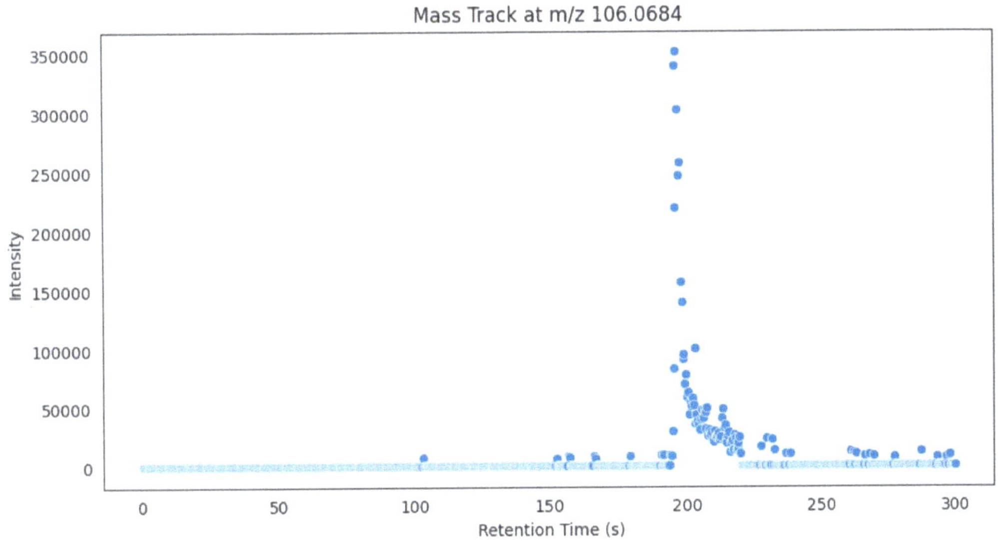

Fig. 3 Plotting a mass track in a Jupyter notebook. The top panel shows the code cell; the graph is the output below

b. In our plots, the samples follow a similar chromatographic profile, indicating good data quality and reproducibility. It also suggests that sample preparation and instrument performance were consistent.

8. Summed ion intensities: We will now evaluate the overall signal abundance by summing ion intensities for each scan within each sample. Execute code cell 14, which sums the ion intensities per scan, applies a log2 transformation, and stores it in a data frame. A boxplot of the summed ion intensities is then generated (Fig. 5—boxplot).

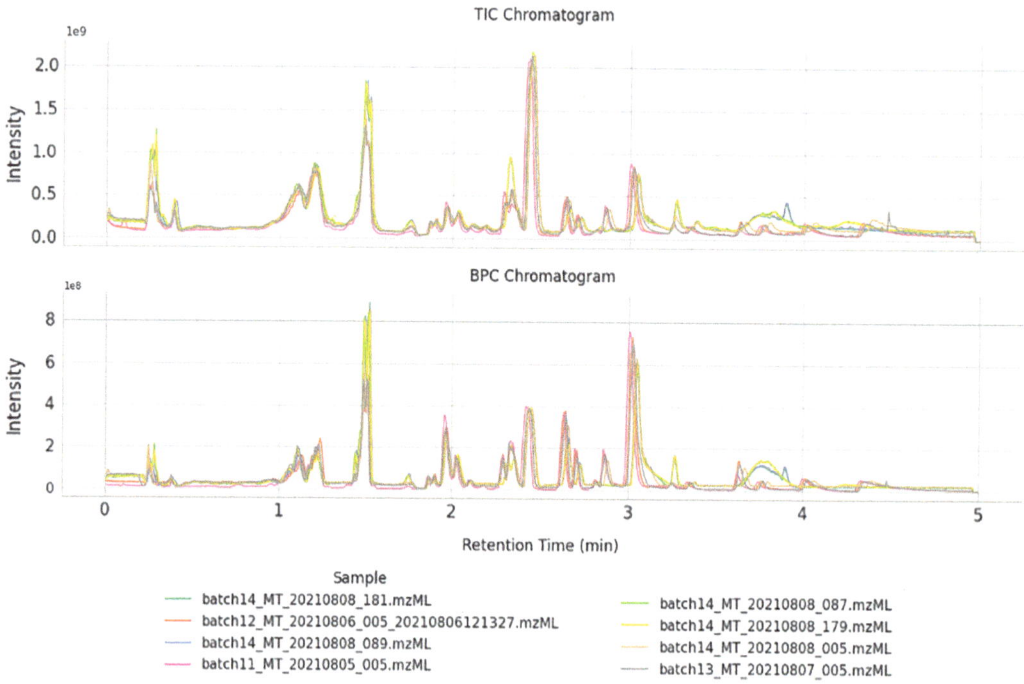

Fig. 4 Plotting total ion chromatograms (top) and base peak chromatograms (bottom) of all samples

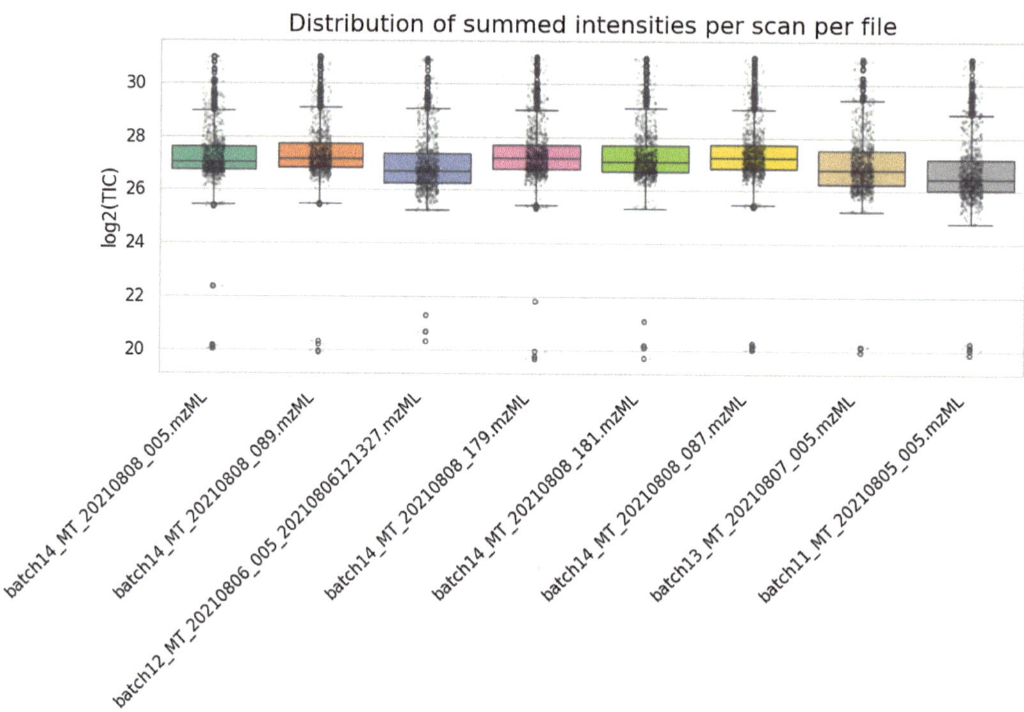

Fig. 5 Boxplots of summed ion intensities per sample

   a. The summed intensities across all files are consistent, with a similar TIC distribution. There are no unexpected surprises here.

9. Feature inspection: The final task is to inspect the data for signals corresponding to known metabolites. In this case, we look for one of our spike-in standards and two common plasma metabolites. Execute code cell 15, which defines a function to extract and plot ion chromatograms for a specified m/z across all samples. Next, run code cells 16–18, which generate ion chromatograms for [$^{13}C_3$]-caffeine, which is a spike-in standard, and L-valine and L-phenylalanine, common plasma metabolites. Since the data were collected in positive ion mode, a proton [H^+] is added to the molecular weights.

   a. Visual inspection can easily tell good peaks from bad peaks. This may also catch retention time drift, meaning the same peak has slightly different retention time across samples. Finally, if there were blank samples, they should not have major peaks other than contaminants.

   b. For caffeine (Fig. 6—caffeine), we observe two peaks. The first peak is our spike-in, matching the known retention time.

   c. For both L-valine and L-phenylalanine, high-quality peaks with matching retention time are observed (figures shown in the notebooks). These can serve as positive controls for data quality.

5 LC–MS Data Analysis—Practical Examples

After raw data are processed to feature tables, annotation takes place to (i) reduce redundant features to metabolites and (ii) identify metabolites. The first is referred to as "pre-annotation" [9]. The second depends on authentic compound libraries, with known m/z, RT, and MS/MS patterns [10, 11]. Annotation is not covered here, but readers can refer to other chapters and literature. In the second half of this protocol, we include two Jupyter notebooks for performing basic statistical analysis and visualization tasks—one in R and one in Python. Both notebooks are available in this chapter's companion GitHub repository and cover similar concepts. Below, we will walk through the R notebook to perform some fundamental data analysis tasks on our processed LC-MSmetabolomics data, including quality control, filtering, normalization, and univariate statistics. The figures are not printed here but shown in the online notebooks.

1. Navigate to the companion GitHub repository for this chapter (https://github.com/shuzhao-li-lab/2025_Chapter_DataScience) and open the notebooks folder. Select the

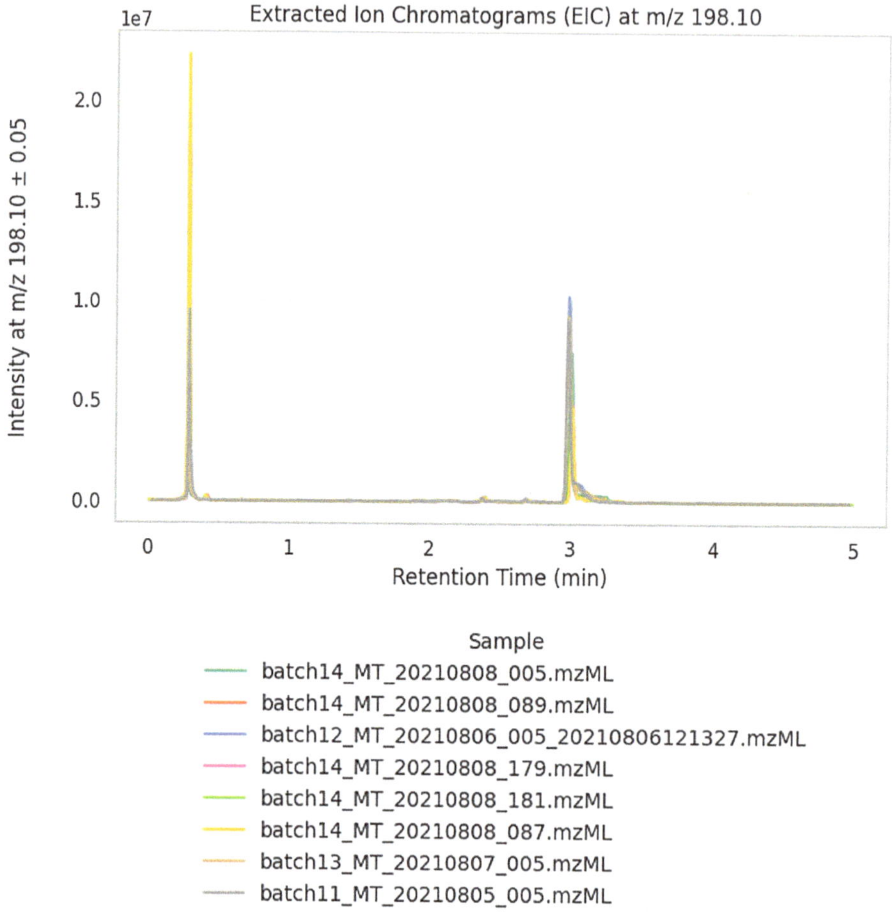

Fig. 6 Inspection of features in a Jupyter notebook.

02_R_for_metabolomics_data_science.ipynb notebook. At the top, you will see a blue button labeled "Open in Colab." Click this button to open the notebook in Google Colab and begin the tutorial.

2. Notebook setup: The R runtime environment in Google Colab comes pre-installed with several common R packages. However, for data science and bioinformatics tasks, we will need to install some additional packages. Generally, R packages come from two major sources:

   - *CRAN*—Repository for general-purpose R packages
   - *Bioconductor*—Repository for bioinformatics-focused R packages

   Execute code cell 1, which will install and load all required R packages. Note that installing R packages can take time, as many dependencies need to be fetched and compiled. This step should take around 5–10 min, depending on your internet connection.

3. Data download: Next, we will download the example feature table to perform data analysis. This is a typical output from data processing tools such as XCMS [12], MzMine [13], MS-DIAL [14], and Asari [7]. The table consists of six *Escherichia coli* (*E. coli*) samples: three grown with regular glucose (labeled as ^{12}C) and three grown with uniformly labeled ^{13}C-glucose. While no biological difference is expected, their metabolomes should reflect the difference from the isotope labeling.
   a. Execute code cell 2, which reads the .tsv file directly from GitHub. The head of the table will be displayed in the output. The first three columns contain information about the metabolic features (e.g., m/z and retention time), while columns 4–9 display feature intensities across all samples.
   b. Next, we will format the data to prepare the table for analysis. Execute code cell 3, which transposes the table so that samples are in rows and features are in columns. It also drops irrelevant columns, removes rows with zero variability across samples (non-informative), and creates a metadata table containing sample names and their assigned group labels (^{12}C, ^{13}C). The heads of both the feature and metadata tables will appear in the output.
4. Global visualization: The first plot we will generate is a Principal Component Analysis (PCA) of all our samples, based on the raw feature intensities. This plot helps us identify extreme outliers and observe how samples cluster together. Additionally, we can see how informative (or not) our features are by examining the variance explained by the principal components.
   a. Execute code cell 4, which loads the custom PCA function into the R environment, where it can be used throughout the notebook. At the bottom of the cell, a PCA plot of the raw feature intensities will be displayed.
   b. In the PCA, samples are colored by their group labels. The first two PCs explain ~70% of the variance in the dataset, suggesting there are clear metabolic differences between the two groups.
5. Data filtering and normalization: Now we will carry out a series of standard steps to ensure our data are suitable for downstream statistical analyses. For brevity, we skip batch effect correction here, but readers can refer to the topic in the relevant literature [15–17].

Step 1: Filter features with excessive missing values. Missing values may result from low-abundance metabolites, inconsistent detection, or technical artifacts. Including too many features with missing data can weaken statistical power, so we will only keep consistently detected features, focusing on robust signals.

Step 2: Imputation. In many cases, an intensity of zero represents a signal below the limit of detection (LoD), rather than a true absence. To avoid zero-inflated data, we will replace zeros with an estimated LoD per feature, calculated as the minimum positive intensity value for each feature divided by 10.

Step 3: Filter features with high variability. We will use RSD (Relative Standard Deviation), which is the standard deviation of a feature divided by its mean, as a measure of variability. Features with high RSD across technical replicates or QC samples are unreliable and may reflect instrument noise rather than the true biological signal. Removing these features improves data reproducibility and reduces the likelihood of false positives downstream. As our data contains only biological replicates, we will calculate RSD per group and use this to filter highly variable features.

Step 4: Data normalization. This final step corrects for unwanted variation in signal intensity due to technical factors (e.g., differences in injection volume, ion suppression). This addresses heteroscedasticity and improves feature comparability across samples. Common normalization methods include auto-scaling, Pareto scaling, and log transformation.

a. Execute code cell 5, which first performs missing value filtering, removing features (columns) that have more than 50% missing intensity values across all samples. Second, it performs LoD replacement of zero values (imputation).

b. Execute code cell 6, which calculates RSD per group and then filters out features with an RSD greater than 30. This is a common threshold in metabolomics.

c. After these steps, we will visually assess the distribution of our data. Execute code cell 7. In the first histogram (of filtered intensity data), we observe a left-skewed distribution, which suggests the data are not normally distributed. This is problematic as many statistical tests assume data normality (e.g., *t*-tests, ANOVA), and skewed data violate these assumptions and may lead to inaccurate results. To address this, we apply a log2 transformation to the intensity data and visually evaluate the distribution with a second histogram. From this plot, we observe a Gaussian-like distribution centered on the middle of the x-axis, suggesting the log2 transformation has successfully normalized the data. If the distribution had remained skewed, one should try other methods until normality is observed. Together, these steps enhance the reliability of our data and give us the confidence to move on to the next steps.

6. Statistical analysis: Now that the data are ready, we will perform univariate analysis to identify metabolic features that differ

significantly across our experimental groups. Each feature will be tested individually to determine whether its intensity varies by group.

We are using *group* as a categorical variable, so the first method we will apply is *t*-tests. A *t*-test compares the means of a continuous variable (e.g., metabolic feature intensity) between two groups.

- Null hypothesis (H0): The group mean intensities are equal.
- Alternative hypothesis (H1): The mean intensity of one group differs from the other.

a. Execute code cell 8, which performs *t*-test for each individual feature in parallel. The resulting *p*-values are corrected by the Benjamini–Hochberg (BH) method [18] to account for multiple testing. In 'omics, thousands of features are tested at the same time, increasing the probability of false positives. Multiple testing correction methods, including BH, control for the false discovery rate (FDR), to ensure our results are reliable. The top of the results should be displayed in the output. We observe that over 800 features are significantly different in one group from the other (adjusted p-value < 0.05).

7. Fold-change: Next, let us look at fold-changes (FC). Fold-changes are a fundamental analysis in 'omics as they provide an easy-to-interpret measure of the magnitude of difference between groups. An FC of 1 means there are no differences, FC > 1 means the feature is more abundant, and an FC < 1 means the feature is less abundant .

   a. Execute code cell 9, which calculates FC per feature and applies a log2 transformation. Features with an absolute log2FC greater than 1 are considered important.

8. Volcano plot: Alone, an FC shows the magnitude of differences between groups, but not the statistical significance. By pairing the *p*-values from the *t*-test with the fold changes, we can build volcano plots. These plots allow us to identify features that are both statistically significant and show substantial change in intensity. Such features are likely the most biologically relevant.

   a. Execute code cell 10, which combines the *t*-test and fold-change results into a single data frame and generates a volcano plot. In the plot, upregulated genes are on the right, downregulated on the left, and most significant at the top. Here, we see there are many features that are statistically significant (above the gray horizontal line), but few that have significant fold-changes. We see four important features (red dots) with an adjusted p-value < 0.05 and an absolute fold-change > 1.

b. Violin plot: Execute code cell 11 to plot the intensity of one of the four significant features. The plot shows that its relative intensity is clearly higher in ^{13}C samples versus ^{12}C samples.

You have now successfully completed a set of common tasks in metabolomicsdata analysis. We have used standard libraries to illustrate the concepts, while readers are encouraged to explore more powerful packages, e.g., MetaboAnalystR [19]. A Python version of this tutorial is included in the chapter's GitHub repository for you to explore at your convenience. We hope these Jupyter notebooks serve as valuable building blocks for your future data science projects (*see* **Note 3**).

6 Notes

1. Saving your Jupyter notebook: When you open or create your own notebook in Google Colab, it is saved by default to your Google Drive. The notebook is accessible from any device, and changes are saved automatically as you work. You can also share your notebook with others from Google Drive or as a standalone file.
2. Troubleshooting session crashes and runtime disconnections: Google Colab sessions can disconnect due to 90 minutes of inactivity, excessive use of computational resources, or runtime errors. When this happens, the runtime environment resets, meaning that all variables, uploaded files, and installed packages are lost. All notebook cells must be rerun from the beginning to restore the environment. Frequent interaction with notebooks will avoid disconnection, and saving notebooks frequently to Google Drive will avoid loss of progress if a session terminates. Being mindful of runtime usage and switching to the CPU instead of the GPU helps to avoid usage limits. Measuring code execution time with the %time and %timeit magic commands helps to identify performance bottlenecks. Delete any unnecessary variables or files to release resources. When finished, disconnect or delete runtimes to free up computational resources.
3. When to (and not to) use Jupyter notebooks: Now that you are familiar with Jupyter notebooks, it is helpful to understand when they should be used. Jupyter notebooks are ideal for interactive and exploratory coding, allowing you to write and run code as you go, make real-time changes, and immediately see the results. They are also a great tool for learning and teaching, as they combine code, output, and text in a single, user-friendly format—just as you experienced in this tutorial!

In the context of academic research, Jupyter notebooks are becoming a standard for open and reproducible data science. It is common to see notebooks included as supplementary files or linked in GitHub repositories of scientific articles. However, they are not ideal for production-scale software development, where code is better managed in standalone scripts or packages.

References

1. Pittard WS, Villaveces CK, Li S (2020) A bioinformatics primer to data science, with examples for metabolomics. Methods Mol Biol 2104:245–263. https://doi.org/10.1007/978-1-0716-0239-3_14
2. Pittard WS, Li S (2020) The essential toolbox of data science: Python, R, Git, and Docker. Methods Mol Biol 2104:265–311. https://doi.org/10.1007/978-1-0716-0239-3_15
3. Perkel JM (2018) Why Jupyter is data scientists' computational notebook of choice. Nature 563(7729):145–146. https://doi.org/10.1038/d41586-018-07196-1
4. Jupyter. https://jupyter.org/install. Accessed 9 Jul 2025
5. Anaconda. https://www.anaconda.com/download. Accessed 9 Jul 2025
6. Code V. https://code.visualstudio.com/docs/datascience/jupyter-notebooks. Accessed 9 Jul 2025
7. Li S, Siddiqa A, Thapa M, Chi Y, Zheng S (2023) Trackable and scalable LC-MS metabolomics data processing using asari. Nat Commun 14(1):4113. https://doi.org/10.1038/s41467-023-39889-1
8. Bald T, Barth J, Niehues A, Specht M, Hippler M, Fufezan C (2012) pymzML-Python module for high-throughput bioinformatics on mass spectrometry data. Bioinformatics 28(7):1052–1053. https://doi.org/10.1093/bioinformatics/bts066
9. Li S, Zheng S (2023) Generalized tree structure to annotate untargeted metabolomics and stable isotope tracing data. Anal Chem 95(15):6212–6217. https://doi.org/10.1021/acs.analchem.2c05810
10. Mitchell JM, Chi Y, Thapa M, Pang Z, Xia J, Li S (2024) Common data models to streamline metabolomics processing and annotation, and implementation in a Python pipeline. PLoS Comput Biol 20(6):e1011912. https://doi.org/10.1371/journal.pcbi.1011912
11. Metz TO, Chang CH, Gautam V, Anjum A, Tian S, Wang F, Colby SM, Nunez JR, Blumer MR, Edison AS, Fiehn O, Jones DP, Li S, Morgan ET, Patti GJ, Ross DH, Shapiro MR, Williams AJ, Wishart DS (2025) Introducing "identification probability" for automated and transferable assessment of metabolite identification confidence in metabolomics and related studies. Anal Chem 97(1):1–11. https://doi.org/10.1021/acs.analchem.4c04060
12. Smith CA, Want EJ, O'Maille G, Abagyan R, Siuzdak G (2006) XCMS: processing mass spectrometry data for metabolite profiling using nonlinear peak alignment, matching, and identification. Anal Chem 78(3):779–787
13. Schmid R, Heuckeroth S, Korf A, Smirnov A, Myers O, Dyrlund TS, Bushuiev R, Murray KJ, Hoffmann N, Lu M, Sarvepalli A, Zhang Z, Fleischauer M, Duhrkop K, Wesner M, Hoogstra SJ, Rudt E, Mokshyna O, Brungs C, Ponomarov K, Mutabdzija L, Damiani T, Pudney CJ, Earll M, Helmer PO, Fallon TR, Schulze T, Rivas-Ubach A, Bilbao A, Richter H, Nothias LF, Wang M, Oresic M, Weng JK, Bocker S, Jeibmann A, Hayen H, Karst U, Dorrestein PC, Petras D, Du X, Pluskal T (2023) Integrative analysis of multimodal mass spectrometry data in MZmine 3. Nat Biotechnol 41(4):447–449. https://doi.org/10.1038/s41587-023-01690-2
14. Tsugawa H, Ikeda K, Takahashi M, Satoh A, Mori Y, Uchino H, Okahashi N, Yamada Y, Tada I, Bonini P, Higashi Y, Okazaki Y, Zhou Z, Zhu ZJ, Koelmel J, Cajka T, Fiehn O, Saito K, Arita M, Arita M (2020) A lipidome atlas in MS-DIAL 4. Nat Biotechnol 38(10):1159–1163. https://doi.org/10.1038/s41587-020-0531-2
15. Fan S, Kind T, Cajka T, Hazen SL, Tang WHW, Kaddurah-Daouk R, Irvin MR, Arnett DK, Barupal DK, Fiehn O (2019) Systematic error removal using random forest for normalizing large-scale untargeted lipidomics data. Anal Chem 91(5):3590–3596. https://doi.org/10.1021/acs.analchem.8b05592
16. Liu Q, Walker D, Uppal K, Liu Z, Ma C, Tran V, Li S, Jones DP, Yu T (2020) Addressing the batch effect issue for LC/MS metabolomics data in data preprocessing. Sci Rep 10(1):13856. https://doi.org/10.1038/s41598-020-70850-0

17. Deng K, Zhao F, Rong Z, Cao L, Zhang L, Li K, Hou Y, Zhu ZJ (2021) WaveICA 2.0: a novel batch effect removal method for untargeted metabolomics data without using batch information. Metabolomics 17(10):87. https://doi.org/10.1007/s11306-021-01839-7
18. Benjamini Y, Hochberg Y (1995) Controlling the false discovery rate: a practical and powerful approach to multiple testing. J R Stat Soc Ser B Stat Methodol 57:289–300
19. Pang Z, Xu L, Viau C, Lu Y, Salavati R, Basu N, Xia J (2024) MetaboAnalystR 4.0: a unified LC-MS workflow for global metabolomics. Nat Commun 15(1):3675. https://doi.org/10.1038/s41467-024-48009-6

Check for updates

Chapter 2

LC–MS Data Processing Using *xcms*

Marilyn De Graeve, Philippine Louail and Johannes Rainer

Abstract

Liquid chromatography–mass spectrometry (LC–MS) is a key technology in metabolomics, enabling high-throughput detection of small molecules across diverse biological samples. However, raw LC–MS data are complex, requiring careful preprocessing to ensure accurate and reproducible feature detection. This chapter introduces a step-by-step protocol for LC–MS data preprocessing using the open-source *xcms* package in R. Designed for users ranging from beginners to experienced analysts, the chapter outlines critical stages including data inspection, peak detection, retention time alignment, correspondence, and result export. Special attention is given to parameter optimization and diagnostic visualization to guide users in making informed decisions tailored to their specific datasets. We demonstrate the approach on human serum samples, using real-life example compounds such as proline to showcase retention time alignment and peak correspondence. By combining practical code snippets with conceptual insights, this chapter empowers researchers to harness *xcms* for robust, reproducible LC–MS workflows in untargeted metabolomics. Whether you are setting up your first analysis or refining an established pipeline, this chapter serves as both a tutorial and a reference for high-quality LC–MS data processing.

Keywords *xcms*, Liquid chromatography, Mass spectrometry, Metabolomics, Data processing, Preprocessing, Peak detection, Retention time correction, Quality control

1 Introduction

Liquid chromatography-mass spectrometry (LC–MS) has become a cornerstone technique in metabolomics,proteomics, and other omics disciplines, such as lipidomics or DNA adductomics, due to its high sensitivity, selectivity, and ability to detect a broad range of compounds in complex biological samples. However, the raw data generated by LC–MS instruments are complex and multidimensional, requiring careful preprocessing to extract meaningful and reproducible quantitative information. Key steps of preprocessing typically include chromatographic peak detection, retention timealignment, correspondence analysis, and gap filling. Preprocessing of LC–MS data is highly influenced by the specific LC setup and MS system used. LC-dependent influences include the type and length of the

Michael Witting (ed.), *Computational Methods and Data Analysis for Metabolomics*, Methods in Molecular Biology, vol. 3063,
https://doi.org/10.1007/978-1-0716-5452-1_2,

chromatography column, particle size, flow rate, gradient composition, and the chemical nature of the mobile phase (solvents, pH, buffer salts) [1]. These parameters affect chromatographic resolution, retention time stability (i.e., batch effects), peak shape, and the degree of co-elution between analytes [2]. Additionally, matrix effects arising from the sample type, such as serum, can influence retention and suppression or enhancement of signals [3]. MS-dependent influences relate to the instrument type (e.g., TOF, Orbitrap), ion source settings, acquisition mode (centroid vs. profile, DDA vs. DIA), mass accuracy, and resolution. These factors affect the sensitivity, dynamic range, and spectral complexity of the acquired data and have a direct impact on the mass deviation tolerance [4]. Therefore, parameters for each of the preprocessing analysis steps need to be adjusted based on the experimental setup as well as instrument configurations. A visual as well as quantitative evaluation of the data before and during the analysis is crucial to ensure high-quality results.

This chapter introduces the *xcms* R package [5], a widely adopted, open-source tool designed for LC–MS data preprocessing and analysis. Developed with a strong foundation in statistical computing and reproducible research principles, *xcms* provides robust and flexible algorithms for key tasks such as peak detection, retention time correction, and feature alignment across samples. Its seamless integration with the broader R ecosystem allows for the entire workflow, from initial inspection to normalization and downstream statistical analysis, to be executed within a single, reproducible environment. Moreover, *xcms* is optimized to handle computationally intensive, big datasets, making it suitable for both exploratory and large-scale studies.

In detail, we will demonstrate a complete, reproducible LC–MS data preprocessingworkflow using *xcms*, illustrated with a small untargeted LC–MS metabolomics dataset. Special emphasis is placed on the best practices for data handling, parameter optimization, and quality control. By the end of this chapter, readers will have a strong foundation in using *xcms* to transform raw LC–MS data into structured, high-quality feature tables suitable for downstream statistical analysis and biological interpretation.

2 Material

This section provides an overview of the dataset, software environment, and workflow used for LC–MS(/MS) data preprocessing with the *xcms* R package. The entire analysis is designed to be reproducible and made publicly available.

2.1 Overview of the Workflow

To make the data processing steps transparent and reproducible, the analysis performed in this chapter is also implemented as a Quarto document (.qmd). Quarto enables seamless integration of code, narrative, and output, allowing readers to execute and modify the workflow directly (*see* **Note 1**). While any code editor may be used, RStudio offers built-in support for both R and Quarto and is a commonly used interface in data science.

As LC–MS preprocessing is strongly influenced by the specific setup used, there is no single set of parameters that fits all configurations; hence, settings for each preprocessing algorithm need to be defined and evaluated on the actual experimental data. This workflow shows examples to derive dataset-specific settings for each preprocessing step and tools to evaluate their impact. Depending on the data, it might be needed to test and evaluate different settings to adapt them to the experiment.

2.2 Dataset

The LC–MS data used in this chapter comefrom a small, publicly available untargeted metabolomics study. It can be found on the MassIVE repository (accession: MSV000099121) and can be downloaded from the archive for use in this tutorial.

Samples were analyzed using HILIC UPLC-QTOF-MS in positive ionization modes, as described in [6]. Raw data files are provided in the open and vendor-independent .mzML format (*see* **Note 2**).

The dataset consists of pooled human serum samples, which represent internal quality control (QC) samples, each spiked with a set of 15 compounds at two concentration levels: high and low. Each sample was measured in the positive ionization mode and as a technical triplicate to assess reproducibility (Table 1).

Additionally, two LC–MS/MS runs were acquired for one of the high-concentration samples using data-dependent acquisition (DDA) at two different collision energies (20 eV and 30 eV). These are included to support downstream MS/MS-based compound annotation (Table 1).

2.3 Software and R Packages

The data analysis was performed using R (version 4.5.0), a statistical programming language available for Windows and Unix (Linux and macOS) systems.

The core LC–MS(/MS) preprocessingworkflow is built upon the following R packages:

1. *Spectra* is used to import and handle the raw mass spectrometry data, providing efficient and flexible access to spectral information [7].
2. *MsExperiment* is employed to manage the experimental design and to link sample metadata with the corresponding raw data files [8].
3. *xcms* performs key preprocessing steps, including chromatographic peak detection, retention timealignment, and correspondence analysis [5].

Table 1
Experimental design of the small untargeted metabolomics dataset

mzML	Polarity	Time	Mode	Collision energy	Concentration	Sample name	Sample origin	Sample type
QC_LowIS_Mix13_1_POS.mzML	POS	2020-01-17 14:33:51	MS1		Low	Low_a	Serum	Pool QC
QC_HighIS_Mix13_3_POS.mzML	POS	2020-01-17 14:51:21	MS1		High	High_c	Serum	Pool QC
QC_LowIS_Mix13_3_POS.mzML	POS	2020-01-17 15:46:56	MS1		Low	Low_c	Serum	Pool QC
QC_HighIS_Mix13_2_POS.mzML	POS	2020-01-17 17:58:34	MS1		High	High_b	Serum	Pool QC
QC_LowIS_Mix13_2_POS.mzML	POS	2020-01-17 20:48:23	MS1		Low	Low_b	Serum	Pool QC
QC_HighIS_Mix13_1_POS.mzML	POS	2020-01-17 22:48:18	MS1		High	High_a	Serum	Pool QC
QC_HighIS_Mix13_CE20_POS.mzML	POS	2020-01-18 19:27:09	DDA	20	High	High_20	Serum	Pool QC
QC_HighIS_Mix13_CE30_POS.mzML	POS	2020-01-18 19:39:02	DDA	30	High	High_30	Serum	Pool QC

Complementary packages that enhance and support mainly metabolomics-specific analyses include:

1. *MetaboCoreUtils* offers core utility functions for metabolomics data processing, including mass-to-charge (*m/z*) and retention time calculations, adductannotation, and common data transformations used across multiple packages [9].
2. *SummarizedExperiment* provides Bioconductor's standardized container for storing assay data (e.g., feature intensities) alongside row (feature) and column (sample) metadata, facilitating consistent data handling and integration in metabolomicsworkflows [10].
3. *MsBackendMgf* implements support for reading and writing mass spectral data in the Mascot Generic Format (MGF), commonly used for storing MS/MS spectra in metabolomics and proteomicsannotation tasks [11].

All these packages are part of Bioconductor [12, 13] and can be installed using the *BiocManager* R package (*see* **Note 3**).

Additional R packages support the workflow by enabling data visualization (*RColorBrewer*, *pheatmap*), reading of metadata and tabular input files (*data.table*), and providing utility functions that enhance reproducibility and automation (*knitr*, *BiocStyle*).

2.4 Reproducible Analysis Workflow

To ensure complete reproducibility [14] of the present analysis, the workflow and the data are publicly available through GitHub and MassIVE, respectively. Users can either manually install all required dependencies or load and run a Docker image that contains the workflow, the data, and all required software, allowing interactive evaluation of the code locally on the user's computer.

Information on the workflow files, data, and installation instructions is available at https://github.com/jorainer/xcms_preprocessing_chapter.

3 Methods

Before the actual data processing, the experimental setup should be defined and the data files/samples organized. The experimental setup is best defined using a spreadsheet with one data file per row and columns containing the technical and phenotypical properties of each data file or sample (as shown in Table 1). Ideally, samples/files should be added in the order in which they were measured. For the present analysis, it is assumed that all MS data files are placed into a *data/mzML* folder, and the spreadsheet with sample annotation is stored in the *data* folder.

3.1 Data Import

In this section, general setup steps are performed, including loading the required R packages, configuring parallel processing, and specifying the data directory. The sample metadata and raw LC–MS data are then loaded and color-coded to facilitate clear and consistent visualization throughout the analysis.

1. Load all required R/Bioconductor packages for the present analysis (*see* **Note 3**).

```
#' Load all R packages
library(xcms)
library(MsExperiment)
library(Spectra)
library(RColorBrewer)
library(pheatmap)
library(MetaboCoreUtils)
library(SummarizedExperiment)
library(MsBackendMgf)
```

2. Configure parallel processing. Configure the number of cores, using the `CORES` variable. All data analysis functions in *xcms* have built-in parallel processing capabilities. In addition, to reduce memory load, *xcms* performs analyses in a chunk-wise manner, i.e., only data from a certain number of samples are loaded and processed at a time. This can be configured with the parameter `chunkSize =` in most data analysis functions. Importantly, the number of CPU cores used should be smaller than or equal to the value for this parameter. Below, we chose to use two CPU cores for parallel processing. R supports the multi-core parallel processing on Unix systems, while on Windows systems, a simple network of workstations (SNOW)-based parallel processing must be used.

```
#' Define the number of cores
CORES <- 2
#' Set up parallel processing using 2 cores
if (.Platform$OS.type == "unix") {
    register(MulticoreParam(CORES))
} else {
    register(SnowParam(CORES))
}
```

3. Load experimental setup/sample information. It is good practice to define the experimental setup using, e.g., a spreadsheet that contains the names of all the experiment's data files (eventually including their paths) and their respective samples, including all phenotypical information and technical/experimental properties. Ideally, the order of the samples should match their injection sequence. The content of the spreadsheet for the present experiment is shown in Table 1. Load

the experiment's spreadsheet using the `read.csv()` function from the data.table R package (*see* **Note 4**).

```
#' Import sample file
sample_data <- read.csv("data/standards_mzml.csv")
```

4. Create a data representation for the experiment. The `readMsExperiment()` function takes the names of the data files (including the path to the directory where they are stored, configured below using the `MZML_PATH` variable) and a `data.frame` with the sample information for the respective data files as input. The resulting `MsExperiment` object (variable `ms_data`) contains a reference to the MS data files and the sample information.

```
#' Define the path to the directory where the data
#' files can be found
MZML_PATH <- file.path("data/mzML/")
#' Linking the raw data to its respective metadata.
ms_data <- readMsExperiment(file.path(MZML_PATH,
                                      sample_data$mzML),
                                    sampleData = sample_data)
```

5. Check if the *data representation* is correct by printing the sample information table (Table 2). Sample information can be extracted using the sampleData() function.

```
#' View the sample information table
sampleData(ms_data)[, c("mzML", "sample_name",
                        "mode", "collision_energy",
                        "concentration")] |>
    kable(format = "pipe")
```

Table 2
Data files and samples

mzML	Sample name	Mode	Collision energy	Concentration
QC_LowIS_Mix13_1_POS.mzML	Low_a	MS1	NA	Low
QC_HighIS_Mix13_3_POS.mzML	High_c	MS1	NA	High
QC_LowIS_Mix13_3_POS.mzML	low_c	MS1	NA	Low
QC_HighIS_Mix13_2_POS.mzML	High_b	MS1	NA	High
QC_LowIS_Mix13_2_POS.mzML	Low_b	MS1	NA	Low
QC_HighIS_Mix13_1_POS.mzML	High_a	MS1	NA	High
QC_HighIS_Mix13_CE20_POS.mzML	High_20	DDA	20	High
QC_HighIS_Mix13_CE30_POS.mzML	High_30	DDA	30	High

6. Define colors to label individual or groups of samples (*see* **Note 5**).

```
#' Define colors by concentration levels: high and
#' low
col_concentration <- c(low = "#FF7F00", high =
"#A65628")
col_sample <- col_concentration
    [sampleData(ms_data)$concentration]
```

3.2 Initial Data Inspection

Before initiating the preprocessing of LC–MS data, it is essential to evaluate the overall data quality and define key parameters. In this section, we perform the visual inspection of the base peak chromatogram (BPC) and extracted-ion chromatograms (EICs), enabling the identification of poorly acquired samples, retention time regions with meaningful signal, and potential contaminants. We also verify whether the data were acquired in centroid mode, a requirement for peak detection, and extract signal characteristics such as chromatographic peak width and *m/z* mass deviation using the example compounds. An EIC visualizes the measured signal (intensity) for a particular ion of a compound along the retention time axis. It allows inspection of the number of available data points and the shape of the elution pattern. While for the present example we extract the EIC only for proline, a metabolite that is known to be present in human serum samples, for real experiments signals for multiple compounds or ions, ideally from retention time regions along the full retention time range, should be extracted evaluated. These initial steps help ensure that the data are suitable for automated processing and allow optimization of parameter settings for robust peak detection and alignment.

3.2.1 Visualize Base Peak Chromatogram

1. Create a base peak chromatogram of the full dataset and plot it (Fig. 1). The BPC is plotted to assess chromatographic drift or shifts, sample quality or outliers, and the presence of systematic noise or chemical contamination (e.g., polymer peaks). With parameter aggregationFun = "max," the highest intensity per spectrum is reported, while setting `aggregationFun = "sum"` would sum all intensities per spectrum, hence generating a total ion chromatogram (TIC).

```
#' Create a BPC
bpc <- chromatogram(ms_data, aggregationFun = "max",
chunkSize = CORES)

#' Plot the BPC
plot(bpc, col = paste0(col_sample, 80), main = "")
grid()
abline(v = c(15, 225), lty = 2)
legend("topright", col = col_concentration, lty = 1,
legend = names(col_concentration))
```

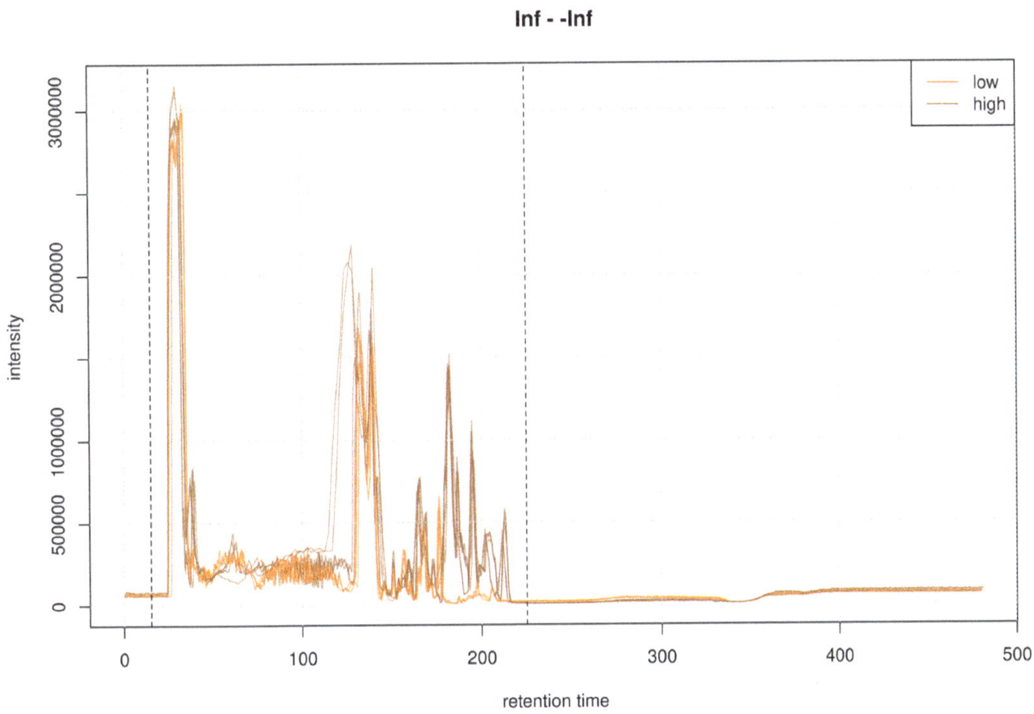

Fig. 1 Base peak chromatogram of the full dataset. Most signals are present between 15 and 225 s

3.2.2 Retention Time Filtering

2. To focus analysis on the region where meaningful compound separation and detection occur, filter the data to a specific retention time window, e.g., for the present dataset, between 15 and 225 s. This can be guided by known elution times of spiked or expected endogenous compounds.

```
#' Filter retention time
ms_data <- filterSpectra(ms_data, filterRt, c(15,
225))
```

3. Create a BPC of the retention time-filtered dataset for further evaluation under Sect. 3.3.2.

```
#' Extract the BPC again - for reference later.
bpc <- chromatogram(ms_data, aggregationFun =
    "max", chunkSize = CORES)
```

3.2.3 Extracted-Ion Chromatogram Selection

4. Determine the *m/z* range for proline. Starting from its chemical formula C5H9NO2, calculate the theoretical isotopic mass and subsequently the theoretical *m/z* for one of its ions. We use `"[M+H]+"` as the most likely adduct of proline in positive polarity, but could also specify any other possible adduct, such

as "[M+Na]+", "[M+NH4]+" etc. Create a range around the theoretical *m/z* by subtracting and adding a value of 0.01 Da. Depending on the accuracy and precision of the MS instrument, this value might be increased or decreased.

```
#' Calculate the mass of proline
proline_formula <- "C5H9NO2"
proline_mass <- calculateMass(proline_formula)
#' Calculate m/z for an adduct of proline
proline_mz <- mass2mz(proline_mass,c("[M+H]+"))[1, ]
proline_mzr <- cbind(mzmin = proline_mz - 0.01,
    mzmax = proline_mz + 0.01)
```

5. Extract and plot the EIC of proline (Fig. 2).

```
#' Extract the EIC for proline
proline_eic <- chromatogram
    (ms_data, mz = proline_mzr, rt = c(160, 200),
    chunkSize = CORES)
#' Plot the EIC for proline
plot(proline_eic, col = paste0(col_sample, 80),
    xlim = c(165, 175))
grid()
```

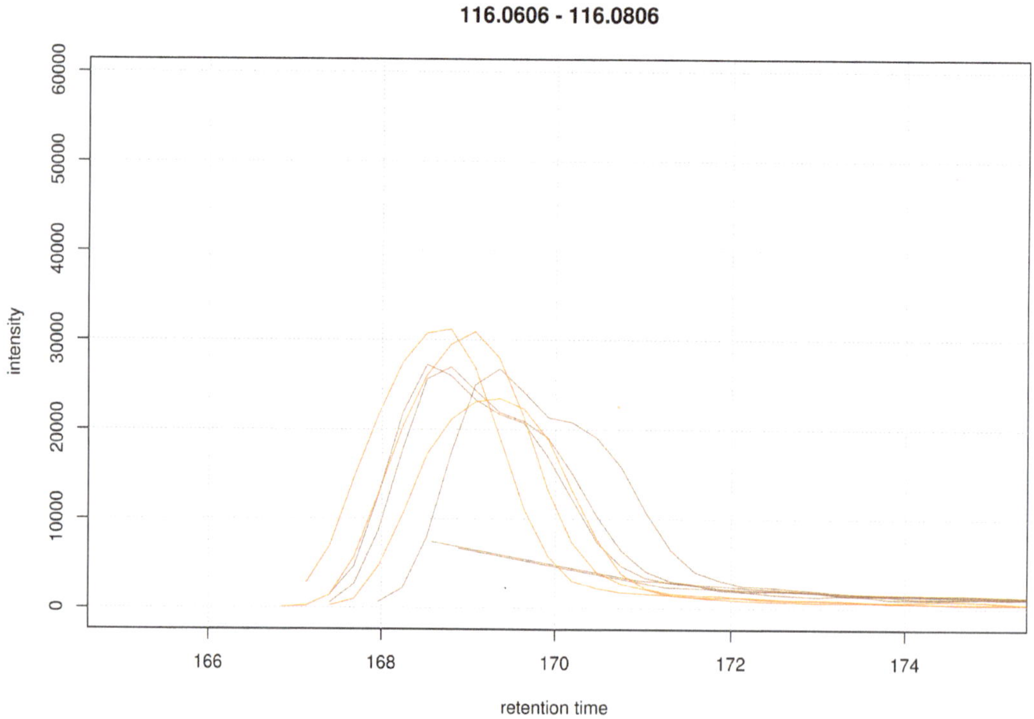

Fig. 2 EIC for proline

6. Repeat this for a minimum of ten compounds and save the EIC objects. Inspect molecules that you have spiked (e.g., internal standards), are known to be endogenously present (e.g., amino acids in serum), or select unknown compounds (*see* **Note 6**). Be aware to select compounds with retention time and *m/z* values representing the full chromatogram and mass spectrum range, respectively. If the expected retention time range is unknown, omit the parameter `rt` of the `chromatogram()` function and extract the ion trace for the full retention time range (Fig. 3).

```
#' Extract the EIC for an unknown compound
unknown_eic <- chromatogram(ms_data, mz =
    c(106.0399, 106.0599), chunkSize = CORES)
#' Plot the full EIC
plot(unknown_eic, col = paste0(col_sample, 80))
grid()
```

3.2.4 Check m/z Data Acquisition Mode

7. Evaluate whether data are acquired in centroid (*see* **Note 7**) mode in the first sample, using proline (Fig. 4). In addition, determine the variance in peaks' *m/z* along the retention time axis for individual ions. For better readability of the code, the R *pipe operator* (|>) is used to concatenate subsequent function calls.

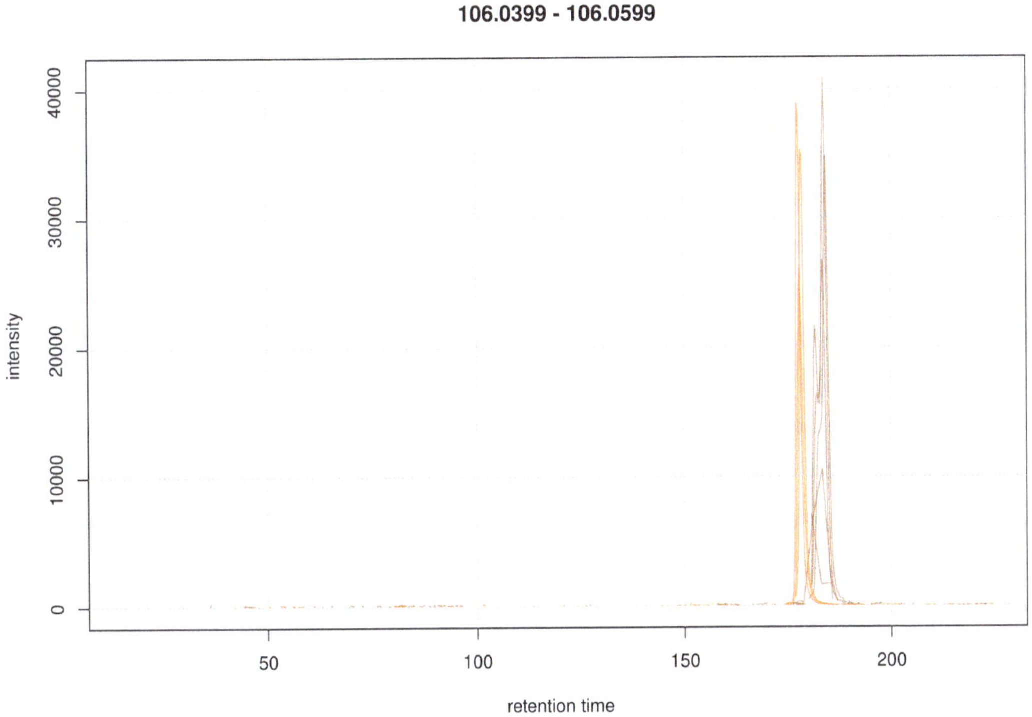

Fig. 3 Signal trace for ions with an *m/z* between 106.039 and 106.059

```
#' Check if data is in centroid mode
ms_data_proline <- ms_data[1] |>
filterSpectra(filterRt, c(167, 172)) |>
filterSpectra(filterMzRange, proline_mzr[1, ])
plot(ms_data_proline)
```

3.3 Data Preprocessing

In this section, we outline the essential steps of LC–MS data preprocessing using the *xcms* package. Chromatographic peak detection identifies relevant features in the raw MS data based on intensity patterns across retention time and mass-to-charge ratio. Retention time shifts between measurement runs are then adjusted in the retention time alignment step. The subsequent correspondence analysis groups chromatographic peaks from different samples that represent the same adduct or isotopic form of the same underlying compound, thus defining the *LC–MS features*. This grouping across samples relies on similarity of the chromatographic peak's *m/z* and retention time values and thus depends on a successful reduction of retention time shifts between the samples. Finally, gap filling recovers missing intensity values where peaks were not initially detected, producing a complete and consistent feature table ready for downstream statistical analysis.

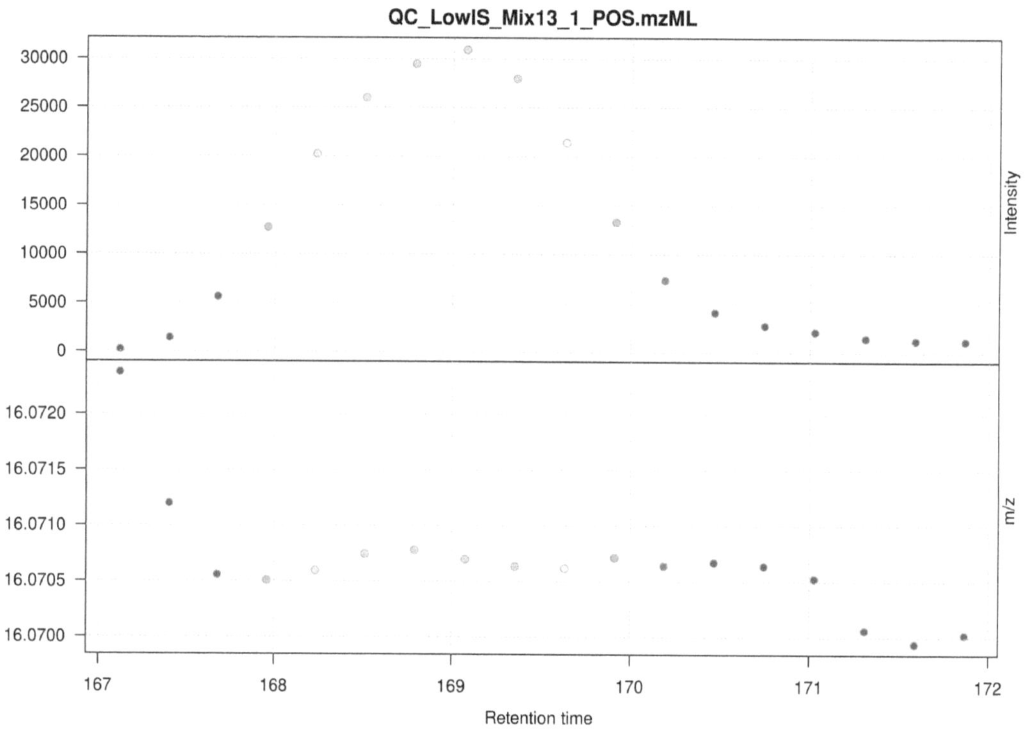

Fig. 4 All mass peaks reported within an *m/z* window from 116.07 to 116.072 and a retention time window from 167 to 172 s in one sample

The *xcms* package provides several different algorithms for each of these steps. In the present example, we use the most robust and commonly used methods. Also, importantly, settings for these algorithms need to be adapted to the analyzed dataset. These are best defined using visual inspection of the data and evaluation of the results, eventually requiring changing and optimizing the parameters in several iterations.

3.3.1 Chromatographic Peak Detection

We will use the *centWave* algorithm [15] for chromatographic peak detection. The most important parameters for this algorithm are `peakwidth`, which defines the expected range of chromatographic peak widths (in retention time dimension) in the dataset, and ppm, which defines the maximal acceptable deviation (in *m/z* dimension) of mass peaks for a certain ion in consecutive spectra. Other parameters for which different values than the default are used are `snthresh` (signal-to-noise threshold) and integrate. `snthresh` defines the trade-off between the number of false positive and false negative peaks. The parameter `integrate` allows to select the method to define the chromatographic peaks boundaries in the retention time dimension. The default `integrate = 1` assumes symmetric, Gaussian-shaped peaks, while `integrate = 2` uses a method that defines peak boundaries through local minima in peak intensities.

1. Define the `peakwidth` parameter. Inspect extracted chromatographic peaks of the ≥ 10 plotted EICs and determine the average observed width of the peaks in the retention time dimension. The rule of thumb is: choose an initial `peakwidth` between half of the observed peak width and up to twice the observed peak width (in seconds). Values 2 and 10 for the lower and upper expected peak width, respectively, seem to be appropriate for the present dataset. Evaluate the impact and result of this choice by applying the chromatographic peak detection on the EICs extracted in Sect. 3.2. The parameters for the *centWave* algorithm are configured with the `CentWaveParam()` function, and peak detection is performed using `findChromPeaks()`. Note that a very low value for `snthresh` is used for peak detection in EIC signals because of the lack of background signal to properly estimate noise. Repeat this analysis for all EICs. Values for the `peakwidth` parameter should be adjusted if the results do not match the expectations.

```
#' Define the peak detection settings
cwp <- CentWaveParam(peakwidth = c(2, 10),
    snthresh = 1,
    integrate = 2)
#' Perform peak detection for the proline EIC
proline_eic <- findChromPeaks(proline_eic, param =
    cwp)
```

2. Plot and visually inspect the chromatographic peak detection results of the EIC (Fig. 5).

```
#' Plot the proline EIC after peak detections
plot(proline_eic, peakCol = paste0(
    col_sample[chromPeaks(proline_eic)[, "column"]], 80),
    peakBg = paste0(
    col_sample[chromPeaks(proline_eic)[, "column"]], 40),
    col = col_sample, xlim = c(165, 175))
```

3. Calculate the average peak width in the retention time dimension. This value across all EICs can be used to evaluate whether the peak width matches the expectation from the LC setup.

```
#' Calculate the mean rt peak width
(chromPeaks(proline_eic)[, "rtmax"] -
chromPeaks(proline_eic)[, "rtmin"]) |>
mean()

[1] 5.744353
```

4. Define the *centWave* ppm parameter. Using the tentative MS signal for proline from Sect. 3.2.4 (i.e., the MS data from its approximate *m/z* and retention time window), the parts-per-million (ppm, *see* **Note 8**) deviation of the mass peaks' *m/z*

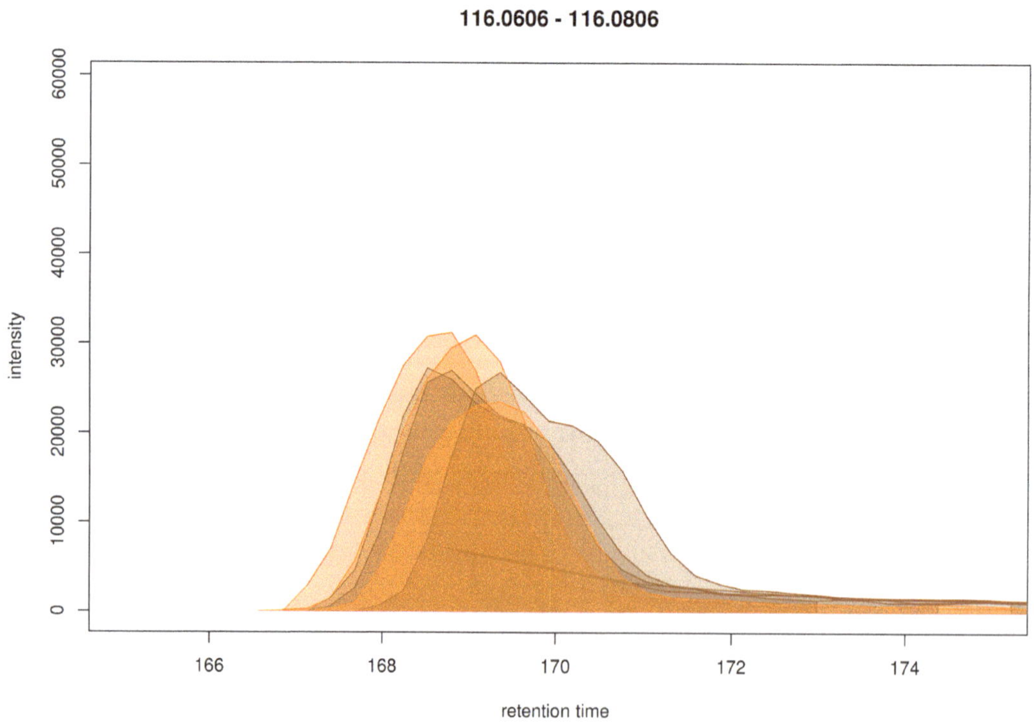

Fig. 5 EIC of proline with the area of identified chromatographic peaks indicated with the sample's color

values is calculated. This should be repeated for each of the ≥ 10 plotted EICs. Use the calculated values to define parameter ppm. Note that it is absolutely acceptable to use larger values for ppm than observed, in particular for data from time-of-flight (TOF) instruments, where *m/z* deviation generally increases with lower intensities. Thus, to ensure that chromatographic peak boundaries are correctly estimated, it can be beneficial to use higher than observed values for the parameter ppm. For the present dataset, we use ppm = 5.

```
#' Estimate m/z deviation for proline
mz_diff <- spectra(ms_data_proline) |>
    mz() |>
    unlist() |>
    diff()
ppm <- max(mz_diff) / proline_mz * 10^6
print(paste0("The expected ppm for proline is: ",
round(ppm, 3)))

[1] "The expected ppm for proline is: 1.279"
```

5. Perform chromatographic peak detection with the defined settings on the whole dataset. Set the parameter `snthresh` depending on whether only high-abundance, high-quality peaks should be detected (i.e., use a value ≥ 10), or also low-abundance peaks, accepting eventually false positive peaks. For `snthresh`, we use a value of 5 (instead of the default 10).

```
#' Define the peak detection settings
cwp <- CentWaveParam(peakwidth = c(2, 10),
    ppm = 5,
    snthresh = 5,
    integrate = 2)
#' Perform peak detection on the full data
ms_data <- findChromPeaks(ms_data, param = cwp,
    chunkSize = CORES)
```

6. Perform peak refinement (optional, but suggested for *centWave*). This step can help correct peak detection artifacts of the *centWave* algorithm. It should, however, be used with caution, as it may also introduce artificial corrections if not properly evaluated. By merging neighboring peaks, closely eluting or overlapping peaks that likely originate from the same compound are combined. This helps reduce redundancy and improves the accuracy of downstream analyses. Parameter `expandRt` should be set to a value close to, or smaller than, half of the average observed retention time width of chromatographic peaks, and for `ppm`, the same value as used for *centWave* peak detection can be used.

```
#' Define the peak refinement settings
mnpp <- MergeNeighboringPeaksParam(expandRt = 3,
    ppm = 5)
#' Perform peak refinement
ms_data <- refineChromPeaks(ms_data, param = mnpp,
    chunkSize = CORES)
Reduced from 70363 to 63236 chromatographic peaks.
```

7. Summaries of identified peaks: determine the number of identified peaks per sample (Table 3). Especially pay attention to the number of peaks per sample and to the quantiles of retention time widths and *m/z* widths.

```
#' View summary of identified peaks
data.frame(sample_name =
    sampleData(ms_data)$sample_name,
    peak_count = as.integer(table(chromPeaks
    (ms_data)[, "sample"]))) |>
    t() |>
    kable(format = "pipe")
```

sample_name	low_a	high_c	low_c	high_b	low_b	high_a	high_20	high_30
peak_count	9270	11950	9056	11847	9029	11724	188	172

8. Summaries of identified peaks: evaluate the *m/z* widths of identified chromatographic peaks. Calculate the distribution (quantiles) of the chromatographic peak widths in *m/z* dimension.

```
#' Distribution of m/z peak widths
(chromPeaks(ms_data)[, "mzmax"] -
```

Table 3
Summary of peak counts per sample

Sample name	Peak count
Low_a	927
High_c	1195
Low_c	905
High_b	1184
Low_b	902
High_a	1172
High_20	188
High_30	172

```
chromPeaks(ms_data)[, "mzmin"]) |>
    quantile()
0% 25% 50% 75% 100%
0.000000000 0.001056088 0.001863562 0.003020259
0.034988519
```

9. Summaries of identified peaks: evaluate the retention time widths of identified chromatographic peaks. Calculate the distribution (quantiles) of the chromatographic peak widths in the retention time dimension.

```
#' Distribution of retention time peak widths
(chromPeaks(ms_data)[, "rtmax"] -
    chromPeaks(ms_data)[, "rtmin"]) |>
    quantile()
0% 25% 50% 75% 100%
0.641 1.674 2.232 3.627 83.423
```

3.3.2 Retention Time Alignment

Retention timealignment aims at adjusting the data for retention time shifts between measurement runs (i.e., samples) [16]. We use the *peak groups* method from *xcms*. This algorithm performs a robust, non-linear alignment based on retention times of signals, the so-called *anchor peaks*, that are present in most samples. This requires an initial, loose grouping of chromatographic peaks across samples. This grouping of chromatographic peaks to features does not need to be perfect yet, i.e., the signal from different ions may be grouped into the same feature, as retention time alignment will correct for systematic shifts. Grouping collects peaks that represent the same compound across samples, even if retention times differ slightly. The *peak groups*alignment method can then select anchor peaks among these initial LC–MS features and align the samples by minimizing the retention time differences between the anchor peaks.

1. Perform an initial correspondence analysis using the *peak density* method. Define the sample groups of the experiment and exclude samples that should not be considered for peak grouping/alignment by assigning NA. In the present example, LC–MS/MS measurements are excluded. Set the parameter `minFraction` depending on the experimental setting. We use `minFraction = 2/3;` thus, to define a feature, a chromatographic peak has to be present in at least two out of the three samples per sample group. Parameters `binSize` and ppm depend on the resolution of the MS instrumentation and proper calibration across measurement runs. Chromatographic peaks (within and between samples) with a difference in *m/z* value smaller than `binSize` + ppm of the peaks' *m/z* will be grouped into a feature. Choices should depend on the maximal expected absolute difference in *m/z* values for signals of the same compound in different sample runs. The parameter ppm

allows to account for the *m/z*-dependent measurement error frequently observed in TOF-based instruments. The value for parameter bw depends on the observed difference in retention times between isomers as well as retention time shifts between samples. For this initial correspondence analysis, it should be set to a larger value to allow/account for retention time differences between samples. A value close to or larger than the average observed retention time width of detected chromatographic peaks could be a good initial choice.

```
#' Define sample groups and set samples that should
#' not be considered to `NA`
grp <- sampleData(ms_data)$concentration
grp[sampleData(ms_data)$mode == "DDA"] <- NA
#' Define the settings for the initial
#' correspondence
pdp <- PeakDensityParam(sampleGroups = grp,
 minFraction = 2 / 3,
 binSize = 0.02,
 ppm = 10,
 bw = 4)
#' Perform the initial correspondence analysis
ms_data <- groupChromPeaks(ms_data, param = pdp)
```

2. Perform retention timealignment to correct systematic shifts between samples using the *peak groups* method. Configure the algorithm using `PeakGroupsParam()`. Parameters: `minFraction`: minimum fraction of samples that must share an anchor peak for alignment. This is relative to the total number of samples, not the number of samples per sample group (as above). Choices around 0.8 (presence of an anchor peak in 80% of samples) work for most experiments. We use a value of 0.9 as the same sample was repeatedly measured. `extraPeaks`: number of extra peaks per feature considered for robust alignment (optimally use a value similar to the number of samples in the experiment). `span`: controls the smoothness of the alignment curve. Must be a value between 0 and 1, smaller values fit more local variation, a value of 1 would perform a constant shift. A value between 0.4 and 0.7 works in most cases. Parameter `subset` allows to configure an alignment that is based only on a `subset` of samples (such as, e.g., quality control (QC) samples). Set subset to the indices of such QC samples or to the indices of the samples on which the initial correspondence was performed.

```
#' Define retention time alignment parameters
pgp <- PeakGroupsParam(minFraction = 0.9,
    extraPeaks = length(ms_data),
    span = 0.4,
```

```
    subset = which(!is.na(grp)))

#' Perform retention time alignment
ms_data <- adjustRtime(ms_data, param = pgp,
    chunkSize = CORES)
Performing retention time correction using 3395
peak groups.


Aligning sample number 7 against subset ... OK
Aligning sample number 8 against subset ... OK
```

3. Evaluate alignment result. Plot the adjusted retention times (x-axis) against the difference between raw and adjusted retention times (y-axis) for each sample as a line (Fig. 6). Deviations observed on the y-axis should meet the expectations from the variance generally observed on the employed LC setup. Data points (i.e., anchor peaks) should span most of the retention time range. If needed, adapt parameters and repeat the initial correspondence and alignment.

```
#' Plot the adjusted rt vs. rt differences before
#' and after correction
```

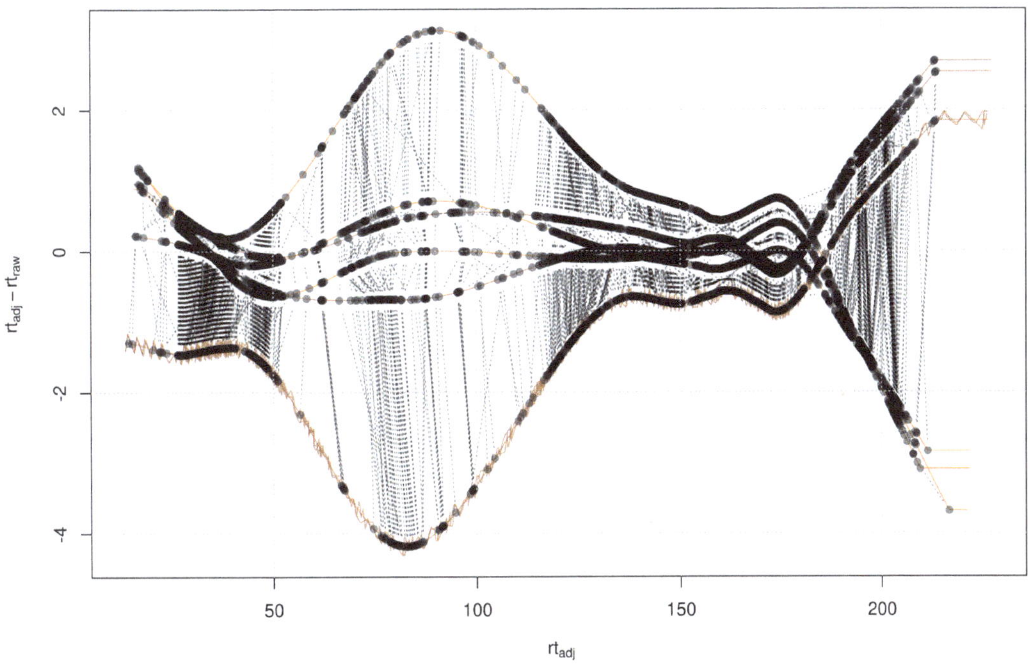

Fig. 6 Visualization of retention time alignment results. The largest retention time differences accounted for by the alignment are approximately 6 s, observed between two samples. For all other samples, differences are below 1 s. Strongest adjustments were performed in the retention time region between 50 and 100 s.

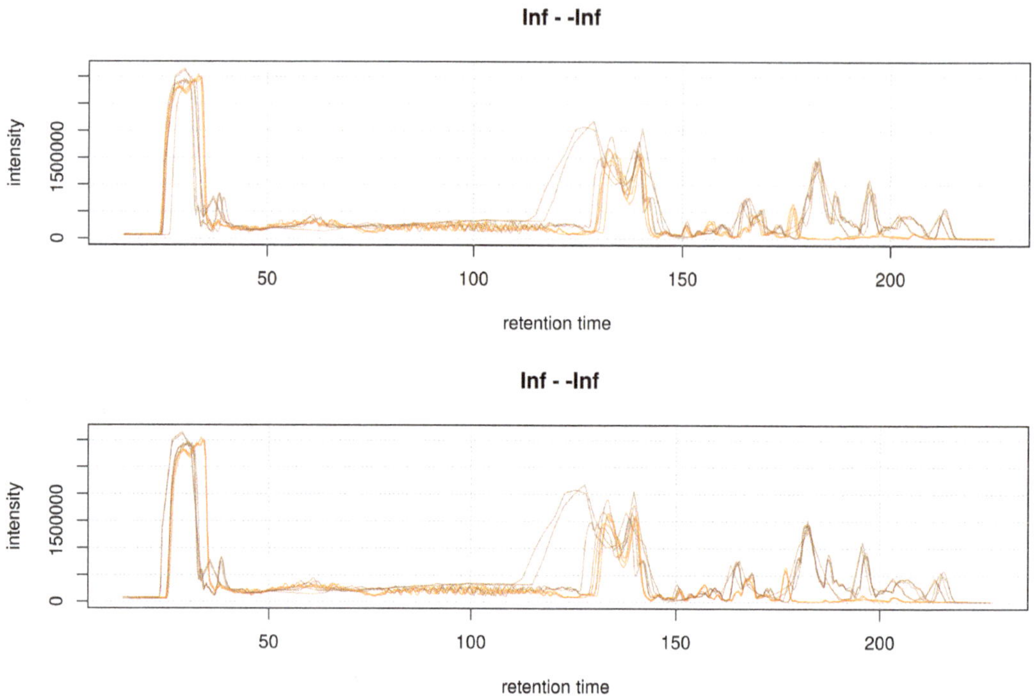

Fig. 7 BPC before and after retention time alignment. A small improvement can be observed

```
plotAdjustedRtime(ms_data, col = paste0(col_sample,
80))
grid()
```

4. Create the BPC after retention timealignment to visually compare with the original BPC (Fig. 7).

```
#' Create the BPC after rt alignment
bpc_adj <- chromatogram(ms_data, chromPeaks =
    "none", aggregationFun = "max",
    chunkSize = CORES)
#' Plot before and after rt alignment
par(mfrow = c(2, 1))
plot(bpc, col = paste0(col_sample, 60), main = "raw
data")
grid()
plot(bpc_adj, col = paste0(col_sample, 60),
    main = "aligned data")
grid()
```

5. Evaluate improved retention timealignment on each EIC (Fig. 8).

```
#' Extract the EIC for proline with corrected rt
```

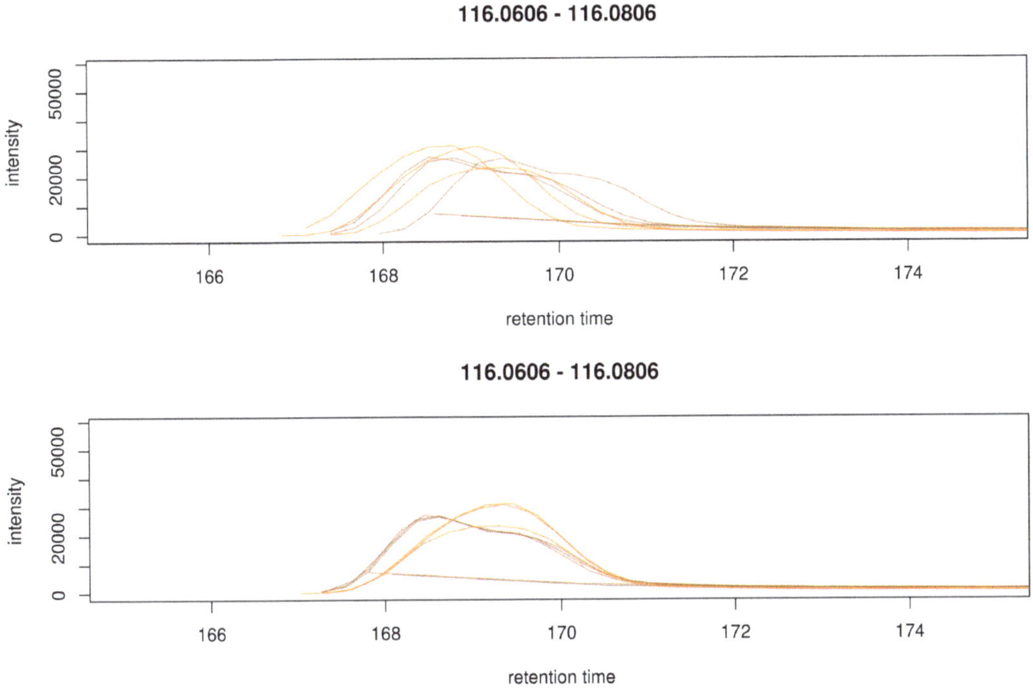

Fig. 8 Alignment results evaluated for signal of proline. Shown is the EIC for proline before (upper panel) and after (lower panel) alignment. Retention time alignment reduced observed retention time shifts considerably

```
proline_eic_adj <- chromatogram(ms_data, mz =
    proline_mzr,
    chunkSize = CORES)
Processing chromatographic peaks
#' Plot before and after rt alignment for proline
#' EIC
par(mfrow = c(2, 1))
plot(proline_eic[1, ], col = paste0(col_sample,
80),
peakType = "none", xlim = c(165, 175))
grid()
plot(proline_eic_adj[1, ], col = paste0(col_sample,
80),
peakType = "none", xlim = c(165, 175))
grid()
```

6. Apply the alignment results to the dataset.

```
#' Apply the corrected rt to data set
ms_data <- applyAdjustedRtime(ms_data)
```

3.3.3 Correspondence Analysis

Correspondence analysis groups signals (chromatographic peaks) across samples that are presumably from the same original ion. This grouping depends on the similarity of the peaks' *m/z* and retention time values.

1. Derive settings for correspondence analysis. Extract the signal trace for the *m/z* of EICs defined in Sect. 3.2.3. Define parameters for the correspondence: values for `binSize` and `ppm` can be adopted from the initial correspondence analysis in Sect. 3.3.2. The parameter `minFraction` should be adapted to match the experimental setup. For `bw`, use the same value as, or a smaller value than, that used for the initial correspondence. Generally, smaller values are used to perform stricter grouping. Plot the resulting EIC and simulate a correspondence analysis with the defined configuration (Fig. 9).

```
#' Extract EIC for the full rt range for EICs' m/z
#' ranges
eic <- chromatogram(ms_data, mz = proline_mzr,
    chunkSize = CORES)
Processing chromatographic peaks
#' Configure correspondence analysis
pdp <- PeakDensityParam(sampleGroups = grp,
    minFraction = 2 / 3,
    binSize = 0.02,
    ppm = 10,
    bw = 2)
#' Plot EIC and simulate correspondence
plotChromPeakDensity(eic, param = pdp, col =
paste0(col_sample, 80))
grid()
```

2. Evaluate the resulting plot. Are chromatographic peaks grouped properly into features? Chromatographic peaks, or more precisely the retention time of their apex, grouped into the same feature are indicated with gray rectangles in the lower part of the plot. Are co- or closely eluting compounds separated into distinct features? Evaluate the results on all EICs defined in Section 3.2.3. Eventually, change one or more parameters of `PeakDensityParam()` and repeat until results are satisfactory. Eventually, reduce the value for bw to avoid grouping of closely eluting, but distinct, compounds into the same features.
3. Perform the correspondence analysis on the full dataset (Fig. 10).

```
#' Final correspondence
ms_data <- groupChromPeaks(ms_data, param = pdp)
```

4. Evaluate results on EICs. Extract ion traces for the m/z ranges of EICs from Sect. 3.2.3 and plot the correspondence results

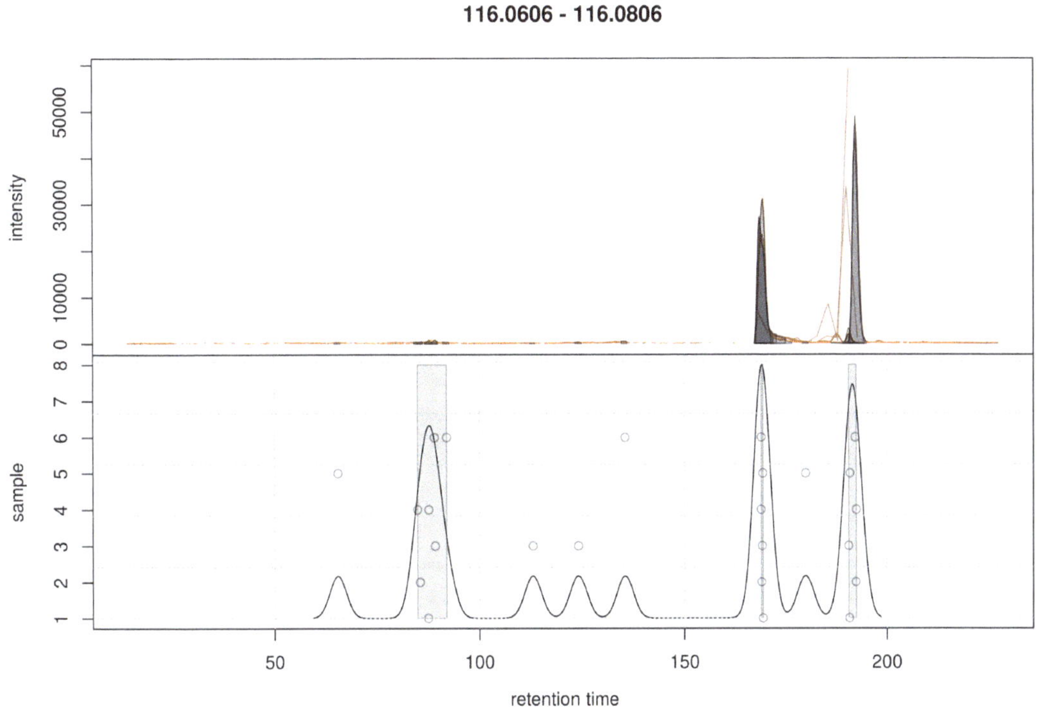

Fig. 9 Visualization of peak grouping using the *peak density* method. Upper panel: signal for ions with an *m/z* between 116.06 and 116.08 along the full retention time range. Lower panel: retention time of identified chromatographic peak apexes (x-axis) in the different samples of the experiment (y-axis). The solid black line represents the *peak density* curve along the retention time. Chromatographic peaks within the same density peak are grouped into a feature. Chromatographic peaks grouped into the same feature are indicated with a gray rectangle

using plotChromPeakDensity(). Use the parameter simulate = FALSE to show the actual correspondence results. Repeat for all m/z ranges from Sect. 3.2.3.

```
#' Extract EIC for the full rt range for EICs' m/z
#' ranges
eic <- chromatogram(ms_data, mz = proline_mzr,
    chunkSize = CORES)
#' Plot EIC and simulate correspondence
plotChromPeakDensity(eic, col = paste0(col_sample,
80),
simulate = FALSE)
grid()
```

5. Visualize the features's *m/z* and retention times in the full *m/z* and retention time space of the dataset (Fig. 11). To avoid loading the *m/z* values of all mass peaks, we manually define the *m/z* range based on the MS instrument setup. Use ylim =

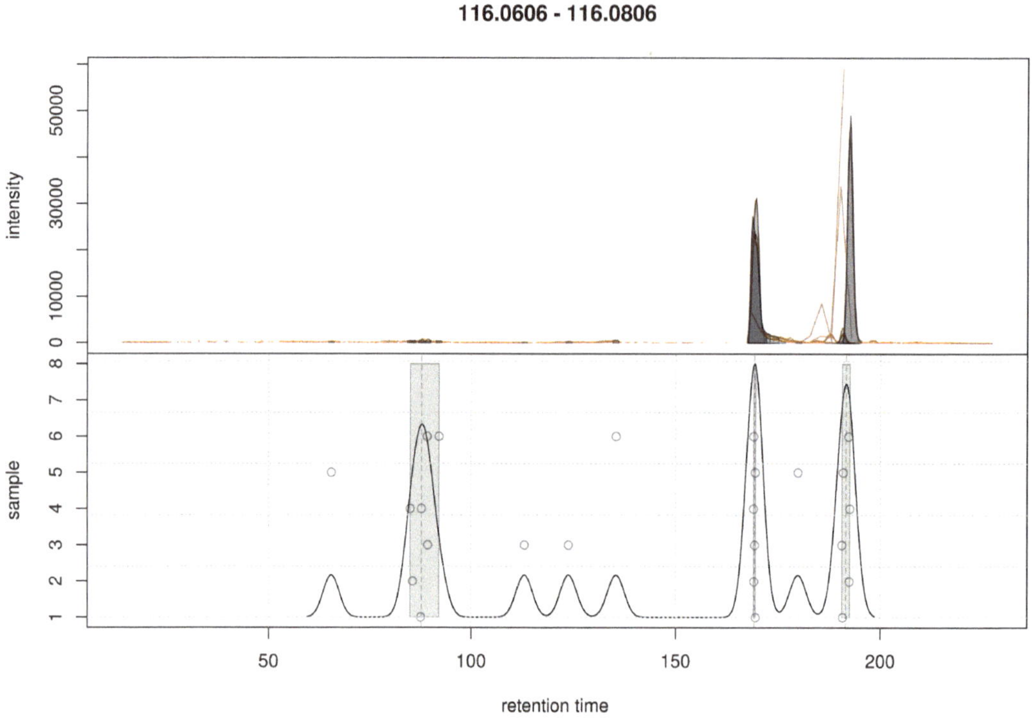

Fig. 10 Correspondence analysis results for the *m/z* window containing signal of proline.

range(unlist(mz(spectra(ms_data)))) to calculate it from the dataset.

```
#' Plot m/z vs. rt for the data set
plot(NA, NA, xlim = range(rtime(spectra(ms_data))),


    ylim = c(0, 1000),
    xlab = "retention time", ylab = "m/z")
grid()
points(featureDefinitions(ms_data)$rtmed,
featureDefinitions(ms_data)$mzmed,
pch = 21, col = "#00000060", bg = "#00000020")
```

3.3.4 Gap Filling

The feature abundance matrix reports values for identified chromatographic peaks. This matrix can contain a considerable number of missing values depending on whether a chromatographic peak was detected for a particular feature in a sample or not. The gap-filling step aims at reducing the number of missing values by integrating all raw MS signals from the feature's expected *m/z* and retention time window in samples in which no chromatographic peak was detected.

1. Determine the percentage of missing values in the feature abundance matrix.

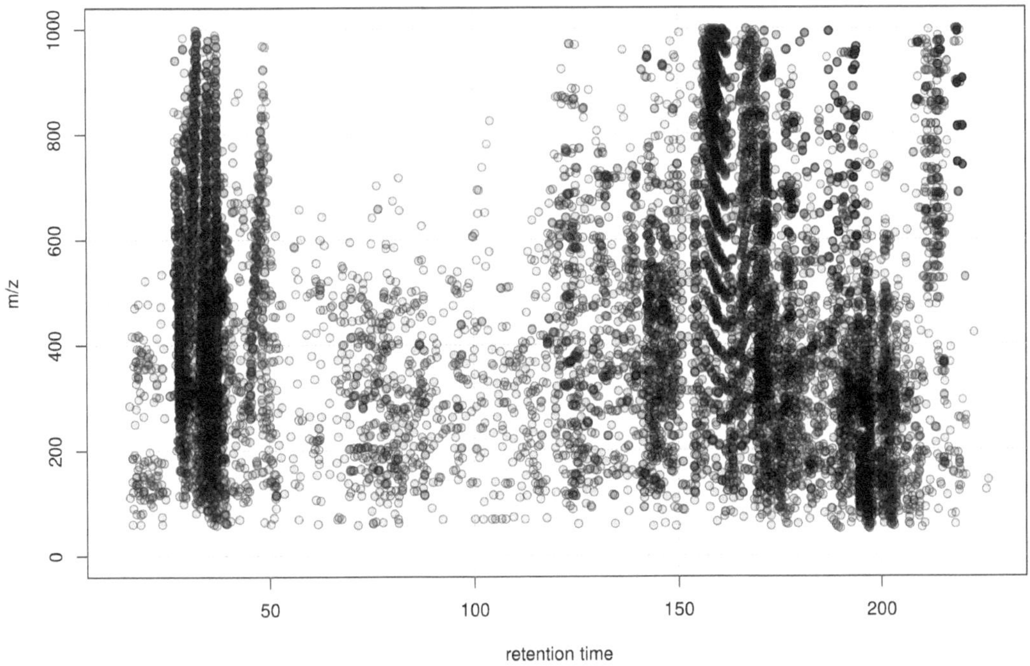

Fig. 11 Dotplot of *m/z* versus retention time values of this dataset.

```
total_feat <- length(featureValues(ms_data))
miss_feat <- sum(is.na(featureValues(ms_data)))
#' Calculate the percentage of missing values in
#' the data set
miss_feat / total_feat * 100
[1] 51.8038
```

2. Perform the gap filling.

```
#' Perform gap filling
ms_data <- fillChromPeaks(ms_data,
    ChromPeakAreaParam(),
    chunkSize = CORES)
```

3. Determine the number of missing values after gap filling.

```
#' Get the number of missing values after gap
#' filling
miss_feat <- sum(is.na(featureValues(ms_data)))
#' Calculate the percentage of missing values in
#' the data set
miss_feat / total_feat * 100
[1] 28.40485
```

3.4 Post-processing and Data Export

Following the preprocessing, the resulting feature table represents the foundation for downstream statistical analysis and biological interpretation. This post-processing phase typically involves several key steps to refine and analyze the data further. These include normalization of feature intensities across samples to correct for systematic variation, removal of outlier features that may represent technical artifacts, and assessment of sample variance to identify trends or inconsistencies.

Subsequently, differential abundance analysis is performed to identify metabolic features that distinguish phenotypes or experimental conditions of interest. Multivariate statistical methods, such as principal component analysis (PCA), hierarchical clustering, (orthogonal) partial least squares discriminant analysis ((O)PLS-DA), linear regression-based approaches, or machine learning algorithms are often employed to visualize global patterns and highlight relevant features.

Finally, metabolite annotation of significant or interesting features is a critical step for biological interpretation, enabling insights into the metabolic pathway or processes involved. The exact post-processing workflow may vary depending on the research question, experimental design, and biological matrix.

For practical code examples and guidance on the post-processing, we refer the reader to the many example workflows of the Metabonaut resource [17], which provides detailed and modular R-based workflows for each component of post-processing.

After completing preprocessing, and optionally some further data filtering and/or normalization, the feature table can be exported for downstream analysis, sharing, or archiving. This step ensures reproducibility and allows interoperability with other software tools for statistical analysis or metabolite identification. Below, we outline key procedures to export the processed data, including both the feature intensity matrix and the features' MS/MS spectra.

1. Export feature table as a `SummarizedExperiment` object (*see* **Note 9**). The parameter `filled = FALSE` extracts only abundances of detected chromatographic peaks, but no gap-filled values.

```
#' Extract feature intensity results as a
SummarizedExperiment
res <- quantify(ms_data, method = "sum", filled =
    FALSE)
#' View the summary of the result object
res
class: SummarizedExperiment
dim: 13132 8
metadata(6): '' '' ... '' ''
assays(1): raw
rownames(13132): FT00001 FT00002 ... FT13131
FT13132
rowData names(10): mzmed mzmin ... low ms_level
```

```
colnames(8): QC_LowIS_Mix13_1_POS.mzML QC_HighIS_
Mix13_3_POS.mzML ...


QC_HighIS_Mix13_CE20_POS.mzML QC_HighIS_Mix13_CE30_
POS.mzML
colData names(11): mzML polarity ... X
spectraOrigin
```

2. Add a second assay containing missing-value-filled data to the `SummarizedExperiment` object. The additional assay contains filled intensity values (i.e., using gap-filling methods for missing peaks), so it can be used in downstream imputation-aware analysis.

```
#' Add a second assay named 'raw_filled' with gap-
#' filled intensities
assays(res)$raw_filled <- featureValues(ms_data,
    method = "sum",
    filled = TRUE )
#' View different assays in the SummarizedExperiment
#' object
assayNames(res)
[1] "raw" "raw_filled"
#' View the top of the filled feature intensity
#' matrix
assay(res, "raw_filled") |> head()
QC_LowIS_Mix13_1_POS.mzML QC_HighIS_Mix13_3_POS.
mzMLFT00001 1875.43619 790.0453
FT00002 22.48469 6331.4319
FT00003 273.61998 151.5588
FT00004 262.42877 NA
FT00005 NA 523.0190
FT00006 6192.42353 6119.8266
QC_LowIS_Mix13_3_POS.mzML QC_HighIS_Mix13_2_POS.
mzML
FT00001 1174.224766 963.4551
FT00002 5.983196 5514.1061
FT00003 191.085146 699.5291
FT00004 350.285451 NA
FT00005 NA 804.6854
FT00006 4453.887713 5858.3308
QC_LowIS_Mix13_2_POS.mzML QC_HighIS_Mix13_1_POS.
mzML
FT00001 736.556377 1624.94089
FT00002 8.051912 2952.14927
FT00003 201.881637 1031.20904
FT00004 313.886706 30.46919
FT00005 NA 260.96907
FT00006 2450.053783 3310.73301
QC_HighIS_Mix13_CE20_POS.mzML QC_HighIS_Mix13_CE30_
POS.mzML
FT00001 NA NA
```

```
FT00002 8947.748 NA
FT00003 4533.783 NA
FT00004 NA NA
FT00005 NA NA
FT00006 NA NA
```

3. Export the feature abundance matrix to a spreadsheet. The ID of the features and their *m/z* and retention time are added as additional columns to this feature abundance matrix. For ease of data sharing or as input into external tools (e.g., Excel, Python), the feature intensity matrix can be exported as a .csv file.

```
#' Compile a data.frame with feature definitions and
#' abundances
fd_data <- data.frame(
    feature_id = rownames(rowData(res)),
    rowData(res)[, c("mzmed", "rtmed")],
    assay(res, "raw_filled")
    )
#' Write the filled feature intensity matrix to a
#' CSV file
write.csv(fd_data,
    file = "data/standards_abundance.csv")
```

4. Export the MS/MS spectra in MGF format. To support metabolite annotation in external tools (e.g., GNPS, MS-DIAL, SIRIUS, or MzMine), the MS/MS spectra associated with detected MS features are exported to the Mascot Generic Format (MGF). The `MsBackendMgf()` backend function is used here to write the compatible files. To allow association between the exported feature abundances and the MS/MS spectra, the MGF file contains a field *feature_id* containing the identifier of the feature for each MS/MS spectrum.

```
#' Extract the MS/MS spectra for each feature
sp <- featureSpectra(ms_data)
#' Export the MS/MS spectra in .mgf format
export(sp, MsBackendMgf(), file = "data/std_spectra.
mgf")
```

3.5 Session Information

1. Print the R session info in the rendered Quarto document (*see* **Note 10**).

```
#' View R session info sessionInfo()
R version 4.5.0 (2025-04-11 ucrt) Platform:
x86_64-w64-mingw32/x64
Running under: Windows 11 x64 (build 22631)
Matrix products: default LAPACK version 3.12.1
locale:
[1] LC_COLLATE=English_United States.utf8
[2] LC_CTYPE=English_United States.utf8
```

```
[3] LC_MONETARY=English_United States.utf8
[4] LC_NUMERIC=C
[5] LC_TIME=English_United States.utf8
time zone: Europe/Rome
tzcode source: internal
attached base packages:
[1] stats4 stats graphics grDevices utils datasets
methods
[8] base
other attached packages:
[1] MsBackendMgf_1.16.0 SummarizedExperiment_1.38.1
[3] Biobase_2.68.0 GenomicRanges_1.60.0
[5] GenomeInfoDb_1.44.0 IRanges_2.42.0
[7] MatrixGenerics_1.20.0 matrixStats_1.5.0
[9] MetaboCoreUtils_1.17.1 pheatmap_1.0.12
[11] RColorBrewer_1.1-3 Spectra_1.18.2
[13] S4Vectors_0.46.0 BiocGenerics_0.54.0
[15] generics_0.1.4 MsExperiment_1.10.0
[17] ProtGenerics_1.40.0 xcms_4.6.0
[19] BiocParallel_1.42.1 BiocStyle_2.36.0
[21] knitr_1.50
loaded via a namespace (and not attached):
[1] DBI_1.2.3 rlang_1.1.6
[3] magrittr_2.0.3 clue_0.3-66
[5] MassSpecWavelet_1.74.0 compiler_4.5.0
[7] vctrs_0.6.5 reshape2_1.4.4
[9] stringr_1.5.1 pkgconfig_2.0.3
[11] crayon_1.5.3 fastmap_1.2.0
[13] XVector_0.48.0 rmarkdown_2.29
[15] UCSC.utils_1.4.0 preprocessCore_1.70.0
[17] purrr_1.0.4 xfun_0.52
[19] MultiAssayExperiment_1.34.0 jsonlite_2.0.0
[21] progress_1.2.3 DelayedArray_0.34.1
[23] prettyunits_1.2.0 parallel_4.5.0
[25] cluster_2.1.8.1 R6_2.6.1
[27] stringi_1.8.7 limma_3.64.1
[29] Rcpp_1.0.14 iterators_1.0.14
[31] snow_0.4-4 BiocBaseUtils_1.10.0
[33] Matrix_1.7-3 igraph_2.1.4
[35] tidyselect_1.2.1 rstudioapi_0.17.1
[37] abind_1.4-8 yaml_2.3.10
[39] doParallel_1.0.17 codetools_0.2-20
[41] affy_1.86.0 lattice_0.22-7
[43] tibble_3.2.1 plyr_1.8.9
[45] evaluate_1.0.3 pillar_1.10.2
[47] affyio_1.78.0 BiocManager_1.30.25
[49] foreach_1.5.2 MSnbase_2.34.1
[51] MALDIquant_1.22.3 ncdf4_1.24
[53] hms_1.1.3 ggplot2_3.5.2
[55] scales_1.4.0 glue_1.8.0
[57] MsFeatures_1.16.0 lazyeval_0.2.2
[59] tools_4.5.0 mzID_1.46.0
[61] QFeatures_1.18.0 vsn_3.76.0
[63] mzR_2.42.0 fs_1.6.6
```

```
[65] XML_3.99-0.18 grid_4.5.0
[67] impute_1.82.0 tidyr_1.3.1
[69] MsCoreUtils_1.20.0 colorspace_2.1-1
[71] GenomeInfoDbData_1.2.14 PSMatch_1.12.0
[73] cli_3.6.5 S4Arrays_1.8.1
[75] dplyr_1.1.4 AnnotationFilter_1.32.0
[77] pcaMethods_2.0.0 gtable_0.3.6
[79] digest_0.6.37 SparseArray_1.8.0
[81] htmlwidgets_1.6.4 farver_2.1.2
[83] htmltools_0.5.8.1 lifecycle_1.0.4
[85] httr_1.4.7 statmod_1.5.0
[87] MASS_7.3-65
```

4 Notes

1. Writing workflow processes using Quarto documents (.qmd) supports literate programming [18] by integrating code, text, and outputs in a single file. Quarto enables the creation of fully reproducible documents by embedding R code directly within narrative text, ensuring that results are dynamically generated from the most up-to-date data. This approach facilitates transparent, data-specific analysis workflows that are easy to audit, share, and update. By combining documentation and computation, Quarto helps streamline reporting, improve collaboration, and enhance the reproducibility of scientific and statistical analyses.
2. The workflow starts from vendor-neutral raw MS files, already converted from vendor-specific proprietary raw file formats (e.g., Thermo Fisher, Agilent, Bruker, Waters) to the open standard for MS data, i.e., the mzML format. Conversion of vendor-specific raw MS files is performed using *MSConvert*, a powerful and flexible tool from *ProteoWizard*. Using mzML as a standard format ensures that your MS data can be analyzed in a reproducible, tool-independent way.
3. R packages can be installed from the Comprehensive R Archive Network (CRAN) or from Bioconductor using the following commands:

```
#' Install CRAN packages
install.packages("knitr")
#' Install Bioconductor packages
install.packages("BiocManager")
BiocManager::install("BiocStyle")
```

4. The user can also read Excel (.xlsx) files using the `read_xlsx()` function from the *readxl* R package. However, the

resulting object should be converted into a `data.frame` using as.`data.frame()` before beginning the analysis. Alternatively, the same tabular data can be created directly in R as a `data.frame`, but it is usually more practical and user-friendly to define and load the spreadsheet using a spreadsheet program.

5. For improved visualization, it is recommended to assign specific colors to individual samples as well as to sample groups. These color settings can be reused in all plots during the analysis. Suggested color-coding schemes include: acquisition batch, injection order, sample type (e.g., QC, spiked, biological), and biological phenotypes of interest (e.g., sex, case-control status). These visual cues aid in assessing data quality and identifying potential systematic variation.
6. When no prior information is available about expected compounds, select high-intensity peaks from the BPC, extract full MS1 scans, and evaluate EICs to derive retention time and *m/z* parameters empirically.
7. Accurate peak detection, in particular with the *centWave* algorithm, requires centroided data. MS instruments allow to acquire data in profile or centroid mode. For profile mode, all measured mass peaks in a spectrum are reported, while in centroid mode, the data are reduced to a single representative mass peak for the signal of one ion in a spectrum. If your data are not centroided, use *MSConvert* or a similar tool to convert profile-mode data to centroid mode before importing it into R.
8. The *m/z* mass deviation is expressed as ppm and not in *m/z* difference (in Da) because the variation scales with the molecular mass of the compound. It is important to estimate the ppm also on large compounds, as the ppm increases with larger compounds. For example, if your highest *m/z* detected is 1000 Da, and you have an estimated error of 5 ppm, the largest possible *m/z* deviation allowed is 1000 ± 0.005 Da (*m/z* mass deviation = $1000 \text{ Da} \times 5 \text{ ppm}/10^6$), ranging from 999.995 to 1000.005 Da.
9. The `SummarizedExperiment` container is a robust Bioconductor class that holds both assay data (e.g., intensity values) and associated metadata (e.g., feature and sample annotations in its `rowData()` and `colData()`, respectively). It provides a standardized format for integrating omics data and simplifies downstream manipulation or analysis.
10. Including a call to `sessionInfo()` at the end of the workflow is important for ensuring reproducibility, as it documents the versions of R and all loaded packages, as well as system-specific details that may influence results.

5 Funding

PL is funded by an European Union HORIZON-MSCA-2021 project, Grant No. 101073062: HUMAN—Harmonizing and Unifying Blood Metabolic Analysis Networks.

References

1. Lindahl A, Sääf S, Lehtiö J et al (2017) Tuning metabolome coverage in reversed phase LC–MS metabolomics of MeOH extracted samples using the reconstitution solvent composition. Anal Chem 89:7356–7364
2. Liu Q, Walker D, Uppal K et al (2020) Addressing the batch effect issue for LC/MS metabolomics data in data preprocessing. Sci Rep 10: 13856
3. Fang N, Yu S, Ronis MJ et al (2015) Matrix effects break the LC behavior rule for analytes in LC-MS/MS analysis of biological samples. Exp Biol Med 240:488–497
4. Vereyken L, Dillen L, Vreeken RJ et al (2019) High-resolution mass spectrometry quantification: impact of differences in data processing of centroid and continuum data. J Am Soc Mass Spectrom 30:203–212
5. Louail P, Brunius C, Garcia-Aloy M et al (2025) xcms in peak form: now anchoring a complete metabolomics data preprocessing and analysis software ecosystem. Anal Chem 97:27639–27645. https://doi.org/10.1021/acs.analchem.5c04338
6. Volani C, Malfertheiner C, Caprioli G et al (2023) VAMS-based blood capillary sampling for mass spectrometry-based human metabolomics studies. Metabolites 13:146
7. Rainer J, Louail P, Stanstrup J et al (2025) Spectra: infrastructure for mass spectrometry data. https://bioconductor.org/packages/Spectra. https://doi.org/10.18129/B9.bioc.Spectra
8. Gatto L, Borman T, Gibb S et al (2025) MsExperiment: infrastructure for mass spectrometry experiments. https://bioconductor.org/packages/MsExperiment. https://doi.org/10.18129/B9.bioc.MsExperiment
9. Rainer J, Vicini A, Salzer L et al (2022) A modular and expandable ecosystem for metabolomics data annotation in R. Metabolites 12:173
10. Morgan M, Obenchain V, Hester J et al (2025) SummarizedExperiment: a container (S4 class) for matrix-like assays. https://bioconductor.org/packages/SummarizedExperiment. https://doi.org/10.18129/B9.bioc.SummarizedExperiment
11. Gatto L, Witting M, Rutz A et al (2025) MsBackendMgf: mass spectrometry data backend for mascot generic format (mgf) files. https://bioconductor.org/packages/MsBackendMgf. https://doi.org/10.18129/B9.bioc.MsBackendMgf
12. Gentleman RC, Carey VJ, Bates DM et al (2004) Bioconductor: open software development for computational biology and bioinformatics. Genome Biol 5: R80
13. Huber W, Carey VJ, Gentleman R et al (2015) Orchestrating high-throughput genomic analysis with Bioconductor. Nat Methods 12:115–121
14. Markowetz F (2015) Five selfish reasons to work reproducibly. Genome Biol 16: 274
15. Tautenhahn R, Böttcher C, Neumann S (2008) Highly sensitive feature detection for high resolution LC/MS. BMC Bioinformatics 9:504
16. Smith R, Ventura D, Prince JT (2013) LC-MS alignment in theory and practice: A comprehensive algorithmic review. Briefings in bioinformatics 16:bbt080–117
17. Louail P, Rainer J (2025) Rformassspectrometry/metabonaut: Metabonaut version 1.4.0. https://doi.org/10.5281/zenodo.15062929
18. Knuth DE (1984) Literate programming. Comput J 27:97–111

Chapter 3

mzmine: Unifying Mass Spectrometry Data Processing

Ansgar Korf, Robin Schmid, Steffen Heuckeroth and Tomáš Pluskal

Abstract

Modern mass spectrometry (MS) experiments generate increasingly complex, multidimensional datasets across diverse analytical modalities, including liquid chromatography (LC)–MS, gas chromatography (GC)–MS, ion mobility spectrometry, and mass spectrometry imaging. Efficient analysis of such heterogeneous data has traditionally required multiple vendor-specific software tools, limiting reproducibility and integration. mzmine 4 addresses this challenge by providing a unified, vendor-neutral, and extensible software platform for comprehensive MS data processing and interpretation. Building on nearly two decades of community-driven development, mzmine 4 introduces substantial improvements in performance, scalability, and usability, enabling the routine analysis of large-scale and multimodal datasets on standard consumer hardware. The software integrates workflows for feature detection, alignment, spectral library matching, molecular networking, small molecule annotation, and advanced MS2 interpretation within a single graphical environment. New capabilities, including interactive molecular networking, guided workflow automation, enhanced visualization, and machine learning-based spectral similarity scoring, further support exploratory and reproducible data analysis. The transition to an enterprise-supported yet open innovation model ensures long-term sustainability while preserving open-source principles. Together, these advances position mzmine 4 as a comprehensive, future-ready platform for untargeted metabolomics and mass spectrometry-based research.

Keywords Mass spectrometry data processing, Vendor-neutral data integration, Multimodal mass spectrometry, Interactive molecular networking, Universal mass spectrometry platform

1 Introduction

Modern mass spectrometry (MS) experiments generate complex, multidimensional datasets that are challenging to analyze without robust software support. Depending on the sample and analytical goals, labs may employ a variety of MS platforms ranging from liquid chromatography (LC)-MS and gas chromatography (GC)-MS to ion mobility spectrometry (IMS)-MS, MS imaging, and more. Historically, each instrument vendor often provided their own software, forcing researchers to juggle multiple tools and data formats. In the field of metabolomics and lipidomics, there has long been a need for an integrated, vendor-neutral software that can handle all these data types in one place.

Michael Witting (ed.), *Computational Methods and Data Analysis for Metabolomics*, Methods in Molecular Biology, vol. 3063,
https://doi.org/10.1007/978-1-0716-5452-1_3,

The mzmine software, formally known as "MZmine," has emerged as one of the most popular open-source frameworks addressing this need. Over nearly two decades of development, the MZmine project has provided a comprehensive suite of modules for MS data processing and visualization (https://doi.org/10.1039/9781788019880). Early versions (MZmine 1 (https://doi.org/10.1093/bioinformatics/btk039) and 2 (https://doi.org/10.1186/1471-2105-11-395)) focused on LC-MSworkflows and introduced features such as chromatographic feature detection, spectral filtering, and library matching. By integrating new algorithms (e.g., ADAP for chromatogram building (https://doi.org/10.1021/acs.analchem.7b00947) and deconvolution) and identification tools such as SIRIUS/CSI:FingerID for structure prediction, and direct links to GNPS (https://doi.org/10.1038/s41592-020-0933-6) for molecular networking, MZmine steadily expanded its capabilities to GC-MS data and advanced metabolite identification (https://doi.org/10.1039/9781788019880). A worldwide community of academic labs contributed to this growth, making MZmine a truly collaborative project (https://doi.org/10.1039/9781788019880). With thousands of users and over 6000 citations in the scientific literature, MZmine became a cornerstone in computational metabolomics.

1.1 *From MZmine 3 to mzmine 4—Evolution and Enterprise Transition*

The third major version, MZmine 3, was a significant milestone, introducing a modernized graphical user interface (GUI) based on JavaFX technology and support for new data modalities such as IMS and MS imaging (https://doi.org/10.1038/s41587-023-01690-2). MZmine 3 demonstrated the power of community-driven development and enabled processing of LC-IMS-MS, GC-MS, and even MS imaging data within a single platform (https://doi.org/10.1038/s41596-024-00996-y). This version laid the groundwork for unified analysis of untargeted metabolomics data across different instrument types, all within one software. MZmine 3's success underscored the value of an open-source, all-in-one tool for MS data, but it also highlighted growing demand for performance, support, and sustainable development.

The development of mzmine 4 represents the next leap in this evolution. It is not only a software upgrade, but also a transition in how the project is managed and delivered. In 2024, the core mzmine developers formed *mzio GmbH* (https://mzio.io/) to provide enterprise-level support and continued innovation for the software. This move from a purely academic, volunteer-driven project to a hybrid academic–industry platform brings several benefits. The company structure secures funding independent of research grants, which accelerates development, while ensuring long-term maintenance and professional user support. Importantly, mzio has embraced an open innovation model. The source code of mzmineprocessing modules remains openly available, and external contributions from the scientific community are actively encouraged (https://github.com/mzmine/mzmine). In other words, mzmine remains an open-source community-driven project at its core, even as an enterprise manages its development. This approach leverages the best of both worlds: academic innovation and industrial robustness.

A key aspect of the enterprise transition is the introduction of an academic program and tiered user packages. mzmine software is distributed in a *Community Edition* that is free of charge for students, scientists, and teachers at universities to ensure they can continue to use the full software without financial barriers. Meanwhile, mzio offers advanced CORE, PLUS, and PRO packages for enterprises, which include added services such as dedicated support, training, and early access to new features (https://mzio.io/mzmine-news/#enterprise-packages). Notably, the free Community edition also includes all frequently released updates and even comes with some bleeding-edge functionalities from the most recent research, e.g., machine learning models such as MS2DeepScore (https://www.biorxiv.org/lookup/doi/10.1101/2024.03.25.586580) and DreaMS (https://www.nature.com/articles/s41587-025-02663-3). By keeping the software itself free for universities and open-source, while providing optional paid support plans for those who need it, the mzmine platform sustains a bridge between academia and industry that enables fast knowledge transfer. This open innovation approach ensures that academic research in metabolomicssoftware can translate into a stable, well-supported tool that can also be adopted by industry.

In summary, the transition to mzmine 4 marks more than just a version update. It signifies mzmine's growth from a successful academic software project into a comprehensive, enterprise-supported data analysis solution. The goal of mzmine is to be one software to do it all for MS data analysis (https://mzio.io/), while fostering an open, collaborative community for continued development. The following sections highlight how mzmine 4 achieves this goal through new features and improvements, and what sets it apart from its predecessor MZmine 3.

2 Materials

mzmine 4 software: Java-based desktop application including modules for raw MS data import, feature detection, alignment, visualization, molecular networking, lipid annotation, spectral library matching, and workflow automation via the mzwizard.

Input mass spectrometry data: Vendor-specific raw data files or open formats (mzML, mzXML, imzML) originating from LC-MS, GC-MS, IMS-MS, and mass spectrometry imaging experiments.

Spectral libraries: Reference MS2 libraries in standard formats (e.g., MGF or MSP) containing precursor m/z values, fragment ion m/z values, and intensities for compound identification.

Computational environment: Standard desktop or workstation computer with a multi-core CPU, sufficient RAM for the dataset size, and an operating system compatible with Java Virtual Machine (Windows, macOS, or Linux).

Optional external software tools: MSConvert installation for the conversion of proprietary vendor formats to open formats.

Internet access: Network connection for optional online database queries and documentation access.

3 Methods

3.1 *One Software for All Your MS Data: Multi-Modal and Vendor Support*

The tagline "a single software for all your MS data" reflects mzmine's broad support for multiple MS modalities and data formats. Researchers no longer need separate tools for different instrument types; mzmine 4 can import, process, and visualize data from LC-MS, GC-MS, IMS-MS, and MS imaging experiments within one unified interface. This expansive compatibility, already initiated in MZmine 3, has been further enhanced in mzmine 4:

- Liquid and Gas Chromatography MS: Full workflows for LC-MS remain a core strength of mzmine. Additionally, GC-MS support has been optimized in version 4. For example, a new GC electron ionization (GC/EI) spectral deconvolution module was introduced to improve peak picking in complex EI fragmentation data (https://github.com/mzmine/mzmine/releases/tag/v4.2.0). This allows users to efficiently deconvolve co-eluting GC features and detect components in complex mixtures. The GC/EI-MS workflow in mzmine 4 has been streamlined end-to-end, from raw data import to feature detection, alignment, as well as downstream data analysis tools, such as statistics and Interactive Molecular Networking (https://github.com/mzmine/mzmine/releases/tag/v4.4.0).
- Ion Mobility Spectrometry: mzmine 4 extends robust support to IMS-enhanced MS instruments, covering major vendors, including Bruker, Waters, Agilent, and MOBILion. The software can handle the additional mobility dimension (making analyses effectively "4D"), and it provides specialized visualization such as Kendrick Mass Defect plots in 4D. The traditional Kendrick Mass Defect plot functionality has been expanded into a "4D feature plot (Kendrick)" tool, which now allows highlighting of features across m/z, retention time, mobility, and intensity (https://mzmine.github.io/mzmine_documentation/visualization_modules/kendrickmass/kendrick_mass_plot.html). This is particularly useful for the analysis of molecules including structural repeating units such as polymers and lipids. Performance improvements in reading and processing large IMS datasets mainly come from improved mzML and vendor-specific raw data import.

- Mass Spectrometry Imaging (MSI): mzmine 4 truly enables untargeted imaging metabolomicsworkflows. The software can directly import imaging MS data (e.g., MALDI-MS or DESI-MS imaging) and build feature lists consisting of all extracted ion images. Additionally, the ability to perform co-localization (https://mzmine.github.io/mzmine_documentation/module_docs/group_imagecorrelate/image-colocalization.html) analysis and Interactive Molecular Networking on imaging data was added (https://github.com/mzmine/mzmine/releases/tag/v4.4.0). In practice, this means one can find clusters of molecules that co-occur in the same tissue regions and visualize those relationships as networks, all within mzmine. A dedicated results viewer lets users inspect co-localized ion images and their spectral correlations. In combination with the Bruker timsTOF fleX instruments, mzmine can perform dataset-dependent MS2 acquisitions and subsequent data analysis directly on tissue (https://www.nature.com/articles/s41467-023-43298-9). Additionally, mzmine 4 added support for MALDI spot analysis (e.g., Bruker timsTOF fleX MALDI spot measurements) via the mzwizard, explained in more detail later in this chapter. This expands mzmine's utility to imaging modalities beyond continuous imaging, catering to both imaging and spatially resolved MS assays.
- Broad Vendor Format Compatibility: To further reduce the need for multiple software tools, mzmine 4 improved support for proprietary vendor file formats. It natively or indirectly supports data from Thermo Fisher (including direct reading of Thermo *.raw files since version 4.8.0), Bruker (.tdf, .tsf, and .baf), Waters (.raw via Masslynx, and via converted open formats for others), SCIEX (.wiff and wiff2), and more. Many of these are enabled via integration with the open-source MSConvert toolkit (https://www.nature.com/articles/nbt.2377): if MSConvert is installed on the user's system, mzmine can seamlessly import additional vendor formats by invoking vendor-specific libraries (https://github.com/mzmine/mzmine/releases/tag/v4.3.0). mzmine 4 also added support for MOBILion IMS data via MSConvert, keeping pace with emerging IMS technologies. In sum, whether the data originate from an Orbitrap, time-of-flight, ion trap, or IMS device, and whether they come from LC, GC, or imaging experiments, mzmine 4 aims to ingest and handle them. This universality fulfills the promise of a "single universal tool" for MS labs, greatly simplifying workflows that previously required multiple software packages.

3.2 Performance and Scalability Enhancements

One of the most striking improvements in mzmine 4 over its predecessor is its drastically improved performance and scalability. As metabolomics studies grow in size, with experiments involving thousands of samples or imaging datasets with millions of spectra, software efficiency becomes critical. mzmine 4 was optimized on multiple fronts to handle large-scale data with speed and reduced resource usage:

- Memory Utilization: The mzio team undertook a significant refactoring of data structures and memory management in version 4. As a result, mzmine 4 uses only a small fraction of the RAM that MZmine 3 required for the same tasks. In internal benchmarks, large projects that previously consumed tens of gigabytes of memory now use roughly one-tenth of that in mzmine 4. For example, processing 45 LC-IMS-MS files that would have needed ~13.7 GB of memory in MZmine 3 reportedly required only ~1.3 GB (~9%) in mzmine 4 after these optimizations. Similar ~9% memory usage was seen for a batch of 500 LC-MS files, dropping from ~42 GB to ~3.9 GB (Fig. 1). Such reductions enable mzmine 4 to run complex analyses on ordinary laboratory computers and greatly reduce the risk of running out of memory on big studies. Improved memory handling also means users can load and compare thousands of MS files simultaneously, unlocking high-throughput comparative metabolomics directly on the desktop.
- Processing Speed: mzmine 4 introduced numerous speed improvements that cumulatively make MSdata processing much faster. Key algorithms were parallelized or optimized in Java, and some computations were offloaded to more efficient libraries. Notably, spectral library matching, a common bottleneck when identifying features against large reference libraries, was accelerated by up to 20× in mzmine 4 (https://github.com/mzmine/mzmine/releases/tag/v4.5.0). This dramatic speed-up means that matching millions of fragmentation spectra to spectral databases (such as GNPS, NIST, or MassBank libraries) is now feasible within seconds. Similarly, the feature alignment step (combining feature lists across multiple samples)

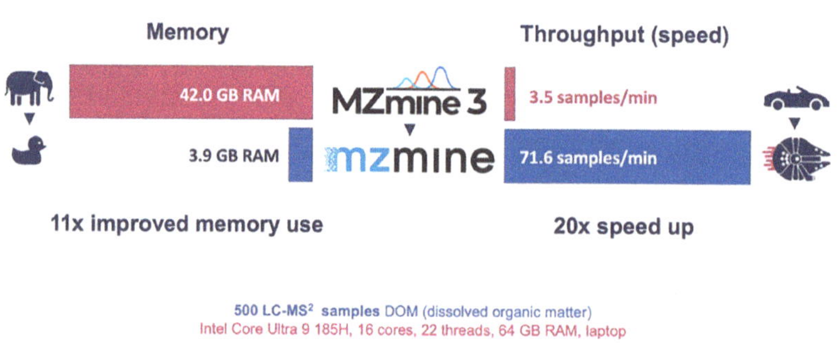

Fig. 1 MZmine 3 vs mzmine 4 memory and speed comparison (modified from mzmine documentation https://www.mzmine.github.io/mzmine_documentation/img/performance/mzmine_performance_both.png, accessed 29 Jan 2026)

was optimized, yielding a roughly 2–3× speed increase in the updated join aligner module (https://github.com/mzmine/mzmine/releases/tag/v4.5.0). The GC-MS aligner also benefited from these improvements, which is especially useful when aligning complex EI fragmentation features. These enhancements allow batch processingpipeline to finish much sooner, even for studies spanning thousands of features across hundreds of samples. For users, this means quicker turnaround from raw data to biological results, facilitating more iterative and exploratory analyses.

- I/O and Data Conversion: The input/output throughput was improved, reducing waiting time when loading data or writing results. Vendor file parsing is faster (e.g., reading native timsTOF.tdf files got a speed boost by a factor of ~4.2), and intermediate data export (such as MGF files, feature tables, or images) was streamlined. mzmine 4 also added options for vendor-based centroiding during import. If raw profile data should be centroided using the vendor's algorithm, this can be done on-the-fly through either MSConvert or native vendor library integration, saving a post-processing step. By harmonizing data import and export (e.g., standardizing the MGF export across different modules for consistency https://github.com/mzmine/mzmine/releases/tag/v4.5.0), mzmine 4 ensures that high-throughput workflows run smoothly with minimal manual intervention.

In practical terms, these performance gains mean that large metabolomics datasets that were borderline impractical in MZmine 3 become routine in mzmine 4. Users can process larger experiments (both in number of files and spectral complexity per file) and still interact with the software in real-time (e.g., filtering or browsing features) thanks to the more efficient resource usage. For MS core facilities, enterprises, and power users, the command-line interface (CLI) in mzmine 4 is also enhanced. It now supports user authentication flags and other options for better integration into automated pipelines, enabling easier deployment on servers or in cloud environments for truly large-scale jobs. Overall, mzmine 4's focus on speed and scalability directly addresses the needs of modern metabolomics and imaging MS, where datasets continue to grow in size.

3.3 New Analytical Features and Capabilities in mzmine4

Beyond improvements in performance, mzmine 4 delivers a rich array of new analytical features that extend its functionality far beyond what MZmine 3 offered. These additions make the software a more powerful platform for data mining, interpretation, and hypothesis generation in metabolomics and related fields. Below, we outline some of the most notable new capabilities.

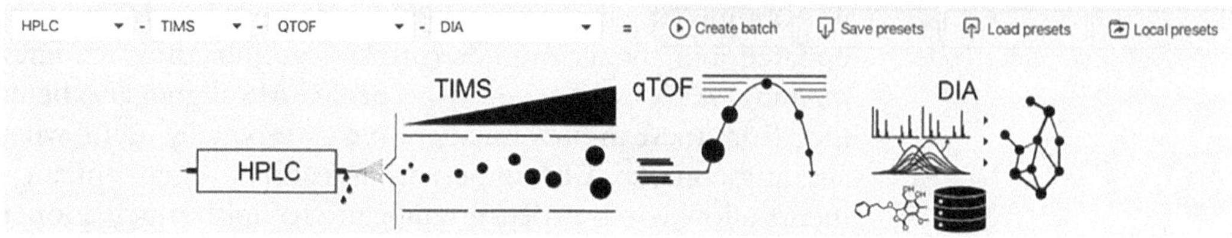

Fig. 2 Screenshot of mzwizard setup

3.3.1 Workflow Automation with "mzwizard"

To help both new and experienced users set up complex data analysispipeline, mzmine 4 introduces the "mzwizard," a guided workflow builder. The mzwizard can automatically generate a sequence of data processing steps optimized for your experimental setup (Fig. 2).

By asking the user a few simple questions (e.g., sampling technology, instrument type, MS2 acquisition mode), the wizard suggests a tailored workflow including appropriate modules for import, feature detection, deconvolution (if GC/EI-MS or DIA), alignment, and identification. This greatly lowers the barrier to entry for novices, allowing them to get started without manually configuring every parameter. In addition, the wizard integrates new tools. For example, it now handles cases such as MALDI spot MSworkflows, which require different handling than LC-MS datasets. By encapsulating best-practice workflows, the mzwizard ensures that even advanced functionalities of mzmine 4 are accessible through an easy step-by-step interface. This is especially valuable for newcomers in metabolomics who may not be familiar with, for instance, how to set up an IMS-enabled feature detection or an imaging co-localization analysis. Expert users can still fine-tune or manually adjust any part of the pipeline after the mzwizard's initial setup. In summary, the mzwizard exemplifies mzmine 4's push for user-friendly automation: it helps users go from raw data to identification with minimal manual intervention, while still leveraging all the advanced modules under the hood.

3.3.2 Interactive Molecular Networking and Annotation

mzmine 4 fully embraces Feature-Based Molecular Networking as a strategy for discovering related molecules in untargeted datasets (https://www.nature.com/articles/s41592-020-0933-6). While MZmine 3 had the ability to submit jobs to the GNPS (Global Natural Products Social Molecular Networking) infrastructure, mzmine 4 makes molecular networking interactive and integrated within the software. Users can now perform Feature-Based Molecular Networking (FBMN) directly in mzmine and visualize the molecular networks in real time, without leaving the platform (https://mzmine.github.io/mzmine_documentation/module_docs/group_spectral_net/molecular_networking.html).

In practical terms, after features are detected and aligned across samples, mzmine 4 can compute spectral similarity networks: each node represents a feature with a corresponding MS2 spectrum (or a consensus spectrum for an aligned feature list row), and edges connect nodes with high spectral similarity, suggesting they are structurally related. These networks can be filtered and explored interactively. Crucially, mzmine 4's networking is coupled to the feature table and annotation views, meaning if you click on a network node, you can see which feature it corresponds to in the feature table, inspect its properties, and vice versa. This direct linkage accelerates the biological interpretation of networks, allowing analysts to quickly pinpoint clusters of analogs (such as groups of lipids, or natural product derivatives) and deep dive into their MS spectra or chromatographic data.

Furthermore, mzmine 4 integrates state-of-the-art spectral similarity metrics powered by machine learning. In addition to classical cosine similarity, users can choose MS2DeepScore (https://www.biorxiv.org/lookup/doi/10.1101/2024.03.25.586580) or DreaMS (https://www.nature.com/articles/s41587-025-02663-3) algorithms for scoring MS2 spectral relatedness. Both are deep learning models that predict molecular similarity from spectra, enabling more sensitive detection of related molecules even when fragment overlap is low. Incorporating these models means mzmine's networking can uncover relationships that might be missed by naive modified cosine comparison. The user can toggle between scoring methods to see how the network connectivity changes, adding a new dimension of analysis.

Molecular networking in mzmine 4 is not limited to LC-MS data. As mentioned earlier, GC/EI-MS data can also be used for networking, which is a relatively new concept compared to ESI-based MS2. Additionally, the software allows networking of MALDI imaging data as well. Users can apply networking on MALDI MS2 data as well as on MS1 data of co-localized ions directly from imaging experiments. This improves untargeted spatial metabolomics analysis, where molecules are related not just by structure but also by their tissue distribution patterns.

Overall, the addition of interactive molecular networking transforms mzmine from a sequential data processing tool into an exploratory analysis platform. Researchers can go beyond lists of features: they can discover new biochemical relationships, gather compound families, and identify unknowns by their network context. This capability, integrated with mzmine's identification tools (spectral libraries, molecular formula prediction, lipid annotation, etc.), accelerates the journey from raw data to biological insight.

3.3.3 Fragmentation Dashboard and In-Depth MS2 Analysis

Another highlight of mzmine 4 is the introduction of a Fragmentation Dashboard, a tool designed for deep interrogation of MS2 data for molecular formula detection. Once features have associated MS2 spectra (through DDA or DIA acquisitions), the Fragmentation Dashboard provides an interactive interface to analyze those spectra in detail. It incorporates algorithms to suggest molecular formulas for features based on their MS2 fragmentation and isotopic patterns.

This dashboard essentially centralizes various fragmentation analysis tasks: users can apply in-silico fragment generation, compare fragments against expected substructures, and evaluate plausible formulas by scoring how well the observed fragments match theoretical ones. The aim is to assist in annotation of unknown compounds when other tools such as spectral library matching yield no results. The tool offers a more guided approach to formula prediction than previous versions. For instance, it might highlight potential molecular formulas and elemental compositions that fit the parent mass and fragmentation pattern and provide a ranking. Users can interactively accept or adjust these suggestions. By linking to the feature table, one can quickly move between features and see how a given formula hypothesis would explain their spectra.

The Fragmentation Dashboard complements external identification tools: while MZmine 3 provided exports for the SIRIUS software to predict structures (https://www.nature.com/articles/s41592-019-0344-8), the new dashboard offers an internal, rapid way to do first-pass molecular formula detection. This also ties in with new additions such as in-software PubChem searches. mzmine enables right-clicking a feature to search its *m/z* or formula directly in PubChem. The combination of formula suggestion and direct online database query can quickly lead to candidate identifications for unknowns. Overall, the Fragmentation Dashboard in mzmine 4 serves as a hub for MS2 interpretation, helping analysts converge on likely formulas and substructures for metabolites in their data.

3.3.4 Enhanced Lipidomics and Polymer Analysis

mzmine 4 brings specialized improvements for particular application domains, notably lipidomics and polymer analysis:

- Lipid Annotation and Quality Control: Building on mzmine's existing lipid identification modules (https://analyticalsciencejournals.onlinelibrary.wiley.com/doi/abs/10.1002/rcm.8117) (https://pubs.acs.org/doi/abs/10.1021/acs.analchem.8b05493), version 4 improves rule-based lipid annotation by incorporating recommendations from the Lipidomics Standard Initiative (LSI). Users can define or use built-in lipid class rules that consider structural hierarchy and shorthand notation, enabling automated annotation of lipids with the correct class names and fatty acyl compositions (https://pubs.acs.org/doi/full/10.1021/acs.analchem.4c05989). Beyond identification, mzmine 4 provides new quality control visualizations for lipid data, including Equivalent Carbon Number (ECN) plots (https://mzmine.github.io/mzmine_documentation/visualization_modules/ecn_plots/ecn_plots.html). ECN plots help in verifying lipid series and checking annotation consistency (e.g., plotting retention time against number of carbons to see whether elution orders are sensible). The updated Kendrick Mass Defect (KMD) plot tool also aids lipidomics by highlighting homologous series of lipids differing by repeating units (e.g., CH_2 or H).

- Polymer and Repeating Unit Analysis: Polymer analysis (and by extension certain natural products or contaminant analyses, e.g., PFAS) benefits from mzmine 4's enhanced Kendrick plot tool. The 4D Kendrick feature plot not only visualizes Kendrick mass defect across features but also includes an automatic repeating unit suggester (https://mzmine.github.io/mzmine_documentation/visualization_modules/kendrickmass/kendrick_mass_plot.html#automatic-repeating-unit-suggestion). Given a cluster of features (e.g., a series of polymer oligomers or homologous compounds), mzmine can suggest the likely repeating unit molecular formula that links them. This is invaluable for identifying polymer additives, oligomeric surfactants, or humic substance series in complex samples. The software can extract regions of interest in the Kendrick plot and list features that form a polymer series. mzmine 4 also computes Remainder of Kendrick Mass (RKM) and allows charge-aware Kendrick analysis, which is important if multiple charge states of polymers appear (https://mzmine.github.io/mzmine_documentation/visualization_modules/kendrickmass/kendrick_mass_plot.html#remainders-of-kendrick-masses-rkm). These capabilities turn mzmine into a useful tool for analysts dealing with synthetic polymers, plastics additives, or even biopolymers, as it automates what was previously a very manual analysis of pattern recognition in mass spectra.

In both lipidomics and polymer analysis contexts, mzmine 4's new features exemplify a theme: interactive visualization coupled with automated suggestions. Rather than just producing static results, the software guides the user to patterns in the data (homologous series, outliers, consistent vs. inconsistent identifications) and provides ways to validate or explore those findings.

3.3.5 Improved Spectral Libraries and Compound Identification

Identifying unknown features by matching them to spectral libraries or databases is a core part of untargeted metabolomics. mzmine 4 makes several advances in this area to increase both speed and flexibility of compound identification:

- Spectral Library Matching Performance: As mentioned earlier, the spectral library search algorithm in mzmine was optimized extensively, yielding up to ~20-fold speed improvements (https://github.com/mzmine/mzmine/releases/tag/v4.5.0). This means that even very large libraries (millions of spectra) can be searched against all features of a large dataset quickly. The accelerated search unlocks the possibility of using multiple libraries simultaneously (e.g., combining GNPS, MoNA, HMDB, and NIST spectral libraries) in an identification workflow without undue time cost.

- Library Generation and Merging: mzmine 4 introduces new ways to generate and curate spectral libraries from user data. Users can now easily create libraries from their own analysis results. One can generate a spectral library from all features annotated as lipids, or from all compound database matches found in a run. This is useful for building project-specific libraries or for sharing curated spectra of novel compounds. The software also supports merging spectral libraries and improved export options, allowing the combination of multiple library files and output in standardized formats for use elsewhere. The spectral library parser was made more flexible to accommodate various formats and to capture more metadata (such as collision energy, instrument type, compound internal IDs, CAS numbers, and acquisition methods) during import. This rich metadata retention means that libraries handled by mzmine 4 preserve important context, which can be used for filtering matches (e.g., preferring library entries acquired on similar instruments or with similar energies as the query data). The library generation capabilities were utilized to build MSnLib, the largest MS^n library available (https://www.nature.com/articles/s41592-025-02813-0).
- Integration with External Tools: mzmine 4 deepens integration with external identification tools. Notably, it provides built-in support for SIRIUS, a popular tool for formula and structure elucidation. Users can send features to SIRIUS/CSI:FingerID directly from mzmine and retrieve the results. In version 4.8, structural insights from SIRIUS, such as predicted formulae or candidate structures, can be viewed within mzmine, streamlining the annotationworkflow for unknowns. Additionally, as previously mentioned, the software can query online databases such as PubChem from within the feature table, accelerating the process of checking whether a given mass or formula might correspond to a known compound.
- Identification of Complex Data (DIA, MS^n): mzmine 4 expands support for more complex acquisition schemes such as Data-Independent Acquisition (DIA) and multi-stage MS^n fragmentation data. For DIA data, improved handling of fragmentation spectra selection and merging is implemented. Users can define advanced rules for how DIA fragment spectra are grouped or filtered to represent compound fragmentation, giving greater control in challenging cases such as overlapping precursors. These enhancements ensure that library matching and networking can be applied to DIA data as effectively as to traditional DDA data. For sequential MS^n, the data model in mzmine 4 and the library search were extended to handle multi-stage fragmentation spectra, in line with the increasing interest in MS^n spectral databases in metabolomics.

All mzmine 4's improvements in identification workflows mean higher confidence and more comprehensive annotation of metabolomic features. By speeding up library searches, supporting user-generated libraries, and integrating with formula prediction tools, mzmine 4 helps researchers move from a list of unknown features to putative compound IDs more efficiently than ever.

3.3.6 User Experience and Workflow Enhancements

In parallel with backend improvements and new analysis modules, considerable attention in mzmine 4 was given to user experience and workflow ergonomics. Recognizing that data analysis is an iterative and often explorative process, the mzio team introduced numerous features to make the software more intuitive, flexible, and user-friendly for both newcomers and power users:

- Refined Graphical User Interface (GUI): The GUI of mzmine 4 was polished in many small ways that enhance usability. For example, metadata coloring was added, allowing users to assign colors to sample groups or metadata categories and have those reflected in visualizations, e.g., colored ion trace shapes in the feature table. This makes it easier to visually distinguish, e.g., case versus control samples. The assigned colors have the biggest effect when used in combination with the newly introduced statistics dashboard (https://mzmine.github.io/mzmine_documentation/visualization_modules/statistics_dashboard/statistics_dashboard.html), another step towards delivering the user an end-to-end data analysis experience. Dialogs and warnings throughout the software were improved to be more descriptive and check for common user errors (e.g., missing parameter settings), thereby preventing mistakes. There is also better linking to documentation: many modules now include direct links to online help, papers, and useful downloads, so users can quickly find guidance. These refinements, though subtle individually, contribute to a smoother learning curve and fewer frustrations during data analysis.
- Workflow Presets: mzmine 4 introduces configurable presets for all modules and workflows (https://github.com/mzmine/mzmine/releases/tag/v4.8.30). Users can save their preferred parameters for a module (e.g., a specific set of noise threshold, chromatographic peak width, etc., for feature detection) as a named preset. These presets can be loaded and applied to new datasets or even shared among colleagues to ensure consistent processing. Not only does this save time, but it also promotes reproducibility. Analysts can report the preset used or share the preset file alongside colleagues or as part of a publication. The software comes with some default presets out of the box for common scenarios, which new users can take advantage of without needing to tweak dozens of parameters from scratch. In addition, entire workflow presets are available in the form of

the mzwizard, which instantly configures a chain of modules for typical analyses (e.g., an LC-MS DDA workflow). This feature is designed for high-throughput environments where standardized processingpipelines are run repeatedly.

- Advanced Filtering and Data Mining: Working with large feature tables is more accessible in mzmine 4 thanks to new feature list row filters. The user can interactively filter features based on a variety of criteria. This includes basic ones like retention time or m/z, but also metadata-driven filters for any annotations, tags, or custom fields (https://mzmine.github.io/mzmine_documentation/module_docs/feature_list_row_filter/feature_list_rows_filter.html). For example, one can quickly filter the feature table to show only features that have a certain substructure, or only lipid features, or those within a certain formula range, etc. This flexible filtering extends to the mzmine batch processing as well, meaning you can incorporate complex filters as part of an automated workflow, e.g., to exclude features with >30% RSD among QC samples via an RSD/CV filter option. The user interface for filters was redesigned to be more intuitive, with a dedicated Filter tab in the batch wizard where multiple filter criteria can be combined. These filtering tools allow users to home in on features of interest quickly and even serve as rudimentary data mining within mzmine, enabling, for instance, a quick search for all features annotated as containing a benzene ring, or all features tagged as "interesting" during a manual review.
- Interactive Visualization and Linking: Several viewers in mzmine's interface were upgraded to be more interactive. A prime example is the pseudo-spectra visualizer for DIA and GC/EI-MS deconvolution results. In mzmine 4, this viewer updates in real time as the user selects different features or rows in the feature table (https://github.com/mzmine/mzmine/releases/tag/v4.7.8). For instance, when reviewing GC/EI deconvolution results, clicking a row in the feature table will refresh the spectrum plot to show that component's spectrum, allowing rapid scanning through processing results. Similarly, in DIA data analysis, selecting a feature will show the assembled fragment spectrum for that feature. This interactive linking between tables and plots significantly speeds up the review and curation of results.
- Batch Processing and Reproducibility: mzmine has long supported batch processing via .mzbatch files (XML files describing a sequence of tasks). mzmine 4 makes this even easier by supporting drag-and-drop loading of mzBatch files to quickly set up a run. It also introduced versioned reporting where users can output a structured report (PDF or HTML) summarizing the processing and results of an analysis, which is extremely useful for Methods sections in papers or for lab records (https://github.com/mzmine/mzmine/releases/tag/v4.8.30). This

automated reporting in PDF/HTML ensures transparency and aids in reproducing experiments.

In essence, the user experience enhancements in mzmine 4 are about making the powerful features more accessible and the workflow more efficient and error-resistant. For an end-user, this translates to less time fiddling with the interface or troubleshooting, and more time focusing on the scientific questions at hand. The comprehensive set of tools is organized in a way that both guides beginners (through the mzwizard, presets, and sensible defaults) and empowers advanced users through interactive exploration. mzmine 4 thus caters to a broad audience in the metabolomics community, from those just starting out with untargetedMS analysis to expert mass spectrometrists looking to push the boundaries of data analysis.

4 Conclusion and Outlook

mzmine 4 represents a pivotal advancement in computational metabolomics and mass spectrometrydata analysis. By building on the strong foundation of MZmine 3 and earlier versions, it realizes the vision of a single, universal software platform for all types of MS data, integrating workflows across LC–MS, GC–MS, IMS, and imaging. The transition to an enterprise-supported model via mzio GmbH has injected resources and sustainability into the project while preserving its open-source, community-driven spirit (https://mzio.io/#openinnovation). This unique synergy ensures that mzmine remains open and innovator-friendly in academia, yet professionally maintained and continually improved with input from industry and academic users alike.

From a technical perspective, mzmine 4's enhancements in performance, scalability, and memory efficiency address the pressing challenge of "big data" in metabolomics, enabling researchers to process larger experiments faster and on consumer hardware. At the same time, its new analytical features, ranging from interactive molecular networking to advanced fragmentation analysis, empower users to extract deeper insights from their data and to do so within a single integrated environment. The software does not just handle data, it helps interpret it, by revealing connections between features, suggesting molecular formulas, and providing rich visual tools for quality control and pattern recognition (Kendrick plots, ECN plots, etc.). This positions mzmine 4 not only as a data processing tool, but also as a platform for scientific discovery, where hypotheses can emerge directly from the analysis interface.

The focus on user experience and automation lowers the entry barrier for newcomers and improves productivity for seasoned users. This democratization of advanced MSdata analysis aligns perfectly with the needs of the metabolomics community, where laboratories of varying computational expertise seek to perform complex analyses reproducibly and efficiently.

Looking forward, the mzmine project exemplifies how open scientific software can evolve and thrive through academia–industry collaboration. The open innovation approach means that cutting-edge algorithms from research, such as new machine learning models for spectral analysis, can be rapidly incorporated, while the enterprise backing ensures rigorous testing, documentation, and user support. mzio GmbH has signaled ongoing improvements, such as continuous integration of new vendor formats, further speed optimizations, and expansion of identification capabilities. Given its modular design and large user base, mzmine is well poised to adapt to future challenges in MS data analysis.

In conclusion, mzmine 4 solidifies its role as an indispensable tool in the metabolomics and mass spectrometry world, combining comprehensive functionality with ease of use. It highlights how far the field has come from patchworks of specialized tools toward unified platforms that accelerate knowledge extraction from complex molecular data. By uniting diverse MS modalities under one roof and fostering a global community of users and developers, mzmine 4 truly delivers "all your MS data" in one software and sets the stage for future innovations in computational metabolomics.

Acknowledgments

T.P. was supported by the Czech Science Foundation (GA CR) grant 21-11563M and by the European Union's Horizon Europe program (ERC, TerpenCode, 101170268).

Check for updates

Chapter 4

LC–MS Data Processing Using OpenMS

Axel Walter, Eftychia E. Kontou, Payam Emami and Oliver Kohlbacher

Abstract

Liquid chromatography–mass spectrometry (LC–MS) is a cornerstone of metabolomics, enabling detailed analysis of complex biological samples. Untargeted approaches, however, face challenges in handling complex data and ensuring scalability. UmetaFlow, built on the OpenMS platform, addresses these challenges through a unique re-quantification step for low-abundance metabolites and a suite of complementary annotation tools. Available in four formats, from command-line tools to an interactive WebApp, UmetaFlow supports both high-throughput automation and manual data exploration. This chapter provides a step-by-step guide to UmetaFlow and its applications.

Keywords OpenMS, UmetaFlow, Workflows, Workflow managers, Untargeted metabolomics

1 Introduction

Liquid chromatography–mass spectrometry (LC–MS) has become a standard in metabolomics research, enabling quantification and annotation of metabolites in biological samples. However, LC–MSdata analysis is nontrivial due to the high complexity of samples as well as the need for reproducibility and automation in large-scale studies [1]. These factors can make LC–MS data processing daunting, especially for bench scientists and non-experts lacking a computational background.

In this chapter, we focus on UmetaFlow, an untargeted metabolomicsworkflow based on OpenMS, a free and open-source toolset for LC–MSdata analysis [1, 2]. UmetaFlow aims to provide data analysis solutions for the above-mentioned challenges by offering a complete pipeline for data-dependent acquisition (DDA) data [1]. Steps include file conversion, preprocessing, an optional re-quantification step, and metabolite annotation via MS2 spectrum identification using in-house libraries and external tools such as SIRIUS [3–5], GNPS [6, 7], and MS2Query [8]. Available as command-line workflows in Snakemake [9] and Nextflow nf-core [10, 11], PythonJupyternotebooks [12], and an OpenMS WebApp [13], UmetaFlow

Michael Witting (ed.), *Computational Methods and Data Analysis for Metabolomics*, Methods in Molecular Biology, vol. 3063,
https://doi.org/10.1007/978-1-0716-5452-1_4,

supports users ranging from experienced bioinformaticians to bench scientists.

This chapter is intended as a comprehensive user guide to UmetaFlow from installation to parameter tuning and includes suggestions for subsequent steps.

1.1 OpenMS

OpenMS is a powerful open-source software platform for LC–MSdata analysis. Its strength lies in its versatile implementation, offering multiple ways to access a collection of algorithms intended for both metabolomics and proteomics. A key advantage of OpenMS is its emphasis on modularity and workflow integration. The platform's architecture allows users to combine different analytical components into customized workflows, providing precise control over each step of the analysis process [2]. At its core, OpenMS offers a robust C++ library, delivering efficient and high-performance computational capabilities. For users who prefer Python, the platform offers pyOpenMS bindings, making it more accessible to the expanding community of Python users in bioinformatics [14]. Researchers favoring command-line interfaces can leverage the TOPP (The OpenMSPipeline) tools, which integrate seamlessly with workflow managers such as Nextflow [10] and Snakemake [9]. Recently, OpenMS further enhanced its accessibility with WebApps, offering an intuitive graphical interface tailored to specialized workflows, broadening its usability for a wider range of users [13].

1.2 Metabolomics Workflows with OpenMS

OpenMS supports two major metabolomicsworkflows, each designed to address specific analytical challenges in metabolomics research: DIAMetAlyzer for targeted analysis of data-independent acquisition (DIA) data [15] and UmetaFlow for untargeted analysis of DDA data [1].

DIAMetalyzer is intended for targeted metabolomics data acquired using data-independent acquisition (DIA), utilizing sequential window acquisition of all theoretical mass spectra (SWATH), where all ions are fragmented regardless of intensity. DDA data of target metabolites are used to generate an assay library, utilizing SIRIUS fragmentation trees for target transitions and their re-rooting for decoy transitions, enabling estimation of the false discovery rate [15]. The OpenSWATH workflow utilizes this library for targeted extraction of metabolite intensities from DIA SWATH data [16]. A major advantage of this approach is data reusability with different assay libraries [15].

UmetaFlow, which will be the primary focus of this chapter, takes a different approach. This workflow emphasizes the discovery and quantification of unknown metabolites across samples, making it particularly well suited for exploratory metabolomics studies where the compounds of interest are not predetermined [1].

2 Challenges in Untargeted Metabolomics

UmetaFlow addresses a number of challenges in untargeted metabolomics:

- **Quantifying low-abundance analytes:** Sensitivity is improved through a re-quantification step that re-inspects raw *m/z* and retention time (RT) coordinates to recover features missed due to intensity thresholds [1].
- **Enhancing annotation coverage:** Due to the chemical diversity of metabolites, UmetaFlow incorporates multiple complementary annotation tools. UmetaFlow offers tools for spectral matching with in-house libraries built from authentic standards measured on the same instrument (MSI level 1) to spectral matching and structure prediction with publicly available databases using GNPS [6, 7] and CSI:FingerID [4] (MSI levels 2–3) as well as compound class prediction and analog search with CANOPUS [5] and MS2Query [8] (MSI level 3).
- **Scalability and automation:** Workflow managers such as Snakemake [9] and Nextflow [10, 11] enable UmetaFlow to support scalable high-throughput processing of data.
- **Reproducibility:** UmetaFlow ensures reproducibility via containerization and environment management using Conda, Docker, and Singularity. These tools encapsulate all software dependencies, enabling deployment across local machines, clusters, or cloud-based environments.

3 Implementation

UmetaFlow offers several implementations for different user needs and skill levels. The options include the command-line interface tools for automated high-throughput analysis (Snakemake and Nextflow), PythonJupyternotebooks for prototyping and customization, and OpenMS WebApps for ease of use and manual investigation of results in an interactive dashboard.

3.1 Snakemake Workflow Manager

The main UmetaFlow implementation is suitable for high-throughput data analysis of large MS datasets or automated routine tasks such as quality control. This version requires a basic understanding of command-line operations. Snakemake provides a robust framework for creating reproducible and scalable data analysis pipelines written in a user-friendly, human-readable, Python-based language [9]. This makes it a great low-barrier-to-entry choice for bioinformaticians already familiar with Python. The UmetaFlow Snakemake implementation is currently the most comprehensive, offering all the features described in Sect. 2.

OS: Linux, macOS, Windows (via WSL)

GitHub: https://github.com/biosustain/snakemake_UmetaFlow

Prerequisites: Conda

Installation: Installation for Linux is simple, as it only requires an existing Conda installation. An additional setup script is provided for macOS. Detailed installation instructions can be found on GitHub.

3.2 OpenMS WebApp

The OpenMS WebApp interface is the most user-friendly option, designed for users such as bench scientists who prefer a graphical interface and have no coding experience [13]. This version requires straightforward installation using the provided Windows executables or no installation when accessed via the online hosted version. WebApps are great for domain experts who can utilize the interactive result dashboards to get valuable insights that might otherwise be overlooked in an automated pipeline.

OS: Linux, macOS, Windows

WebApp: https://www.openms.de/webapps/umetaflow/

GitHub: https://github.com/axelwalter/umetaflow-gui

Prerequisites: None

For a local installation on Windows, simply download the installer file from our online-hosted WebApp (a download button on the front page) and follow the guided installation. The WebApp will appear as a native application in your Windows application menu. For Linux and macOS, please follow the instructions on GitHub.

Comment: The UmetaFlow WebApp does not support integration of GNPS FBMN results.

Deployment: Hosting a UmetaFlow WebApp is possible using Docker Compose. Please refer to this GitHub repository for hosting OpenMSWebApps.

3.3 Jupyter Notebooks Using pyOpenMS

Jupyternotebooks offer an interactive Python coding environment [12] and synergize very well with pyOpenMS and its seamless integration in the Python data science ecosystem by exporting OpenMS MS data structures into a tabular format using Pandas DataFrames [2, 14, 17]. This implementation provides great flexibility for customizing and extending UmetaFlow while enabling the inspection and visualization of intermediate results.

OS: Linux, macOS, Windows

GitHub: https://github.com/biosustain/pyOpenMS_UmetaFlow

Prerequisites: Python 3.11, Conda (optional but recommended)

Installation: Installation is straightforward using Conda and the provided environment file. However, all packages can be installed via pip as well.

3.4 Nextflow as Part of Metaboigniter 2.0

Metaboigniter 2.0 is a suitable choice for users who prefer Nextflow, a workflow system for creating scalable, portable, and reproducible workflows, for their high-throughput data analysis tasks [10]. Metaboigniter also supports flexible experimental design, including full MS2 on all samples and MS2 on pooled samples. As part of the nf-core collection of curated pipelines [11], Metaboigniter 2.0 is well documented and reliable.

OS: Linux

nf-core: https://nf-co.re/metaboigniter/2.0.0/

GitHub: https://github.com/nf-core/metaboigniter

Prerequisites: Nextflow, nf-core (optional)

Comment: Besides the GNPS integration, Metaboigniter 2.0 offers the complete UmetaFlow pipeline.

4 Workflow Overview

The UmetaFlowpipeline comprises three main steps (see Fig. 1 for details):

Preprocessing (see Sect. 4.2)

This initial step transforms raw data into a feature quantification matrix (FeatureMatrix) containing metabolite intensities across all samples. This includes steps such as RTalignment, adduct detection, and mapping of MS2 spectra to features for subsequent annotation.

Re-quantification (optional, see Sect. 4.3)

Unique to UmetaFlow, this optional step is designed to detect low-abundance features that may be missed during preprocessing (e.g., features below the noise-threshold intensity). By re-evaluating the data for subtle signals, this step enhances sensitivity, improving the quantification of low-abundant metabolites.

Annotation (optional, see Sect. 4.4)

UmetaFlow offers complementary annotation methods to maximize meaningful biological interpretation. The various annotation approaches are feature identification with in-house libraries generated from authentic standards, followed by querying publicly available spectral databases for metabolite annotation, e.g., GNPS [6], and structure prediction by CSI:FingerID [4]. If no specific compound was found, analog search by MS2Query [8] and compound class prediction by CANOPUS [5] can still provide valuable insights.

While all intermediate results are accessible to the user, the most important result file is a FeatureMatrix containing intensity values and annotations for consensus features across all samples (see Table 1).

4.1 LC–MS Data Input

UmetaFlow accepts input files in the open mzML format, developed by the Proteomics Standards Initiative (HUPO-PSI) [18]. The UmetaFlow Snakemake implementation allows Thermo RAW files as well, which will be converted to mzML format internally utilizing

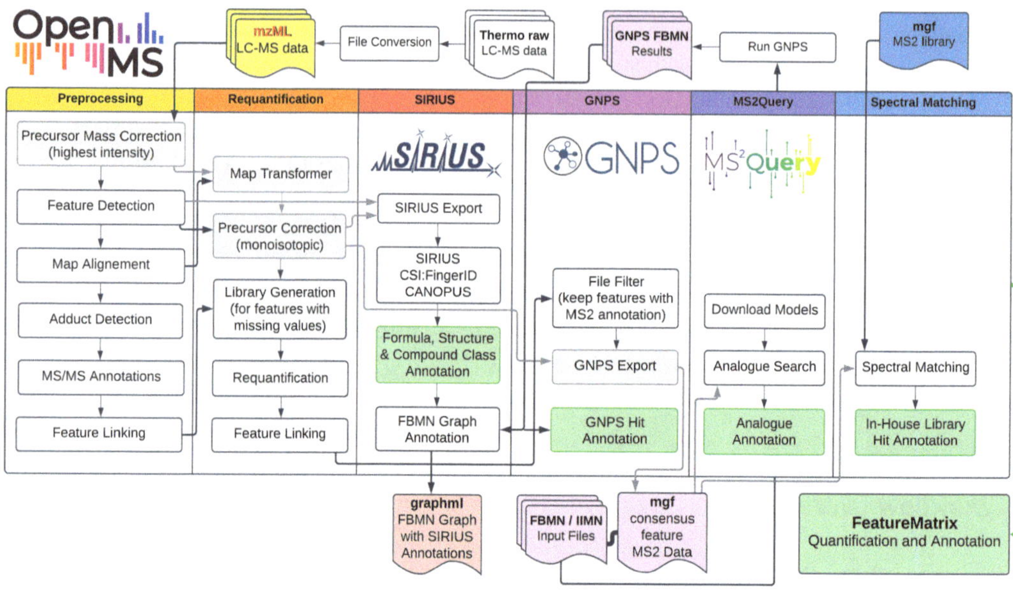

Fig. 1 UmetaFlow overview. UmetaFlow accepts raw data files (MS1 or MS1/MS2 for annotation), either in Thermo RAW file format (which will be automatically converted to mzML) or directly as mzML files (manually converted, e.g., using ProteoWizard's msconvert). Preprocessing detects metabolic features within the raw spectra, aligns them along the RT dimension, detects adducts, annotates MS2 spectra to features, and generates a list of consensus features. Consensus features containing missing values can be used to build an MS1-level library (with RT and *m/z*) for an optional re-quantification step to ensure that low-abundance features are not missed. The consensus feature map from preprocessing or re-quantification (if enabled) is exported to a FeatureMatrix in TSV file format. The spectrum processing pipeline, highlighted in gray, includes precursor *m/z* correction to the highest-intensity peak (if possible, within the monoisotopic trace of a feature), map transformation based on feature map alignment, and finally, export of all consensus feature MS2 spectra to an MGF file by GNPSExport, which also generates input files for GNPS FBMN and IIMN. FeatureMatrix annotation steps are highlighted in green. MS2Query and spectral matching annotate the FeatureMatrix using the MS2 MGF file with chemical analogs and in-house spectral library matches, respectively. SIRIUS predicts chemical formulas, structures (CSI:FingerID), and compound classes (CANOPUS) for features on a sample level, which are propagated to the FeatureMatrix. Finally, input files are generated for GNPS FBMN and IIMN. Running FBMN manually enables the annotation of GNPS library hits to the FeatureMatrix and the propagation of SIRIUS results to the FBMN network graph

ThermoRawFileParser [19]. For other vendor formats, it is recommended to convert the data to mzML format using the ProteoWizard msconvert tool [20]. Mass spectrometers do not have perfect precision. Consequently, peak shapes spread in the *m/z* dimension in a Gaussian distribution. Therefore, peak picking (centroiding) should be applied to data acquired in profile mode. Centroiding is the process of converting raw profile-mode data into single-centroided peaks (see Fig. 2). This process simplifies the data for downstream analysis while retaining essential information about the detected ions.

Table 1
FeatureMatrix columns with descriptions

Column name	Description
metabolite	Unique identifier for each consensus feature with the format m/z@ RT@adduct (e.g., 234.20568@400.72@[M+H]+)
RT (s)	Retention time in seconds
m/z	Mass-to-charge
adduct	Most likely adduct configuration
charge	Consensus feature charge
ending with .mzML	Feature intensity values for each sample
starting with SIRIUS	SIRIUS chemical formula and explained feature peak intensity score
starting with CSI:FingerID	Predicted structure, chemical formula, SMILES, and InChI key
starting with CANOPUS	Predicted compound pathway and class
GNPS	GNPS library hits
starting with MS2Query	Analog name, SMILES, InChI key, pathway, and compound class
starting with SpectralMatch	Spectral match compound name and SMILES

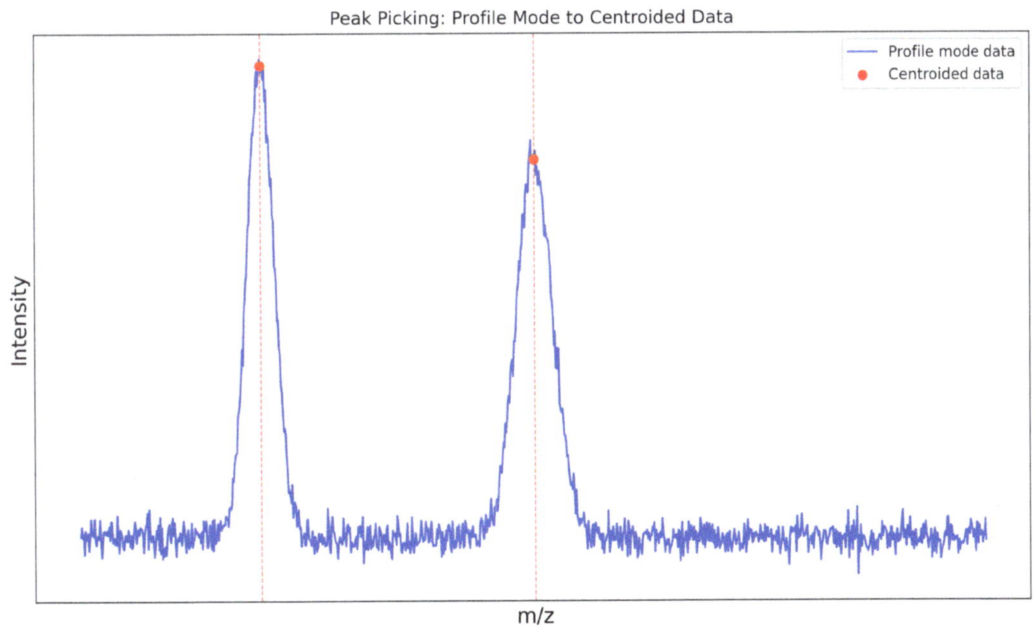

Fig. 2 Peak picking. Conversion of profile mode to centroided data. Data complexity is reduced significantly by picking a single ion peak within a distribution

4.2 Preprocessing

The preprocessing step is crucial in every untargeted metabolomicsworkflow and the only non-optional step in UmetaFlow (see Figs. 1 and 3).

Precursor mass correction (optional, for annotation)

Each MS2 spectrum acquired in DDA mode has an annotated precursor mass (the ion that was fragmented). For annotation of MS2 spectra, it is necessary to correct precursor masses in two ways:

Correction to the highest intensity peak: Ideally, fragmentation is triggered at the MS1 peak apex to maximize the number of ions entering the fragmentation process. In reality, however, this is not always the case, and MS2 spectra can be linked to a precursor MS1 spectrum far from the peak center. This presents two main issues. First, the MS2 spectrum should be linked to an MS1 precursor spectrum with high intensity to increase the chance of association with a metabolic feature. Second, the MS2 spectrum RT should be as close as possible to the feature's peak apex to improve the chances of spectral matching. The only relevant setting here is the mass tolerance defined by the MS instrument accuracy (see Table 2).

Correction to the monoisotopic mass trace: In DDA data, where only the highest intensity ions are fragmented, it is common to place already fragmented masses on an exclusion list. As a result, the second-highest intensity isotopic trace of an abundant metabolite may be selected for fragmentation. To match MS2 spectra to detected features (which are annotated with the monoisotopic *m/z*), the precursor mass must be corrected to match the monoisotopic trace. No settings need to be modified for this step.

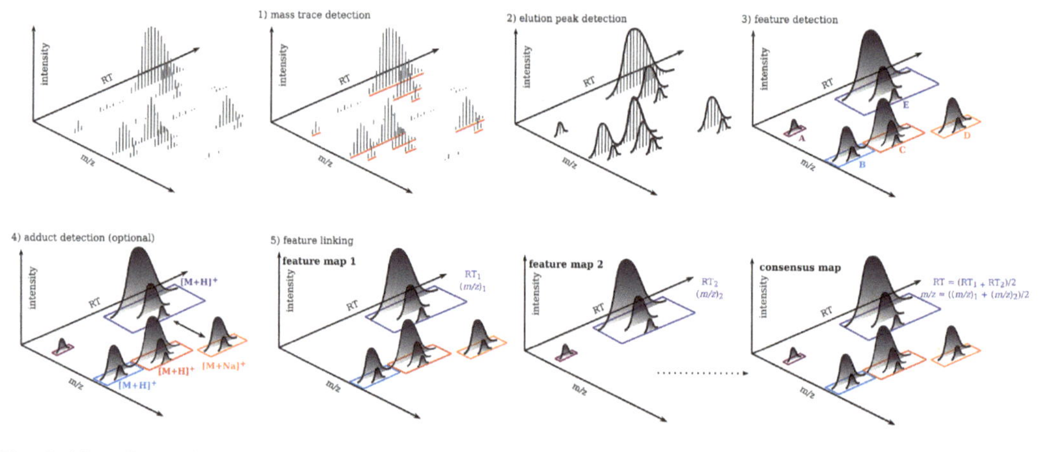

Fig. 3 LC–MS raw data processing main steps. (1) Mass trace detection groups peaks with similar *m/z*. (2) Elution peak detection identifies peaks across retention times within each mass trace, capturing ion signals that elute as chromatographic peaks. (3) Feature detection, elution peaks are combined into individual metabolic features with similar retention times and *m/z* of a metabolic isotope pattern. (4) Adduct detection (optional) annotates these features with potential ion species (e.g., [M+H]+, [M+Na]+). (5) Finally, feature linking aligns and matches features across multiple samples

Table 2
Preprocessing parameters and OpenMS TOPP tools

Name	Description	Default
Precursor mass correction (HighResPrecursorMassCorrector)		
highest_intensity_peak_mz_tolerance	The precursor mass tolerance to find the highest intensity MS1 peak in ppm	10.0
Feature detection (FeatureFinderMetabo [21])		
noise_threshold_intensity	Intensity threshold below which peaks are regarded as noise	1000.0
chrom_peak_snr	Minimum signal-to-noise ratio a mass trace should have	3.0
chrom_fwhm	Expected chromatographic peak width (in seconds)	5.0
mz_tolerance	Allowed mass deviation (in ppm)	10.0
remove_single_traces	Remove unassembled traces (single traces)	true
Adduct detection (MetaboliteAdductDecharger [22])		
ion_mode	MS instrument ion mode	positive
adducts_pos	Potential adducts in positive mode. Format: adduct:charge:probability	H:+:0.6 Na:+:0.1 NH4:+:0.1 H-1O-1:+:0.1 H-3O-2:+:0.1
adducts_neg	Potential adducts in negative mode. Format: adduct:charge:probability	H-1:-:1 H-2O-1:0:0.05 CH2O2:0:0.5
Map alignment (MapAlignerPoseClustering [23])		
mz_tolerance	Never pair features with larger *m/z* distance (in ppm)	10.0
Feature linking (FeatureLinkerUnlabeledKD [24])		
rt_tolerance	RT tolerance (s) for two features to be considered consensus	100.0
mz_tolerance	*m/z* tolerance (ppm) for two features to be considered consensus	8.0

Set according to your LC–MS instrument setup with respect to mass accuracy (in ppm), ion mode, and chromatography

Feature detection

Given a raw 3D peak map with RT, *m/z*, and intensity dimensions, data preprocessing aims to filter out noise signals and group peaks in a meaningful way, e.g., match peaks that are putative, likely to be derived from real metabolites. These metabolic features cover the chromatographic elution profile in the RT dimension and all isotopic traces for a specific charge state (usually one or two charges in metabolomics) in the *m/z* dimension.

The OpenMS TOPP tool FeatureFinderMetabo performs untargeted extraction of metabolic features from a raw peak map [21]. The first algorithm in feature detection is **mass trace detection** (Fig. 3-1). The process begins by identifying peaks with the highest intensity as seed points. These are extended in both directions along the retention time axis to form mass traces. Mass traces are extended until a given number of outlier spectra do not contain a peak within the given mass. Chromatographic peak shapes are elucidated during the **elution peak detection** step, which is important to separate metabolites with similar *m/z* and slightly different RT into two overlapping chromatographic peaks. This allows splitting the mass trace at the local minimum into two distinct elution profiles (Fig. 3-2). In the final, actual **feature detection step**, individual mass traces are grouped into putative metabolic features spanning an area on the peak map, the RT range defined by its elution profile, and the *m/z* range defined by its isotopic pattern (Fig. 3-3). In order to be matched together as a feature, mass traces need to fulfill some criteria. Since mass traces from the same metabolite pass the chromatographic column together, they are co-eluting, which means their peak apex RT and peak shape (in regard to tailing or fronting) should be similar. The *m/z* distance between two mass traces has to match a distinct charge state (e.g., 1 Da distance for single-charged ions, 0.5 Da for double-charged ions). The isotopic trace intensity ratios should be explainable by a metabolite with the given *m/z*. Metabolites are typically small (< 1000 Da). Therefore, the monoisotopic trace is always the most abundant. Each possible combination of mass traces is scored based on these criteria, leading to the assembly of the highest-scoring mass trace combinations into metabolic features. They are defined by their monoisotopic *m/z*, charge, centroid RT, and intensity, which is calculated from the area under the monoisotopic trace.

The most important settings in the feature detection step are related to the mass accuracy of the MS instrument and the chromatographic peak width (see Table 2).

Adduct detection (optional)

A metabolite can potentially be detected with multiple adduct configurations (e.g., with proton or sodium adduct). Hence, each feature can be assigned a specific adduct configuration (Fig. 3-4:) by the OpenMS TOPP tool MetaboliteAdductDecharger [22]. Given the adduct configuration, the neutral mass of a metabolite can be elucidated, and multiple features can be assigned to one metabolite, reducing overall complexity. Probabilities for certain adduct configurations are mainly dependent on the LC buffer and ion mode. Consequently, the adduct detection settings are highly specific to the individual experiment (see Table 2), and the default settings have to be adjusted in most cases.

Map alignment (optional)

RT shifts can occur during data acquisition due to slight variations in instrument conditions, sample preparation, chromatography, and even trivial external factors such as room temperature. These shifts can make it challenging to correctly associate metabolic features across multiple samples. To ensure accurate downstream analysis (e.g., feature linking), feature RTs across samples should be aligned by the OpenMS TOPP tool MapAlignerPoseClustering. First, feature maps from multiple samples are aligned by clustering features across maps using *m/z* and RT coordinates. The feature map with the highest number of features serves as a reference, based on which feature RTs in the other samples are transformed with the goal of minimizing the overall time shifts of clustered features [23] (Fig. 4). Second, the raw peak maps are transformed by the OpenMS TOPP tool MapRTTransformer accordingly as well, based on the feature map alignment transformation description. The most important setting is the maximum *m/z* distance for a feature to the reference feature in order to get aligned (see Table 2).

Mapping MS2 spectra to features (optional, for annotation)

To map annotations to specific features, MS2 spectra must be linked to the corresponding features. This step does not require setting any parameters manually.

Feature linking

Feature linking is the process of associating corresponding features across multiple samples in a dataset. Features are matched based on their *m/z* values and RTs within manually defined tolerance window settings (see Table 2) by the OpenMS TOPP tool FeatureLinkerUnlabeledKD [24]. If map alignment was performed, the RT tolerance can be set narrow. This step consolidates features that represent the same metabolite, enabling consistent comparisons across samples. In

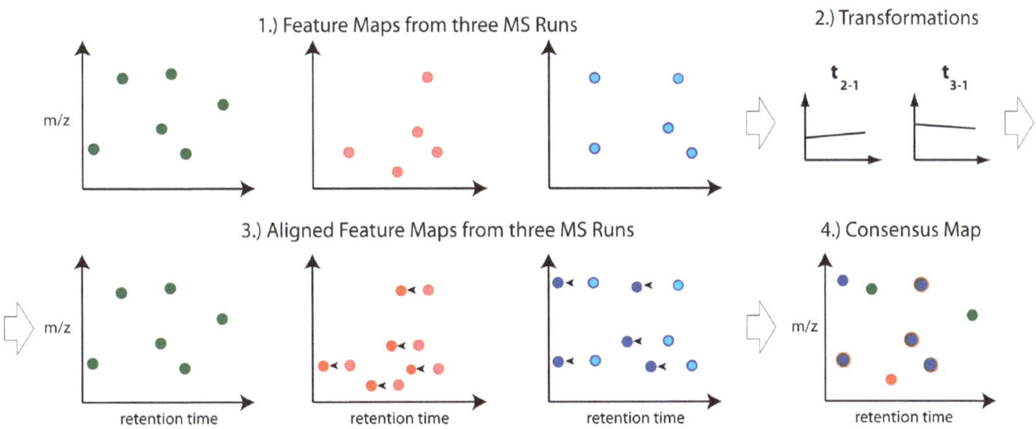

Fig. 4 Feature map alignment. (1) Input feature maps from different runs share some consensus features, which need to be aligned. (2) Transformation of RT based on a reference map (here map 1, since it has the highest number of features). (3) Features in maps 2 and 3 are aligned with the reference map. (4) In a consensus map, they will have the exact same RT

UmetaFlow, the preprocessed consensus features get exported to a FeatureMatrix in TSV file format, including a unique identifier, *m/z*, RT, charge, adduct, and intensity values for each sample, at this point without any annotations (see Table 1, orange section).

4.3 Re-quantification

One of the unique features of UmetaFlow is its re-quantification step, which aims to improve the accuracy of metabolite quantification, particularly for low-abundance metabolites that may not have been detected in the initial feature detection phase during preprocessing. While re-quantification is an optional step, it is recommended if the data analysis focuses on quantification. For metabolite annotation, however, this step can be omitted.

In the first feature detection step, UmetaFlow utilizes an untargeted feature detection tool, FeatureFinderMetabo, which applies a noise-intensity threshold to filter out low-abundance signals that are likely noise. While this reduces processing time and minimizes false positives, it can also mean that some biologically relevant metabolites present at low concentrations might not be detected. Therefore, UmetaFlow introduces a re-quantification step for any consensus feature that contains at least one missing value. This process can be thought of as a form of missing-value imputation, but instead of using statistical models to estimate missing values, it re-quantifies the features using actual experimental data, leading to more accurate and reliable results.

The re-quantification process (see Fig. 5) is performed by FeatureFinderMetaboIdent, a tool for targeted feature quantification at the MS1 level (see parameters in Table 3). This tool takes a target library as input, which contains *m/z* and RT values generated from the consensus feature centroid *m/z* and RT values obtained in the initial feature detection step. Re-quantification for a consensus feature is applied to all samples, even if a quantification value from preprocessing exists already, to make sure intensity values for a consensus feature are derived from the same algorithm and are therefore comparable. The final result is a more complete and accurate FeatureMatrix, which now includes data for low-abundance metabolites that were initially missed. This improved quantification is especially valuable in studies where low-abundance metabolites play a significant biological role and subtle metabolic changes across conditions have a great impact.

4.4 Annotation

UmetaFlow supports four distinct annotation methods, each offering complementary approaches to metabolite identification (see Fig. 6). These methods will annotate the FeatureMatrix with molecular identifications, structures, compound classes, or analogs, enhancing the biological interpretability of the dataset.

SIRIUS, CSI:FingerID, and CANOPUS

SIRIUS, CSI:FingerID, and CANOPUS are powerful tools for molecular annotation from high-resolution tandem mass spectrometry (MS/MS) data. SIRIUS focuses on molecular formula

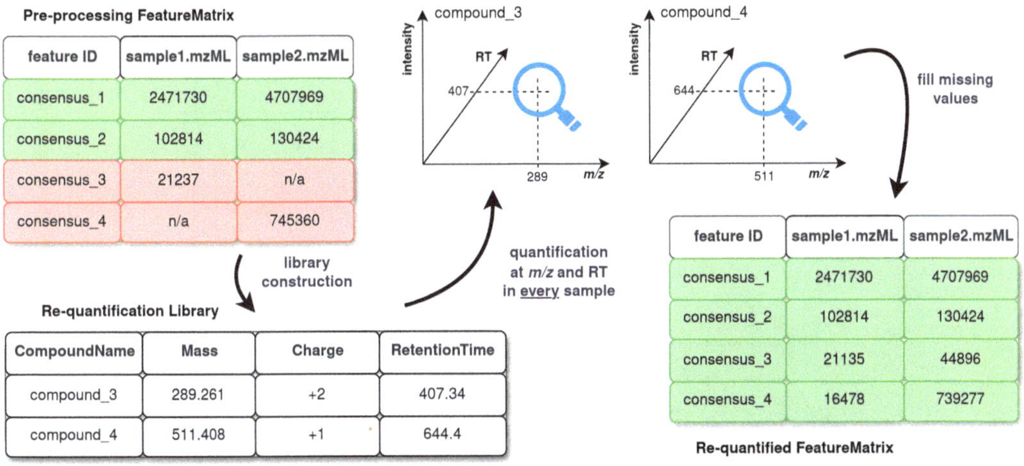

Fig. 5 Re-quantification. A library with mass, charge, and RT values is constructed from all consensus features that have at least one missing quantification value. The specific *m/z* and RT coordinates are inspected (an area of defined RT and *m/z* tolerance windows) within each sample. The resulting quantification values fill the gaps in the FeatureMatrix

Table 3
Re-quantification parameters

Name	Description	Default
Re-quantification (FeatureFinderMetaboIdent)		
mz_window	*m/z* window size for chromatogram extraction (in ppm)	10.0
rt_window	RT window size (in seconds) for chromatogram extraction	30.0

Set according to your LC chromatogram peak width and MS instrument accuracy

elucidation. It uses isotope patterns and fragmentation spectra to construct fragmentation trees, which explain how precursor ions break down into smaller fragments [3, 25]. These trees are scored against theoretical formulas in order to determine the best-matching formula. CSI:FingerID predicts molecular structures based on molecular fingerprints derived from the fragmentation trees, which encode chemical properties such as functional groups. These fingerprints are matched against large chemical databases (e.g., PubChem) for compound identification [4]. CANOPUS complements these tools by predicting compound classes directly from MS2 spectra and does not require a database. CANOPUS classifies molecules into hierarchical categories (e.g.,

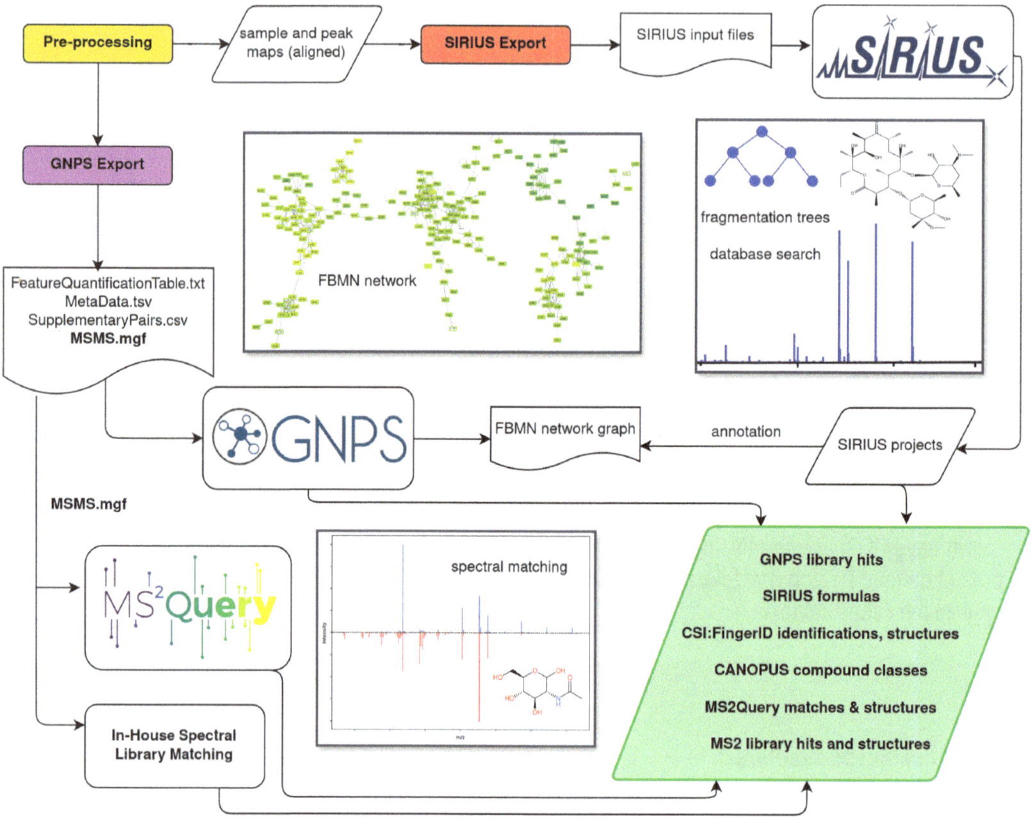

Fig. 6 Annotation. UmetaFlow integrates four annotation approaches to enhance metabolite identification. SIRIUS, CSI:FingerID, and CANOPUS: These tools utilize fragmentation trees and database search to perform molecular formula prediction, molecular structure prediction, compound identification, and compound class classification. GNPS FBMN/IIMN: UmetaFlow exports files for GNPS workflows, which construct molecular networks linking similar spectra. MS2Query: Machine learning predicts exact matches and analogs with high chemical similarity. MS2 spectral matching: Matches experimental spectra to in-house libraries for identification, ideal for private or instrument-specific libraries

lipids, flavonoids), which provides valuable biological context, even when features remain unidentified by CSI:FingerID [5].

UmetaFlow integrates SIRIUS for direct execution, annotating the FeatureMatrix with its top-ranked hits. For more detailed analysis, UmetaFlow also generates SIRIUS input files, allowing users to run SIRIUS externally with greater control over the settings (see parameters in Table 4).

GNPS

GNPS (Global Natural Products Social Molecular Networking) is an open-access platform designed for the analysis, sharing, and annotation of MS2 spectra, particularly in natural products research. By leveraging community-contributed spectral libraries, GNPS enables the identification of metabolites and the exploration of related

Table 4
SIRIUS parameters

Name	Description	Default
Export SIRIUS input files (SiriusExport)		
export_only	Only export input files for SIRIUS, without executing SIRIUS	TRUE
SIRIUS		
predict_structure_and_class	Run CSI:FingerID and CANOPUS as well to predict compound structures and classes. Can increase run time significantly	FALSE
combine_annotations	Whether to combine annotations (e.g., SIRIUS_molecularFormula) from all files into a single column separated by "##" or keep a separate column for each file (e.g., sample1_SIRIUS_molecularFormula, sample2_SIRIUS_molecularFormula, ...)	TRUE
max_mz	The highest feature *m/z* to be considered and annotated. Higher values significantly increase the SIRIUS run time	500
instrument	Set this option according to your instrument. Valid options: default, qtof, orbitrap, ftic	orbitrap
pos_ions_considered	Adduct configurations (positive mode) to consider	[M+H]+, [M-H2O+H]+, [M+Na]+ ,[M+NH4]+
neg_ions_considered	Adduct configurations (negative mode) to consider	[M-H]-, [M-H2O-H]-, [M-HCOOH]-
elements_considered	Set the allowed elements for rare element detection	SBrClBSe
elements_enforced	Enforce elements for molecular formula prediction	CHNOP
mz_tolerance	Maximum allowed mass deviation in ppm for decomposing masses (in ppm)	10
mz_tolerance_ms2	Maximum allowed mass deviation in ppm for decomposing masses in MS2 (in ppm)	10
formula_database	Search sum formulas in database(s). Check SIRIUS documentation for available options	none
structure_database	Search structures in database(s). Check SIRIUS documentation for available options	BIO

molecules in networks [6, 7]. Feature-based molecular networking (FBMN) builds networks where nodes represent molecular features and edges indicate spectral similarity. This approach provides a visual and interactive way to explore the relationships between metabolites, aiding in the identification of unknown compounds that have identified neighbors [6]. If adduct detection was enabled during preprocessing, ion identity molecular networking (IIMN) can reduce the complexity in FBMN by collapsing different ion species of the same metabolite, which would have otherwise appeared as separate nodes [7].

UmetaFlow exports all necessary files for GNPS FBMN and IIMN, but these tools must be executed externally. GNPS result files, including a table of library hits and the network graph, can then be used to annotate the FeatureMatrix with library hits and enrich the FBMN network graph with SIRIUS results.

MS2Query

Unlike traditional library search methods that focus exclusively on exact matches, MS2Query identifies related spectra of analogs with high chemical similarity. It employs machine learning-based chemical similarity predictors trained on existing spectral libraries [8]. UmetaFlow automatically downloads the default models, which are trained on the GNPS library. The FeatureMatrix is annotated with MS2Query hits, including compound names, structures, and classes.

For interpretation, two specific columns in the FeatureMatrix are crucial. Since every consensus feature is annotated, a *MS2Query_ms2query_model_prediction* score of at least 0.6 is recommended for confident results. While MS2Query does not distinguish between exact matches and chemical analogs, a *MS2Query_precursor_mz_difference* value close to zero suggests an exact match.

Aside from the ion mode, which is set during the preprocessingadduct detection step (see Table 2), no additional parameters need to be configured by the user. MS2Query relies on the consensus feature MS2 data file provided by the GNPS Export step.

MS2 spectral matching

This method enables researchers to utilize proprietary or specialized in-house spectral libraries to identify unique compounds not found in public databases, addressing scenarios where data privacy is a concern. Additionally, these spectral libraries can be customized for specific LC–MS instrument setups, significantly improving identification success rates. UmetaFlow facilitates seamless integration by matching experimental MS2 spectra to entries in these libraries. The in-house library must be supplied in MGF file format. The FeatureMatrix is then annotated with the matched compound names and corresponding structures in SMILES format (if available). Like MS2Query, the only relevant parameter is the ion mode, set for adduct detection in preprocessing (see Table 2).

5 Statistical Analysis

For statistical analysis of untargeted metabolomics data processed with UmetaFlow, we recommend using the comprehensive FBMN Stats Guide, including methods for data cleaning (e.g., blank filtering, missing value imputation), principal component analysis, univariate and multivariate statistics, and hierarchical clustering. The guide supports both a JupyterNotebook and a web application interface with an interactive dashboard for data exploration [26]. The final UmetaFlow FeatureMatrix (Table 1) serves as a direct input for the statistical workflow. In addition, a meta-value table is required, containing biologically and experimentally relevant metadata (e.g., blank samples, treatments, extraction methods, or growth media). A template for this meta-value table is included in the GNPS FBMN input files generated by UmetaFlow.

GitHub: https://github.com/Functional-Metabolomics-Lab/FBMN-STATS

Web application: https://fbmn-statsguide.gnps2.org

References

1. Kontou EE, Walter A, Alka O, Pfeuffer J, Sachsenberg T, Mohite OS et al (2023) UmetaFlow: an untargeted metabolomics workflow for high-throughput data processing and analysis. J Cheminform 15:52. https://doi.org/10.1186/s13321-023-00724-w
2. Pfeuffer J, Bielow C, Wein S, Jeong K, Netz E, Walter A et al (2024) OpenMS 3 enables reproducible analysis of large-scale mass spectrometry data. Nat Methods 21:365–367. https://doi.org/10.1038/s41592-024-02197-7.
3. Dührkop K, Fleischauer M, Ludwig M, Aksenov AA, Melnik AV, Meusel M et al (2019) SIRIUS 4: a rapid tool for turning tandem mass spectra into metabolite structure information. Nat Methods 16:299–302. https://doi.org/10.1038/s41592-019-0344-8
4. Dührkop K, Shen H, Meusel M, Rousu J, Böcker S (2015) Searching molecular structure databases with tandem mass spectra using CSI:FingerID. Proc Natl Acad Sci USA 112:12580–12585. https://doi.org/10.1073/pnas.1509788112
5. Dührkop K, Nothias L-F, Fleischauer M, Reher R, Ludwig M, Hoffmann MA et al (2021) Systematic classification of unknown metabolites using high-resolution fragmentation mass spectra. Nat Biotechnol 39:462–471. https://doi.org/10.1038/s41587-020-0740-8
6. Nothias L-F, Petras D, Schmid R, Dührkop K, Rainer J, Sarvepalli A et al (2020) Feature-based molecular networking in the GNPS analysis environment. Nat Methods 17:905–908. https://doi.org/10.1038/s41592-020-0933-6
7. Schmid R, Petras D, Nothias L-F, Wang M, Aron AT, Jagels A et al (2021) Ion identity molecular networking for mass spectrometry-based metabolomics in the GNPS environment. Nat Commun 12:3832. https://doi.org/10.1038/s41467-021-23953-9
8. de Jonge NF, Louwen JJR, Chekmeneva E, Camuzeaux S, Vermeir FJ, Jansen RS et al (2023) MS2Query: reliable and scalable MS2 mass spectra-based analogue search. Nat Commun 14:1752. https://doi.org/10.1038/s41467-023-37446-4
9. Köster J, Rahmann S (2012) Snakemake—a scalable bioinformatics workflow engine. Bioinformatics 28:2520–2522. https://doi.org/10.1093/bioinformatics/bts480
10. Di Tommaso P, Chatzou M, Floden EW, Barja PP, Palumbo E, Notredame C (2017) Nextflow enables reproducible computational workflows. Nat Biotechnol 35:316–319. https://doi.org/10.1038/nbt.3820
11. Ewels PA, Peltzer A, Fillinger S, Patel H, Alneberg J, Wilm A et al (2020) The nf-core framework for community-curated bioinformatics pipelines. Nat Biotechnol 38:276–278. https://doi.org/10.1038/s41587-020-0439-x
12. Kluyver T, Ragan-Kelley B, Pérez F, Granger B, Bussonnier M et al (2016) Jupyter Notebooks

– a publishing format for reproducible computational workflows. In: Positioning and power in academic publishing: players, agents and agendas. IOS Press, pp 87–90. https://doi.org/10.3233/978-1-61499-649-1-87
13. Müller TD, Siraj A, Walter A, Kim J, Wein S, von Kleist J et al (2025) OpenMS WebApps: building user-friendly solutions for MS analysis. J Proteome Res 24(2):940–948. https://doi.org/10.1021/acs.jproteome.4c00872
14. Röst HL, Schmitt U, Aebersold R, Malmström L (2014) pyOpenMS: a Python-based interface to the OpenMS mass-spectrometry algorithm library. Proteomics 14:74–77. https://doi.org/10.1002/pmic.201300246
15. Alka O, Shanthamoorthy P, Witting M, Kleigrewe K, Kohlbacher O, Röst HL (2022) DIAMetAlyzer allows automated false-discovery rate-controlled analysis for data-independent acquisition in metabolomics. Nat Commun 13:1347. https://doi.org/10.1038/s41467-022-29006-z
16. Röst HL, Rosenberger G, Navarro P, Gillet L, Miladinović SM, Schubert OT et al (2014) OpenSWATH enables automated, targeted analysis of data-independent acquisition MS data. Nat Biotechnol 32:219–223. https://doi.org/10.1038/nbt.2841
17. Pandas Development Team (2024) pandas-dev/pandas. https://doi.org/10.5281/zenodo.13819579
18. Martens L, Chambers M, Sturm M, Kessner D, Levander F, Shofstahl J et al (2011) mzML-a community standard for mass spectrometry data. Mol Cell Proteomics 10:R110.000133. https://doi.org/10.1074/mcp.R110.000133
19. Hulstaert N, Shofstahl J, Sachsenberg T, Walzer M, Barsnes H, Martens L et al (2020) ThermoRawFileParser: modular, scalable, and cross-platform RAW file conversion. J Proteome Res 19:537–542. https://doi.org/10.1021/acs.jproteome.9b00328
20. Adusumilli R, Mallick P (2017) Data conversion with ProteoWizard msConvert. Methods Mol Biol 1550:339–368. https://doi.org/10.1007/978-1-4939-6747-6_23
21. Kenar E, Franken H, Forcisi S, Wörmann K, Häring H-U, Lehmann R et al (2014) Automated label-free quantification of metabolites from liquid chromatography–mass spectrometry data. Mol Cell Proteomics 13:348–359. https://doi.org/10.1074/mcp.M113.031278
22. Bielow C, Ruzek S, Huber CG, Reinert K (2010) Optimal decharging and clustering of charge ladders generated in ESI–MS. J Proteome Res 9:2688–2695. https://doi.org/10.1021/pr100177k
23. Lange E, Gröpl C, Schulz-Trieglaff O, Leinenbach A, Huber C, Reinert K (2007) A geometric approach for the alignment of liquid chromatography-mass spectrometry data. Bioinformatics 23:i273–i281. https://doi.org/10.1093/bioinformatics/btm209
24. Weisser H, Nahnsen S, Grossmann J, Nilse L, Quandt A, Brauer H et al (2013) An automated pipeline for high-throughput label-free quantitative proteomics. J Proteome Res 12:1628–1644. https://doi.org/10.1021/pr300992u
25. Böcker S, Dührkop K (2016) Fragmentation trees reloaded. J Cheminform 8:5. https://doi.org/10.1186/s13321-016-0116-8
26. Pakkir Shah AK, Walter A, Ottosson F, Russo F, Navarro-Diaz M, Boldt J et al (2024) Statistical analysis of feature-based molecular networking results from non-targeted metabolomics data. Nat Protoc 20:92–162. https://doi.org/10.1038/s41596-024-01046-3

Check for updates

Chapter 5

Metabolomics Data Processing Using the Asari Toolkit

Joshua Mitchell, Jasmine Chong and Shuzhao Li

Abstract

Asari is a software tool for metabolomics data processing, which is designed from the ground up to address issues in reproducibility, performance and interoperability by introducing new algorithms and data structures. It is significantly faster than other tools and offers qualities that are suitable for large-scale metabolomic and exposomic analyses. The reusable data structures and modular libraries enabled the rapid development of a full-scale pipeline that includes QA/QC, MS/MS integration, annotation, and extension to untargeted stable isotope tracing data. This chapter provides an overview of the key concepts and designs in Asari, step-by-step applications to metabolomics data processing and annotation, and resources related to both LC-MS and GC-MS data.

Keywords Asari, Pipeline, Metabolomics, Data processing

1 Introduction

Data processing is the foundation of computational metabolomics, which transforms raw instrument output into quantitative information of features in biological samples. Many software tools have been developed for this purpose [1–7]. However, the data are highly dependent on methods and platforms, and significant complexity arises from the volume and dimension of data, as well as numerous chemical reactions that occur during sample analysis. These lead to confusion in data reproducibility [8, 9] and even the applicability of metabolomics [10]. The root of many software tools is in instrument analysis and signal processing, while metabolomics demands the data to meet the needs of statistics and data science. The excessive parameterization of software tools, requiring specialized expertise, causes frequent issues when they are used in broader communities. Additional issues were identified by Li et al., including high correspondence error rates in large datasets and spurious low-quality peaks [11]. Asari was designed from the ground up to overcome these limitations. Acknowledging the wide variability in mass spectrometryworkflows and instrumentation, it adopts a flexible, modular architecture. Its algorithms

Michael Witting (ed.), *Computational Methods and Data Analysis for Metabolomics*, Methods in Molecular Biology, vol. 3063,
https://doi.org/10.1007/978-1-0716-5452-1_5,

are implemented in a suite of interoperable Python libraries—collectively referred to as the Asari toolkit—supporting seamless integration across cheminformatics, metabolomics, and biomedical domains. This toolkit includes `asari`, `mass2chem`, `metDataModel`, `khipu`, and `JMS`, each playing a specific role and serving as foundational infrastructure (*see* **Note 4**). Though independently usable, these libraries often operate in concert to enable efficient and scalable data processing, which is also required by the rapid progress of instrumentation. The Python-Centric Pipeline for Metabolomics, `pcpfm`, bundles these components into an extensible, end-to-end solution for processing, annotation, and data standardization [12]. All components are open source and available on GitHub.

The strengths of this toolkit are already evident on several fronts. Asari is substantially faster than other major tools in runtime, up to multiple orders of magnitude [11]. The defined data structure is a key factor for data assembly, for example, it supported the construction of a consensus serum metabolome using over 100,000 samples [13]. The extensibility of the toolkit has enabled untargeted stable isotope tracing analyses [14]. This chapter provides an overview of the Asari toolkit's key concepts and step-by-step workflows for untargeted metabolomics—encompassing processing, pre-annotation, and annotation.

1.1 Algorithms and Terminology in the Asari Toolkit

To effectively use these tools, it is helpful to understand the underlying data structures and concepts (Table 1). We start with LC-MS (liquid chromatography coupled mass spectrometry) data processing. Taking advantage of the high resolution of modern mass spectrometry, Asari prioritizes mass alignment (Fig. 1A). In other tools, elution peaks are typically detected per sample before features are aligned across samples. In Asari, alignment is performed first in *m/z* (mass-to-charge ratio) space, followed by retention time (RT). This decomposes a difficult two-dimensional alignment problem into two simpler one-dimensional problems. Feature detection is then enhanced by recurrent peak patterns by superimposing signals from all samples.

A fundamental unit in Asari is the **mass track**—a zero-filled extracted ion chromatogram (XIC) indexed by scan number. Because *m/z* values that differ by less than the instrument's mass accuracy are indistinguishable, these can be reduced to a set of consensus *m/z* values across samples. A mass track may have zero to multiple elution peaks. Even if no elution peak is detected on a mass track, it still helps *m/z* alignment across samples. When multiple elution peaks are present on a mass track, the mass track facilitates their correct correspondence across samples.

As chromatography may shift from sample to sample, RTalignment is necessary. A set of highly resolved elution peaks is selected and mapped to their counterparts in a reference sample using LOWESS

Table 1
Operational terminology in the Asari toolkit

Name	Operational definition
MS spectrum	A list of m/z values and associated intensities, typically from a scan on a mass spectrometer.
Mass track	An extracted ion chromatogram of consensus m/z, spanning the full retention time.
Elution peak	Peak of intensity values along the axis of chromatography.
Feature	A set of peaks that are aligned across samples, specific to an experiment and method.
Empirical compound	A group of associated features, typically isotopes and adducts, that belong to the same tentative compound.
Compound	A metabolite or a chemical of xenobiotic origin, including contaminants.
Biochemical reaction	A biochemical process that interconverts one or more compounds, often catalyzed by an enzyme.
In-source reaction	Formation of adducts, fragments, and conjugates during mass spectrometry analysis.
Enzyme	A protein that catalyzes a biochemical reaction.
Gene	An inheritable sequence of nucleotides, some of which code for proteins.
Metabolic pathway	A series of linked reactions that typically involve structurally related compounds, usually defined by knowledge.
Metabolic network	A set of reactions connected by shared compounds. Not limited by pathway definitions.
Metabolic model	A collection of metabolic reactions with associated metabolites, enzymes, and genes. May include flux rates.
Study	A collection of experiments on a set of related samples.
Experiment	A set of acquisitions collected on a set of samples using consistent methods.
Method	The approach and parameters used for data collection (e.g., chromatography and ionization).
Sample	A biological or control sample analyzed in one or more experiments. A data file is referred to as an acquisition.

[1] Modified from Mitchell et al. [12]

regression (a local polynomial smoothing technique). This creates a mapping dictionary of scans between samples, as mass tracks are scan-indexed. After establishing alignment mappings for both m/z and RT, per-sample mass tracks are co-indexed and summed to form

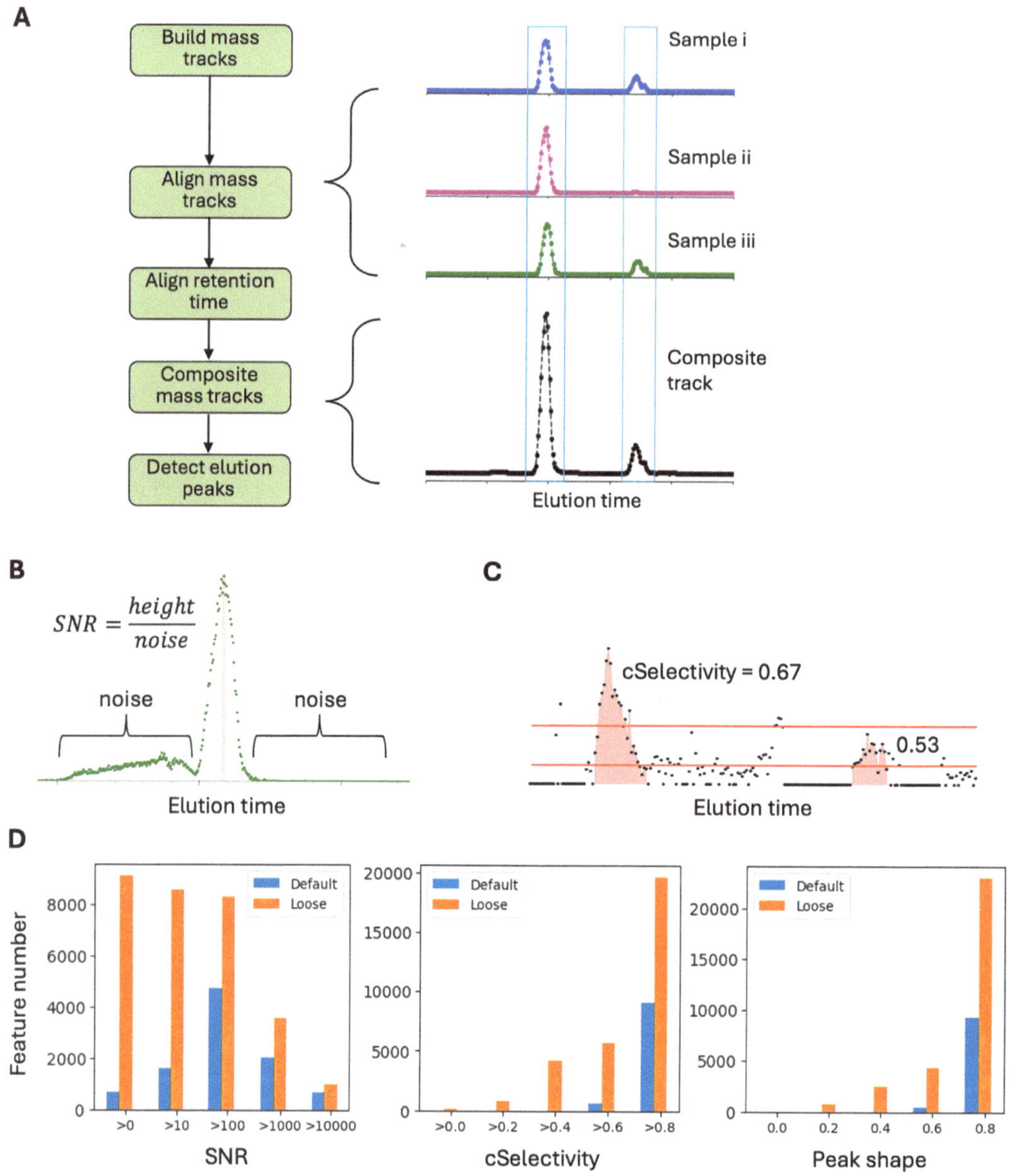

Fig. 1 Overview of Asari algorithm and quality control metrics. (**A**) The mass-first approach in Asari. Individual mass tracks are shown in colors, and the composite mass track is shown in black (right). (**B**) Illustration of SNR (signal-to-noise ratio). (**C**) Illustration of cSelectivity, the ratio between valid data points in features and spurious data points above half of the peak height. (**D**) A high number of low-quality features can be introduced by "loose" parameters. Input data are the 85 samples analyzed by negative ionization, reverse phase, in Mitchell et al. [12]. The default parameters in Asari are gaussian_shape: 0.5; min_peak_height: 100,000; signal_noise_ratio: 2; min_timepoints: 6. The "loose parameters" are 0.3, 50,000, 1.2, and 4, respectively. SNR is typically the most important feature quality metric

composite mass tracks, which represent the entire experiment (Fig. 1A).

Instead of performing peak detection per sample, Asari detects peaks only once on composite tracks. Each detected peak in the composite track defines a single feature. By computationally back-tracking through the mass and retention time alignments, each feature at the composite level can be mapped back to the peaks it represents in the samples. Integrating this per-sample peak yields a per-sample intensity for each feature. This strategy not only improves computational efficiency but also enhances consistency and reduces noise-driven variability in downstream analyses. The formulation of mass tracks and their alignment allows them to be cleanly summed across samples—an additive structure that lends itself naturally to composite representation. As the composite track consists of the sum of individual mass tracks, and because mass tracks are zero-filled, even a peak existing in only a single sample will be detected, as samples without the peak contribute zero intensity. This approach improves scalability by amortizing peak picking and enhances sensitivity by aggregating weak signals across all samples.

1.2 Quality Metrics in Asari Feature Detection

For each detected feature, Asari reports quality metrics such as signal-to-noise ratio (SNR, Fig. 1B) and peak shape (goodness-of-fitting to the Gaussian model), both calculated from the composite peak. It also introduces two novel metrics: mass selectivity (mSelectivity) and chromatographic selectivity (cSelectivity, Fig. 1C) [11]. These values quantify how well a given feature is resolved in *m/z* and retention time, respectively. Values near 1 indicate well-resolved features; values closer to 0 suggest ambiguity. The mSelectivity values are not reported explicitly because the Asari algorithm ensures high mSelectivity in its output results. The cSelectivity values allow users to gauge the quality of chromatography per feature without looking up the raw data.

It is a meaningless statement of how many features are detected by a software tool without their quality assessment. Because one can extract more peaks by lowering the threshold for peak definition. To illustrate this point, we process the same dataset using lowered parameter values (SNR, peak shape, minimum peak height, and required scan numbers) and compare the results to those using default parameter values (Fig. 1D). The "loose" parameters lead to many low-quality features, which often have a negative impact on biomedical projects, as false leads waste time and resources.

1.3 Feature Annotation

Annotation and feature detection are conceptually separate in the Asari toolkit, and we divide annotation into pre-annotation and compound identification. Pre-annotation refers to the grouping of isotopologues, adducts, in-source fragments, and conjugates that belong to the same compounds. This is performed by `khipu`, which models the ions in a tree structure [15]. Khipu differs from previous tools, such as CAMERA [16], Mz.unity [17], xMSannotator [18], MS-FLO [19], Binner [20], and NetID [21], as its tree structure is computable and chainable, and neutral mass is inferred agnostic of ion assumption. With pre-annotation, only the top ion or neutral mass is used for downstream search, greatly reducing false matches. Compound identification depends on libraries of chemical standards and spectral databases. This is illustrated in the PCPFM pipeline, which uses JMS for database and library search, and a third-party tool, MatchMS, for MS/MS matches. The PCPFM pipeline also performs quality control and data standardization [12]. Annotation results are centered on the JSON format; therefore, advanced users can take them or Asari feature tables to perform customized annotations.

2 Materials

Asari is typically run in a terminal using command lines (Fig. 2A, B), which is cloud-friendly. Interactive analysis can be carried out in a Jupyter notebook environment (Fig. 2C). A web-based, user-friendly application is available (Fig. 2D).

2.1 Installing the Asari Toolkit

The performance of Asari allows processing large studies on consumer hardware. A minimum of 4 cores and 16 GB of memory is recommended. The software is typically limited by disk access: faster hard drives help; space of approximately twice the size of the input data should be available to accommodate intermediate files and outputs. Any operating system supporting Python 3.11 or newer is compatible, though Linux and macOS are the most tested environments (*see* **Note 1**).

To install the complete pipeline in command line:

```
pip install pcpfm
```

To install only the core Asari library:

```
pip install asari-metabolomics
```

Asari requires input files in the `.mzML` format. They can be converted from vendor-specific formats by `msConvert` [22] or other format-specific converters [23]. The PCPFM pipeline includes limited conversion capabilities (e.g., Thermo `.raw` to `.mzML`) but does not provide a comprehensive solution.

The PCPFM was designed using a command-line-first philosophy [12]. Each operation is invoked via a specific verb, reflecting a functional programming model where clearly defined inputs are transformed into traceable outputs. To support reproducibility, every command invocation is recorded in the internal `experiment .json`

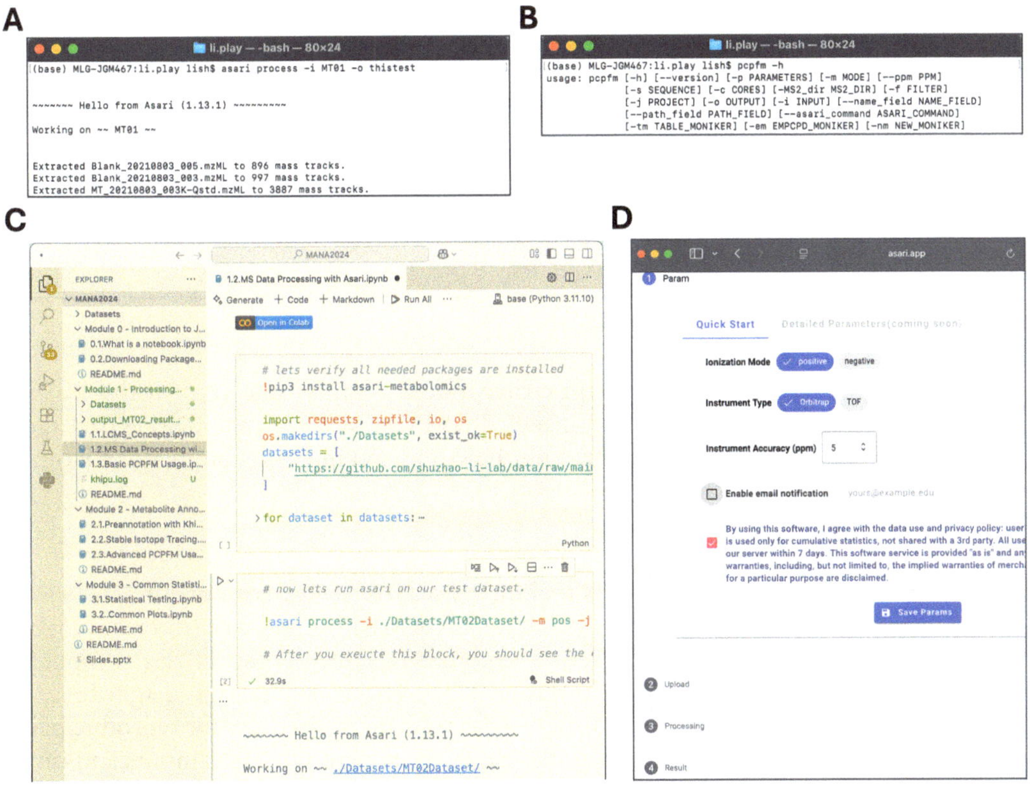

Fig. 2 Running Asari via command lines and via the web app. The terminal is the primary environment for running Asari (**A**) and the PCPFM pipeline (**B**). Many tutorials are available as Jupyternotebooks, which can be run in VS Code (**C**). A free web app allows processing of small datasets without using the command line (**D**)

file, enabling users to replay or audit analyses with exact parameters. While this design increases the number of intermediate files generated throughout the pipeline, it provides complete transparency into each step of the analysis. Commands are provided to compile a final output set, including a processed feature table, a co-indexed annotation table for mapped compounds, and a sample annotation table reflecting the metadata used to inform the experiment. This structure emphasizes reproducibility, modularity, and transparency. User interfaces and workflow engines can be layered atop this core for added usability without sacrificing traceability.

2.2 Additional Data

For annotation, external MS1 and tandem-MS reference libraries are required. They are not included during the default installation process. To download them, run:

```
pcpfm download_extras
```

And agree to the license terms. The annotation sources downloaded using this command include the Human Metabolome Database [24], the LIPID MAPS [25], and the MassBank of North America.

2.3 Tutorial Instructions

Detailed tutorials on Asari and the pipeline are available at https://github.com/shuzhao-li-lab/asari_pcpfm_tutorials/. Example datasets can be found therein (via GitHub Releases and embedded in documentation/notebooks). A series of Jupyternotebooks is provided on GitHub as tutorials. These notebooks can be downloaded and run locally (Fig. 2C) or in the cloud using tools such as Google Colab (see Chap. 1. *See* **Note 2**).

3 Processing LC–MS Data Using Asari

Asari can be run either independently or as part of the PCPFM pipeline. When installed via PCPFM or as a standalone package, it is accessible through the command line. All commands follow the syntax:

```
asari <verb>
```

Core verbs include `process` (feature detection), `analyze` (basic information about an mzML file), and `viz` (interactive visualization). The `process` command converts a set of `.mzML` input files into a processed feature table. These include logs, diagnostics, and quality metrics, all stored in a project directory for reproducibility.

As discussed in Sect. 2, Asari expects `.mzML` inputs and can be installed via pip. In combination with PCPFM and other tools, the Asari toolkit supports full end-to-end workflows including annotation, quality control, and report generation. Outputs from Asari are compatible with many existing workflows that anticipate feature tables as inputs. Asari can be run standalone or as part of a pipeline workflow. The following sections walk through the command-line usage of Asari before advancing into pipeline usage for QA/QC and annotation.

3.1 Command Line Use of Asari

Asari requires centroided `.mzML` files as input. All files in an analysis should be generated using comparable methods—for example, the same chromatographic column, gradient, ionization mode, and instrument configuration. With these files organized into a single directory (e.g., `<input_files>`), Asari can be run using (e.g., Fig. 2A):

```
asari process -i <input_files>
```

The default parameters in Asari are positive ionization mode and a mass accuracy of 5 ppm. Although all parameters are configurable, routine use generally requires only setting the ionization mode and ppm tolerance. For instance, to process a negative mode dataset acquired at 10 ppm mass accuracy:

```
asari process -i <input_files> -m neg --ppm 10
```

Running `asari process` creates a new output directory containing the generated feature tables and intermediate files (*see* **Note 3**). The `preferred_Feature_table.tsv` includes high-confidence features, while `export/full_Feature_table.tsv` contains all detected peaks. The results are placed in a directory named based on a project name specified using an optional `-j` flag. The project directory will be located in the present working directory or an alternative location specified by `-o`.

Fig. 3 Screenshot of part of the Asari dashboard. Results from a project can be visualized and navigated. The top panel contains quality and data summaries. The feature browser here shows one of the three features detected on a mass track. A mass track viewer is also provided. All peak data in Asari are plotted as raw scatters without artificial "beautification"

The results can be visualized interactively using:

```
asari viz -i <result_directory>
```

The interactive dashboard includes peak traces, intensity profiles, and alignment visualizations, allowing for rapid inspection of processing quality (Fig. 3). This dashboard serves as the recommended entry point for both troubleshooting and exploratory analysis.

3.2 Web-Based Processing Using Asari.app

Asari.app is a web-based implementation of Asari, accessible via https://asari.app, providing a lightweight option for processing small datasets without local installation or using command line. The interface allows users to upload `.mzML` files via drag-and-drop and specify key processing parameters such as ionization mode and mass accuracy. Upon submission, users are issued a unique retrieval ID. All uploaded data and results are deleted automatically after a retention period. As a free service, the web application is limited by hardware resources. While security measures are taken at the server end, the users should use discretion on data management policies, as the data are uploaded to an internet web server.

4 Pipeline Usage

The PCPFM integrates tools from the Asari ecosystem into a full-featured pipeline for metabolomicsdata processing. While Asari alone supports feature detection and visualization, the PCPFM adds structured support for MS/MS annotation, QA/QC operations, and reproducible report generation. Each command in PCPFM is invoked using a "verb" and produces clearly named outputs, supporting transparent, modular workflows (Fig. 2B). All intermediate results are organized in a user-defined project directory for transparency and traceability.

This functional approach ensures reproducibility: each input, output, and transformation is tracked. Experimentation is encouraged, as each verb will output an error message detailing its required and optional parameters. A `reset` verb is also available to delete all intermediates and return the pipeline to its post-Asari state to clean up after experimentation and before running the finalized analysis. The final output can include: (1) a processed feature table, (2) a mapped annotation table co-indexed by feature ID, and (3) a metadata-informed sample annotation table.

4.1 Basic Pipeline Usage

Download the example data and prepare the metadata CSV file to begin the analysis. This file should list each `.mzML` file's absolute file path on the system under "Filepath," its sample type (e.g., "Blank," "Unknown") as "Sample Type," and the name of the sample under

"File Name." Example code is provided in the notebooks for generating this automatically. Once prepared:

```
pcpfm assemble -o. -j <project_name> -s <metadata.
csv>
```

This command uses the same flags as Asari to produce a result directory in a specified location with a given name. The resulting directory appears to contain a copy of the mzML data but by default only symlinks them to prevent additional copies. Once assembled, note the project_path and use it to indicate which project the pcpfm should act upon. Once assembled, run Asari:

```
pcpfm asari -i <project_path>
```

Outputs include Asari's standard feature tables and all processing intermediates. Note: Asari prior to version 1.17 did not support mixed MS1/MS2 acquisitions within the same file. MS2 data are assumed to be DDA-like and provided as separate acquisitions. This will yield an asari folder in the project directory as if Asari were run in standalone mode. This table can be used for further processing manually; however, the pipeline provides an integrated solution.

4.2 Annotation with the Pipeline

Annotation here follows a modified Schymanski taxonomy [26]. Level 4 annotations are based on *m/z* similarity alone, while Levels 2 and 1 require MS/MS and optional retention time matches. Pre-annotation—grouping features into empirical compounds —is performed first using:

```
pcpfm build_empCpds -i <experiment_directory> -tm
<feature_table> -nm <empirical_compounds>
```

The default adducts and isotopologues (up to m+13C3) can be modified. Once empirical compounds are built, Level 4 annotations are added using:

```
pcpfm l4_annotate -i <experiment_directory> -em <
empcpds> -nm <annotated_empcpds>
```

To associate MS2 spectra with compounds:

```
pcpfm map_ms2 -i <experiment_directory> -em <
annotated_empcpds> -nm <mapped_empcpds>
```

Note: simply update the input compounds if Level 4 annotation was skipped. Optionally include additional MS2 files via the optional flag `--ms2_dir`. By default, MS2 data are queried against the default libraries using this command:

```
pcpfm l2_annotate -em <mapped_output>
```

Alternatively, "`--msp_files`" may be passed to enable custom library search.

Library searches currently do not permit ionization energy or retention time matching. For the highest confidence annotations, custom libraries are required with retention time information. mzVault exports are compatible with this tool and can be constructed using the Thermo software or made custom via scripting. For in-house libraries with retention time matching:

```
pcpfm l1a_annotate -em <input_empcpds> -nm <
output_empcpds>
```

Alternatively, Level 1b annotations (*m/z* and RT, no MS2) can be added using:

```
pcpfm l1b_annotate -em <input_empcpds> -nm <
output_empcpds>
```

The allowed tolerance in *m/z* and retention time matching can be configured through user-provided options provided in the documentation.

4.3 QA/QC and Report Generation

The PCPFM includes commands for data wrangling:

- Imputing missing features
- Filtering infrequent features
- Total ion current (TIC) normalization
- Outlier sample removal
- Blank masking
- Log transformation
- Batch correction
- Manual curation of samples

These operations act on feature tables and produce new, versioned output data. Unlike annotation, QA/QC strategies often depend on the study's goals. For example, in personalized medicine studies, rare features may be retained as they may reflect individual-specific biology. Sample types must be defined in metadata. The default label "Unknown" designates study samples to match Thermo naming conventions in their software. A typical QA/QC pipeline might involve:

```
pcpfm blank_masking -i <exp> -tm preferred
-nm preferred_blank_masked --blank_value Blank
--sample_value Unknown --query_field "Sample Type"
pcpfm drop_samples -i <exp> --table_moniker
preferred_blank_masked --new_moniker cleaned
--drop_value Unknown --drop_field "Sample Type"
--drop_others True
pcpfm normalize -i <exp> --table_moniker cleaned
--new_moniker normalized
```

```
pcpfm impute -i <exp> --table_moniker normalized
--new_moniker normalized_imputed
```

This sequentially performs blank masking, drops non-study samples, then normalizes the resulting table before imputing it. The full documentation provides details on configuring the cutoffs for each verb; however, each has reasonable defaults. Examples are provided in the notebooks of various workflows, while the published works include the workflows needed for reproduction.

Blank masking is used to remove features that are likely background or contaminants. By default, it removes features whose mean intensity in study samples is not at least three times greater than in the specified blank type. In studies using multiple types of blanks (e.g., process blanks vs. solvent blanks), each blank type should be masked separately. This ensures that masking is context-appropriate and avoids suppressing features that are only present in specific control conditions.

TIC normalization is performed by first identifying features detected in the majority of samples (≥90% by default). For each sample, the sum of intensities across these features is used to calculate a TIC value. Each sample's intensities are then scaled to match the median TIC of all study samples. This minimizes the effect of technical variability such as injection volume differences or instrument sensitivity drift, while reducing the influence of rare or inconsistently detected features. In this example, we first dropped samples that were not study samples to minimize any bias due to inconsistent matrices.

Reports are generated with:

```
pcpfm report -i <exp> --color_by='["Sample Type"]'
--text_by='["File Name"]'
```

By default, reports visualize key experiment metrics, including TIC distributions, feature missingness, and unsupervised clustering (PCA and heatmaps). However, the report can also serve broader purposes—for example, exploratory analysis—by adjusting `--color_by` and `--text_by` fields to highlight experimental variables (e.g., genotype, treatment). Reports also include a log of all executed pipeline commands and the versions of all libraries used, ensuring full reproducibility.

4.4 Combined Pipeline Output

To consolidate results into three tables—features, annotations, and sample metadata—use:

```
pcpfm generate_output -i <exp> -tm <feature_table>
-em <empirical_compounds>
```

Users must specify the feature table (`--tm`) and empirical compound set (`--em`) to be merged. Annotations are co-indexed by feature ID, supporting one-to-many mappings. The resulting output can be used directly for statistical analysis, visualization, or exported to downstream tools.

5 Asari-GC and Other Workflows

Asari and the related tools continue to evolve to support a broader range of analytical modalities. Users should refer to online documentation for accurate version description. Starting in Asari v1.17, support for GC–MS and other workflow types was introduced via a new `--workflow` flag.

For example, LC–MS/MS data can be processed using:

```
asari process --workflow LCMSMS -i <mzML_path>
```

The processing of GC data is the same as LC-MS data, but annotation is different. Both GC and MS/MS results are searched against spectral databases. Annotation of GC result also utilizes a retention index standard file.

Annotation is supported using any standard-compatible .msp file, while internal indexing is often used to speed up searching. Clustering of MS/MS spectra reduces data redundancy. Asari applies multiple approaches to group co-eluting features for EI-MS (electron ionization mass spectrometry) along spectral similarity search. This hard ionization workflow shares several unified modules with the soft ionization framework.

In addition, isotope enrichment detection is supported without requiring full annotation. Using the `detect_labelling` function in Khipu, users can evaluate isotopic labeling within empirical compounds. This method was used in Mitchell et al. to classify compounds into labeling confidence levels based on intensity shifts across isotopologues [14].

The above is a subset of examples on how to use the Asari toolkit. The design is to support flexible reuse of the tools and software libraries, so that solutions can be provided and customized for different projects and work environments.

6 Notes

1. Installation using a Python virtual environment is recommended but not required. For users working in containerized environments, Dockerfiles are available for Asari and the pipeline at their respective GitHub repositories.
2. Most packages have tutorials provided as Jupyternotebooks. We recommend using Visual Studio Code, with Jupyter and package management support, for running these notebooks (Fig. 2C).
3. With `--anno True`, Asari performs pre-annotation using Khipu to identify empirical compounds and infer the neutral masses. These neutral masses are then used for annotation by

comparing them against a default metabolite database. This pre-annotation uses common adducts based on the provided ionization mode, while isotopologues up to m+13C12 are considered.

4. Each library within the Asari toolkit is designed as a standalone module with a clearly defined scope. `mass2chem` provides low-level functions for manipulating molecular formulas, chemical descriptors, functional groups, and reaction rules. `metDataModel` serves as a shared repository of standardized data models that ensure consistency across tools [12]. The `asari` package supports feature detection and alignment for LC-MS and GC-MS data. `JMS` focuses on downstream annotation tasks, including integration with authentic compound libraries, database searches, and interfacing with genome-scale metabolic models. `Khipu` performs pre-annotation grouping of isotopologues, adducts, in-source fragments, and conjugates. All software packages are open-source and available at https://github.com/shuzhao-li-**lab/**.

References

1. Smith CA, Want EJ, O'Maille G, Abagyan R, Siuzdak G (2006) XCMS: processing mass spectrometry data for metabolite profiling using nonlinear peak alignment, matching, and identification. Anal Chem 78(3):779–787
2. Tautenhahn R, Böttcher C, Neumann S (2008) Highly sensitive feature detection for high resolution LC/MS. BMC Bioinformatics 9(1):1–16
3. Melamud E, Vastag L, Rabinowitz JD (2010) Metabolomic analysis and visualization engine for LC– MS data. Anal Chem 82(23):9818–9826
4. Rurik M, Alka O, Aicheler F, Kohlbacher O (2020) Metabolomics data processing using OpenMS. In: Li S (Ed.) Computational methods and data analysis for metabolomics. Humana Press, 49–60
5. Tsugawa H, Ikeda K, Takahashi M, Satoh A, Mori Y, Uchino H, Okahashi N, Yamada Y, Tada I, Bonini P, Higashi Y, Okazaki Y, Zhou Z, Zhu ZJ, Koelmel J, Cajka T, Fiehn O, Saito K, Arita M, Arita M (2020) A lipidome atlas in MS-DIAL 4. Nat Biotechnol 38(10):1159–1163. https://doi.org/10.1038/s41587-020-0531-2
6. Yu T, Park Y, Li S, Jones DP (2013) Hybrid feature detection and information accumulation using high-resolution LC–MS metabolomics data. J Proteome Res 12(3):1419–1427
7. Schmid R, Heuckeroth S, Korf A, Smirnov A, Myers O, Dyrlund TS, Bushuiev R, Murray KJ, Hoffmann N, Lu M, Sarvepalli A, Zhang Z, Fleischauer M, Duhrkop K, Wesner M, Hoogstra SJ, Rudt E, Mokshyna O, Brungs C, Ponomarov K, Mutabdzija L, Damiani T, Pudney CJ, Earll M, Helmer PO, Fallon TR, Schulze T, Rivas-Ubach A, Bilbao A, Richter H, Nothias LF, Wang M, Oresic M, Weng JK, Bocker S, Jeibmann A, Hayen H, Karst U, Dorrestein PC, Petras D, Du X, Pluskal T (2023) Integrative analysis of multimodal mass spectrometry data in MZmine 3. Nat Biotechnol 41(4):447–449. https://doi.org/10.1038/s41587-023-01690-2
8. Myers OD, Sumner SJ, Li S, Barnes S, Du X (2017) Detailed investigation and comparison of the XCMS and MZmine 2 chromatogram construction and chromatographic peak detection methods for preprocessing mass spectrometry metabolomics data. Anal Chem 89(17):8689–8695
9. Delabriere A, Warmer P, Brennsteiner V, Zamboni N (2021) SLAW: a scalable and self-optimizing processing workflow for untargeted LC-MS. Anal Chem 93(45):15024–15032. https://doi.org/10.1021/acs.analchem.1c02687
10. Sindelar M, Patti GJ (2020) Chemical discovery in the era of metabolomics. J Am Chem Soc 142(20):9097–9105. https://doi.org/10.1021/jacs.9b13198
11. Li S, Siddiqa A, Thapa M, Chi Y, Zheng S (2023) Trackable and scalable LC-MS metabolomics data processing using asari. Nat Commun

14(1):4113. https://doi.org/10.1038/s41467-023-39889-1
12. Mitchell JM, Chi Y, Thapa M, Pang Z, Xia J, Li S (2024) Common data models to streamline metabolomics processing and annotation, and implementation in a Python pipeline. PLoS Comput Biol 20(6):e1011912. https://doi.org/10.1371/journal.pcbi.1011912
13. Chi Y, Mitchell JM, Thapa M, Zheng S, Frohock Z, Li Y, Smirnov A, Du X, Li S (2025) Constructing a consensus serum metabolome. bioRxiv:2025.2005.2007.652782https://doi.org/10.1101/2025.05.07.652782
14. Mitchell JM, Chi Y, Zheng S, Thapa M, Wang E, Li S (2024) Annotation of metabolites in stable isotope tracing untargeted metabolomics via Khipu-web. J Am Soc Mass Spectrom 35(12):2824–2835. https://doi.org/10.1021/jasms.4c00175
15. Li S, Zheng S (2023) Generalized tree structure to annotate untargeted metabolomics and stable isotope tracing data. Anal Chem 95(15):6212–6217. https://doi.org/10.1021/acs.analchem.2c05810
16. Kuhl C, Tautenhahn R, Bottcher C, Larson TR, Neumann S (2012) CAMERA: an integrated strategy for compound spectra extraction and annotation of liquid chromatography/mass spectrometry data sets. Anal Chem 84(1):283–289. https://doi.org/10.1021/ac202450g
17. Mahieu NG, Spalding JL, Gelman SJ, Patti GJ (2016) Defining and detecting complex peak relationships in mass spectral data: the Mz.unity algorithm. Anal Chem 88(18):9037–9046. https://doi.org/10.1021/acs.analchem.6b01702
18. Uppal K, Walker DI, Jones DP (2017) xMSannotator: an R package for network-based annotation of high-resolution metabolomics data. Anal Chem 89(2):1063–1067. https://doi.org/10.1021/acs.analchem.6b01214
19. DeFelice BC, Mehta SS, Samra S, Cajka T, Wancewicz B, Fahrmann JF, Fiehn O (2017) Mass Spectral Feature List Optimizer (MS-FLO): a tool to minimize false positive peak reports in untargeted liquid chromatography-mass spectroscopy (LC-MS) data processing. Anal Chem 89(6):3250–3255. https://doi.org/10.1021/acs.analchem.6b04372
20. Kachman M, Habra H, Duren W, Wigginton J, Sajjakulnukit P, Michailidis G, Burant C, Karnovsky A (2020) Deep annotation of untargeted LC-MS metabolomics data with Binner. Bioinformatics 36(6):1801–1806. https://doi.org/10.1093/bioinformatics/btz798
21. Chen L, Lu W, Wang L, Xing X, Chen Z, Teng X, Zeng X, Muscarella AD, Shen Y, Cowan A, McReynolds MR, Kennedy BJ, Lato AM, Campagna SR, Singh M, Rabinowitz JD (2021) Metabolite discovery through global annotation of untargeted metabolomics data. Nat Methods 18(11):1377–1385. https://doi.org/10.1038/s41592-021-01303-3
22. Adusumilli R, Mallick P (2017) Data conversion with ProteoWizard msConvert. Methods Mol Biol 1550:339–368. https://doi.org/10.1007/978-1-4939-6747-6_23
23. Hulstaert N, Shofstahl J, Sachsenberg T, Walzer M, Barsnes H, Martens L, Perez-Riverol Y (2020) ThermoRawFileParser: modular, scalable, and cross-platform RAW file conversion. J Proteome Res 19(1):537–542. https://doi.org/10.1021/acs.jproteome.9b00328
24. Wishart DS, Guo A, Oler E, Wang F, Anjum A, Peters H, Dizon R, Sayeeda Z, Tian S, Lee BL, Berjanskii M, Mah R, Yamamoto M, Jovel J, Torres-Calzada C, Hiebert-Giesbrecht M, Lui VW, Varshavi D, Varshavi D, Allen D, Arndt D, Khetarpal N, Sivakumaran A, Harford K, Sanford S, Yee K, Cao X, Budinski Z, Liigand J, Zhang L, Zheng J, Mandal R, Karu N, Dambrova M, Schiöth HB, Greiner R, Gautam V (2022) HMDB 5.0: the Human Metabolome Database for 2022. Nucleic Acids Res 50(D1):D622–d631. https://doi.org/10.1093/nar/gkab1062
25. Conroy MJ, Andrews RM, Andrews S, Cockayne L, Dennis EA, Fahy E, Gaud C, Griffiths WJ, Jukes G, Kolchin M, Mendivelso K, Lopez-Clavijo AF, Ready C, Subramaniam S, O'Donnell VB (2024) LIPID MAPS: update to databases and tools for the lipidomics community. Nucleic Acids Res 52(D1):D1677–D1682. https://doi.org/10.1093/nar/gkad896
26. Schymanski EL, Jeon J, Gulde R, Fenner K, Ruff M, Singer HP, Hollender J (2014) Identifying small molecules via high resolution mass spectrometry: communicating confidence. Environ Sci Technol 48(4):2097–2098. https://doi.org/10.1021/es5002105

Chapter 6

LC–MS/MS-Based Untargeted Lipidomics Using MS-DIAL

Bujinlkham Buyantogtokh, Yoshimasa Todoroki and Hiroshi Tsugawa

Abstract

Mass spectrometry (MS) has become the cornerstone of metabolomics and lipidomics because of its high sensitivity, specificity, and ability to analyze complex biological samples. However, the data complexity necessitates the use of robust computational tools. MS-DIAL is a versatile and user-friendly software for untargeted metabolomics and lipidomic workflows. It was launched as a universal program for untargeted metabolomics and supports multiple instruments (GC/MS, GC/MS/MS, LC/MS, and LC/MS/MS) and MS vendors (Agilent, Bruker, LECO, Sciex, Shimadzu, Thermo, and Waters). Common data formats such as netCDF (AIA) and mzML can also be managed. In addition, several MSP files, including EI and MS/MS spectra, were included as a "start-up kit." Moreover, MS-DIAL has an internal version of the Fiehn lab's GC/MS database (oriented by the FAME RI index), in silico retention time, and MS/MS database for LC/MS/MS-based lipidomics. Isotope-labeled tracking can be executed using an LC/MS project. This chapter provides two detailed protocols using LC–MS/MS for hydrophilic metabolome and lipidome analyses using MS-DIAL. Each protocol covers the key steps, including raw data import, peak detection, alignment, annotation, and export. Practical tips, troubleshooting strategies, and case studies are included to enhance the reproducibility and interpretation of the results.

Keywords LC–MS/MS, Lipidomics, MS-DIAL, Annotations, Curations

1 Introduction

Liquid chromatography–mass spectrometry (LC–MS) is an indispensable tool in metabolomics that enables the separation, detection, and quantification of small molecules in complex biological matrices. However, the data generated from LC–MS are highly multidimensional, comprising variables such as mass-to-charge ratio (*m/z*), retention time (RT), and ion intensity.

MS-DIAL (Mass Spectrometry-Data Independent Analysis) is a comprehensive software designed to process this complex data for both targeted and untargetedworkflows. It is widely used in metabolomics, lipidomics, and proteomics owing to its ability to manage various types of MS data, including data-independent acquisition (DIA) and data-dependent acquisition (DDA).

Michael Witting (ed.), *Computational Methods and Data Analysis for Metabolomics*, Methods in Molecular Biology, vol. 3063,
https://doi.org/10.1007/978-1-0716-5452-1_6,

The latest version of MS-DIAL offers an improved user experience with a new graphical interface. Unlike before, users no longer need to convert raw MS data from various vendor formats into the common ABF format, which is still supported but no longer mandatory. Now, data can be imported directly, streamlining the workflow for untargeted metabolomics, lipidomics, and proteomics analyses (Fig. 1).

The latest version of MS-DIAL is the most versatile tool for multiomics data analysis. Therefore, it supports direct injection, many separation methods (GC, LC, CE, and SFC), ionization (EI, ESI, and MALDI), fragmentation (CID, HCD, ETD, ECD, EIEIO, EID, and OAD), and MS/MS (DDA, SWATH, AIF, and ZT scan). In addition, technological advances in fragmentation mechanisms have also been considered, making it possible to work with these types of data to offer a comprehensive structural identification of metabolites, lipids, and proteins (Fig. 2).

After data processing, including peak picking, deconvolution, compound identification, and peak alignment, MS-DIAL provides several normalization methods (including LOWESS, internal standard, SPLASH, and TIC), multivariate analysis by principal

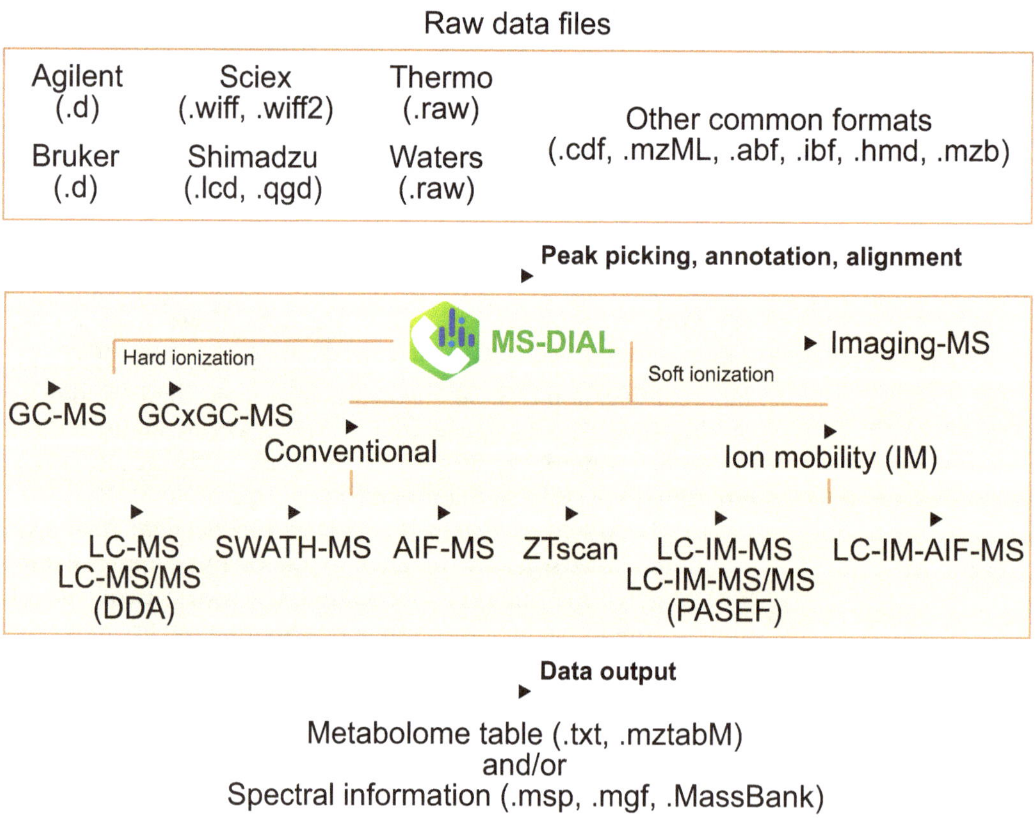

Fig. 1 Overview of supported file formats and analytical techniques

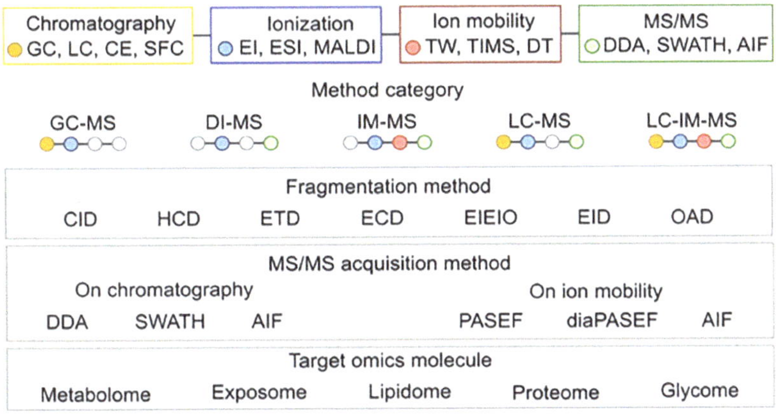

Fig. 2 Summary of analytical techniques compatible with MS-DIAL 5

component analysis (PCA), Notame (a two-parameter processing tool that requires more than five quality control (QC) samples and two parameters), and molecular spectrum networking.

Advantages of MS-DIAL

- **Multi-format compatibility**: This supports data formats such as mzML, mzXML, and vendor-specific raw files (e.g., *.raw*, *.d*, .lcd).
- **DIA, DDA, and AIF acquisition type support**: Optimized for untargetedworkflows using these acquisition methods.
- **Target omics**: Supports metabolomics, lipidomics, and proteomics data.
- **Visualization tools**: Offering chromatogram views, peak tables, and an alignment table for data exploration.
- **Batch processing**: Efficiently processes multiple datasets, making it suitable for large-scale studies.

2 Materials

Download and use:

1. **Download MS-DIAL**:
   - Visit the official MS-DIAL website and download the binary program (https://systemsomicslab.github.io/compms/msdial/main.html).
   - The latest version is recommended.
2. Use:
   - Extract the downloaded folder and run the application.
   - Ensure that the computer meets the minimum requirements (i.e., Windows 11 or later, ≥32 GB RAM).

*Demo data used here can be downloaded from "MS-DIALLC–MSlipidomics demo data" on Zenodo repository (https://zenodo.org/records/15780607/files/MSDIAL%20LCMS%20demo%20data.zip).

3 Methods

3.1 Preprocessing in Raw Data Viewer (Separate Application)

1. Browse raw data.
2. Set your method type and ion mode. Click "Load" to load your data.
3. When your raw data file name appears on "Loaded files:", click it (if you have more than one data file here, select one to open).
4. Click "Show" to see histograms (Fig. 3).
5. Use histograms to set MS and MS/MS cutoff values to delete noise spectra (Fig. 4).

3.2 Starting up your project

1. **Project parameters:**
   - Click "New project" to create a project (or "Load project" to load a previous project).
   - Browse a path to save the project.
   - Browse the raw data files. Raw files (.abf, .hmd, .mzb, .mzml, .cdf, .ibf, .wiff, .wiff2, .raw, .lcd, .qgd, .lrp) can be browsed directly, and an Agilent raw file (.d folder) can be imported by drag-and-drop).

   We highly recommend using raw data without any converter! (Fig. 5).
2. **Raw measurement files**:
   - Further assign identifiers to measurements by sample type (Sample, Standard, QC, Blank) and class ID (according to your experimental setup).
   - Acquisition type can be set one by one or for all raw data by using the "Set to all" button.
   - Set the batch (if multiple batches are analyzed), analytical order, and dilution factor, or exclude some samples from further data processing, if needed (Fig. 6).

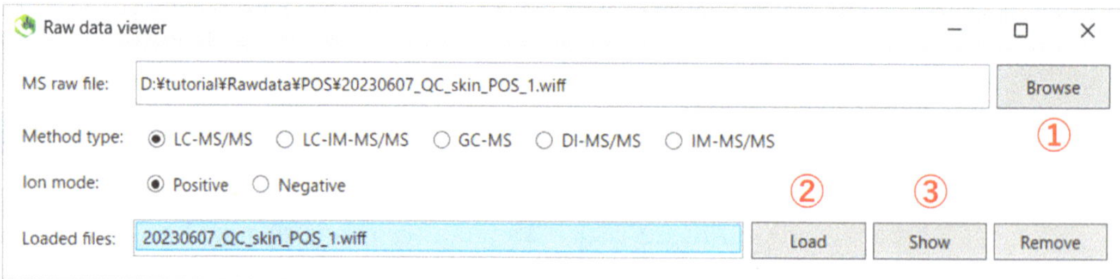

Fig. 3 User interface of the Raw data viewer

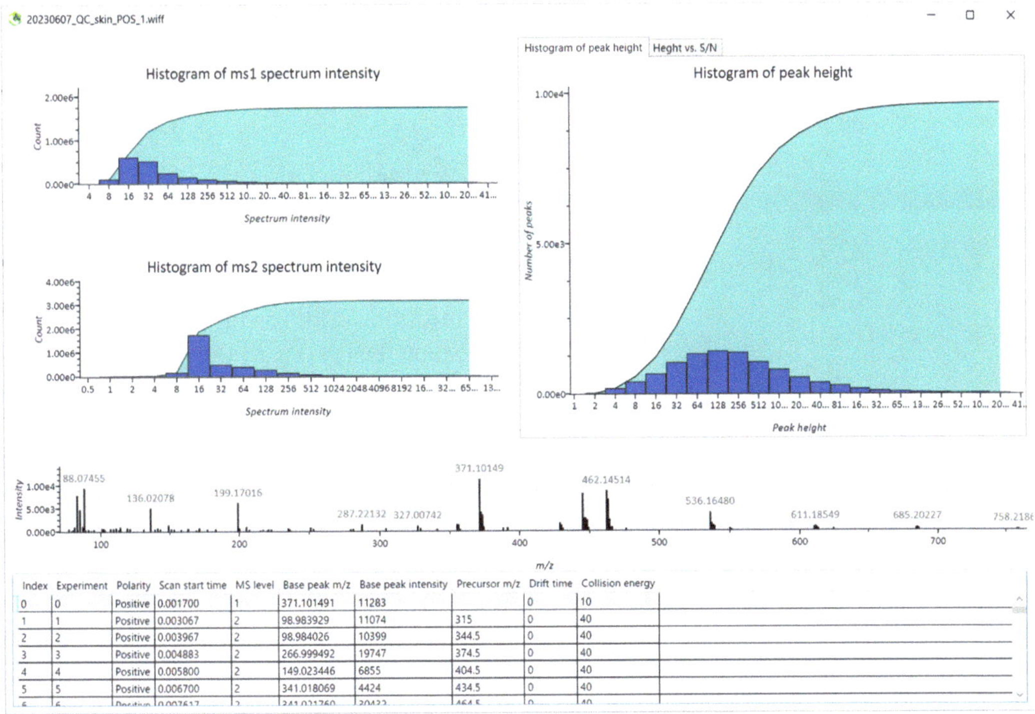

Fig. 4 Result window of the Raw data viewer

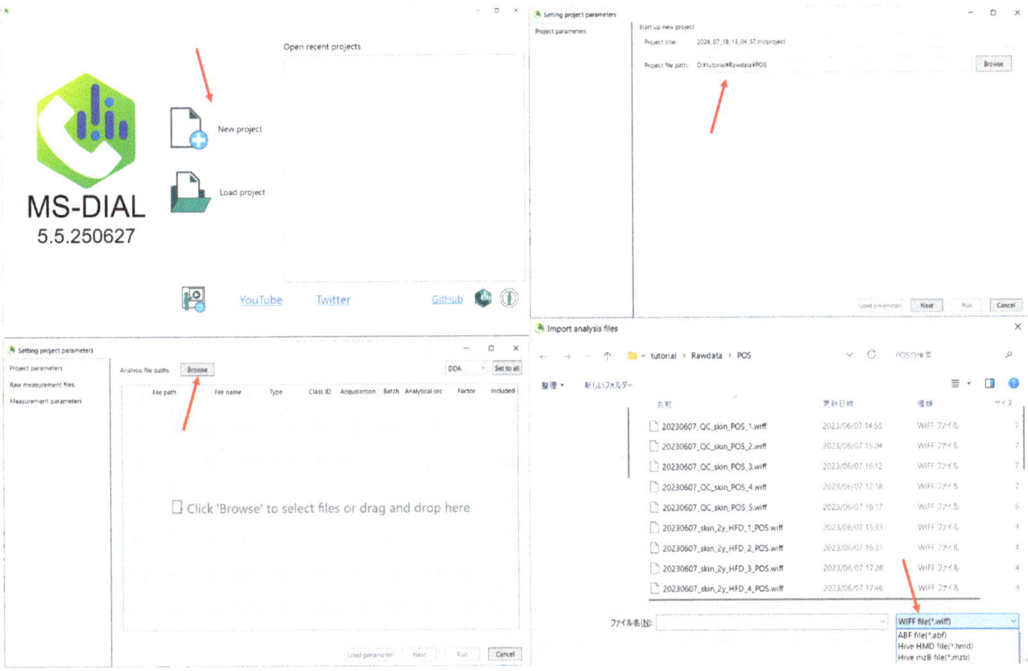

Fig. 5 Create a new project in MS-DIAL 5

3. **Measurement parameters**:

- **Project name**: The project name can be changed to distinguish the created files, such as "date_sample_ionMode.mddata."
- **Ionization type**: Set the ionization type (select soft ionization when analyzing ESI data or hard ionization when analyzing EI data).
- **Separation type**: Select chromatography (for LC, GC, CE, or SFC) or direct infusion. Additionally, you can check the ion mobility box coupled with one of the previous options.
- **Collision type**: Select the corresponding collision type according to your analysis.
- **Data type (MS1)** or **(MS/MS)**:

Centroid or Profile?

In the previous version of MS-DIAL, users were required to define the type of imported data (centroid or profile). However, in the new version (above MS-DIAL 5 [1]), the centroid data can be chosen by default. MS-DIAL retrieves data as a centroid, even for vendor formats that were previously considered as profiles (e.g., SCIEX or Thermo). The only exception is the Agilent data, which can be stored as a centroid, a profile, or both. To make sure what type of

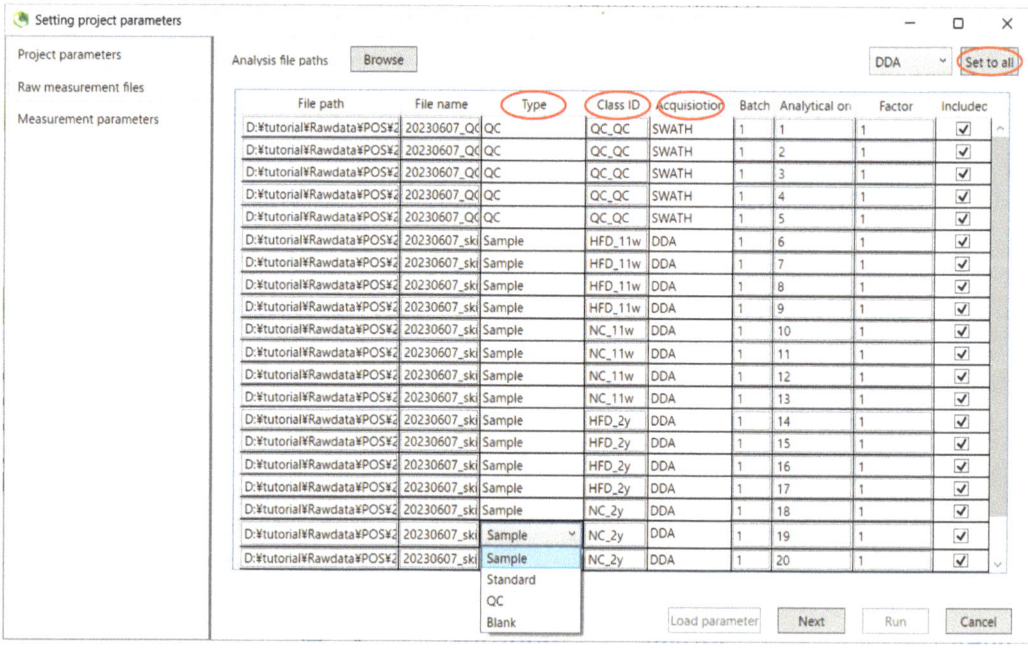

Fig. 6 Raw measurement file parameter settings in MS-DIAL 5

data you have, we recommend using our program "raw_data_viewer.exe" (which is available in the MS-DIAL base folder). The Raw data viewer shows the spectrum retrieved by the MS-DIAL backend (here, you can determine the data type), and users can additionally determine the threshold values for the amplitude cutoff by looking at the statistics of peak height versus frequency.

Your data profile changes when a converter, such as an ABF converter, is used. For instance, ThermoFisher raw mass spectrometry data contains profile data, so after conversion to ABF format, the data remain in profile mode, requiring you to select "Profile" in MS-DIAL. However, if you use ThermoFisher raw data directly, you have the option to choose "Centroid."

In conclusion, your data type changes when using a converter. Therefore, we recommend the use of raw data without converters.

- **Ion mode**: Select the corresponding ion mode according to your analytical setup.
- **Target omics**: Select the type of omics data that you want to perform. If you select the "lipidomics" project, you do not have to prepare the library since MS-DIAL internally contains the library and theoretical MS/MS spectra of lipids (and actually, the fragment ions are further evaluated by the decision tree algorithm to provide the proper lipid structure representation). On the other hand, when you select the "metabolomics" project, your own library will be required for compound identification.
- **Further metadata**: Provide more information regarding your experiment to include the information in MassBank, MoNA, and NIST(MSP) format exports (Fig. 7).

4. **Data collection**

- **Mass accuracy (centroid parameter)**: After initial peak detection in MS-DIAL, mass tolerance defines the acceptable range of mass deviation for both MS1 and MS/MS data. By default, mass tolerance values are set to ±0.01 Da for MS1 and ±0.025 Da for MS/MS. This indicates that the peaks within these ranges around the detected mass are integrated and considered for further analysis. Importantly, MS/MS tolerance also influences the construction of MS/MS chromatograms, affecting how closely the detected fragment ions must match the theoretical masses to be included in the MS2Dec deconvolution process.
- **Data collection parameters**: In MS-DIAL, the analysis ranges can be defined as per the analysis of retention time (RT), MS1, and MS/MS axis to focus on specific parts of the data.
- **Execute retention time correction**: This setting adjusts for any shifts or drifts in retention times that may occur during the analysis, ensuring that the peaks are aligned

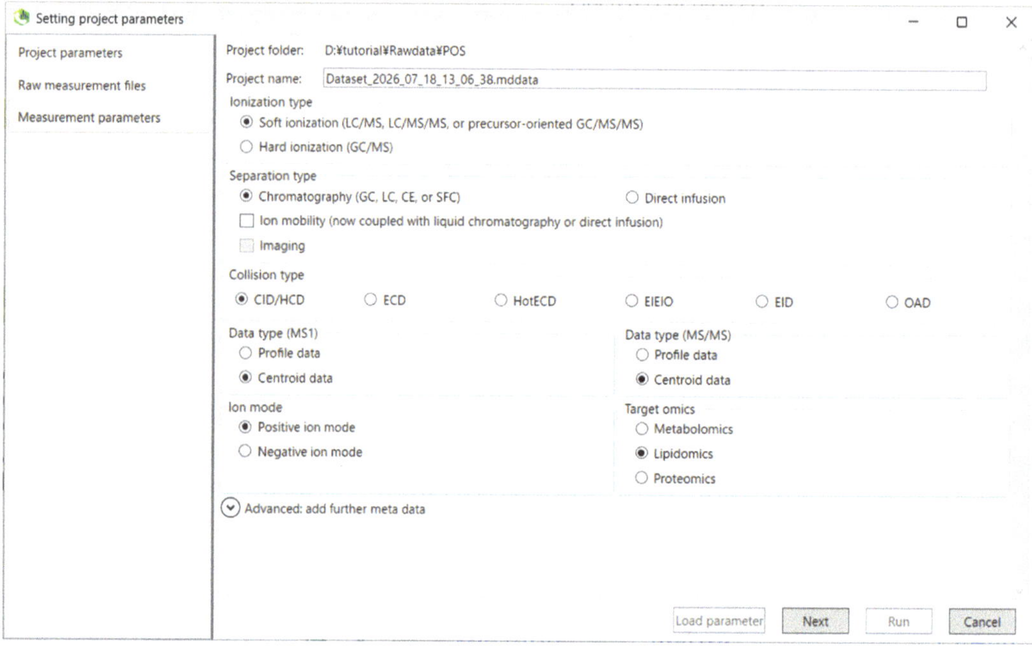

Fig. 7 Measurement parameter settings in MS-DIAL 5

accurately across different samples or runs. Use this function when analyzing data under different elution conditions (e.g., different columns). For further information, please refer to our tutorial (https://www.youtube.com/watch?v=Phzd7ucD5C4&t=2s).

- **Isotope recognition**: Isotope recognition is used to identify and handle isotopic peaks that correspond to the same molecular ion but with different masses owing to natural isotopic variations (e.g., [M+1] or [M+2]). This process helps accurately quantify the main molecular ion by distinguishing it from its isotopic counterparts. Focusing on small-molecule research (less than 2000 Da), the maximum number of charges can be set to 2. However, the parameters can be changed to 8 or more to process the proteome or snRNA research data.
- **Consider halogen elements**: If the samples may contain chlorine (Cl) or bromine (Br), enable this option. These elements exhibit unique isotopic distribution patterns, which help the software identify and quantify compounds by recognizing the specific isotopic peaks created by Cl and Br.
- **Multithreading**: Please set the number of threads you want to use. You can check the maximum available thread count in the resource monitor (Task Manager -> Resource Monitor). Each dataset uses two threads, so if you set the

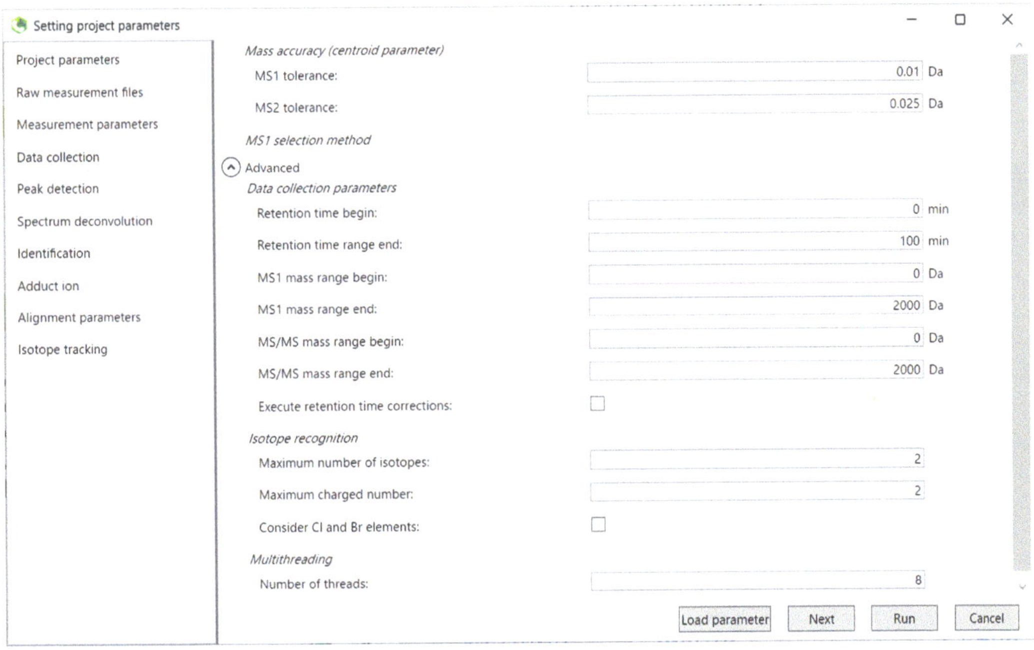

Fig. 8 Data collection parameter settings in MS-DIAL 5

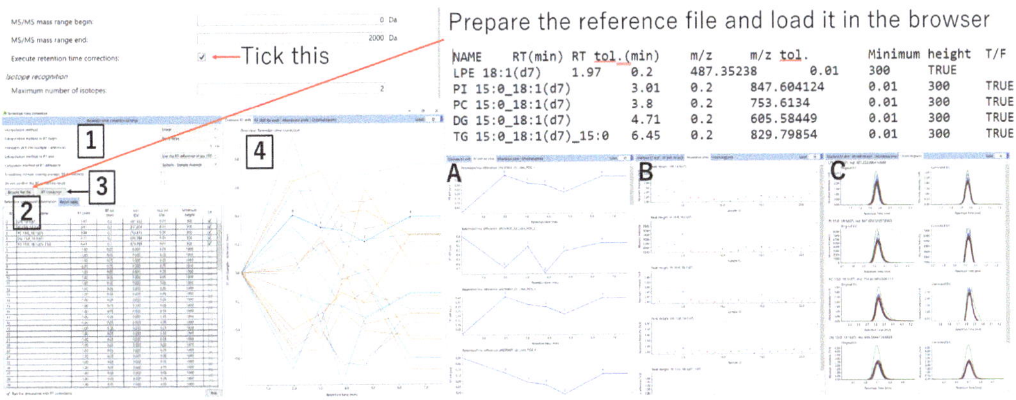

Fig. 9 Retention time correction tutorial

count to 8, four raw datasets will be analyzed simultaneously (Figs. 8 and 9)

5. **Peak detection**

- **Minimum peak height**: Set the minimum intensity level required for a peak to be detected in the data. Peaks with heights below this threshold are ignored, which helps in filtering out noise and focusing only on significant signals for accurate analysis.
- **Mass slice width**: Define the range of mass-to-charge ratios (m/z) around each detected peak that will be considered

during data analysis. This parameter helps group ions within a specified *m/z* range, improving peak detection and quantification by accounting for slight variations in mass measurements. In this case, we are using 0.1 Da. Based on our experience, 0.1 Da is suitable for Q-TOF, while 0.05 Da works well for Orbitrap instruments.

- **Smoothing method**: This method is used as a default for peak detection to accurately determine the left and right edges of the peak.
- **Smoothing level**: The recommended smoothing level is 1–3. This represents the number of points on each side of the peak.
- **Minimum peak width**: A threshold for filtering peaks based on their width. For example, if the minimum peak width is set to 5 scans, any peaks spanning fewer than 5 scans will be excluded from the results. Only peaks with a width of 5 scans or greater will be included.
- **Exclusion mass list**: If you already know of unwanted *m/z* peaks caused by column or solvent contaminants, you can specify them here to exclude those peaks from the analysis (Fig. 10).

6. **Spectrum deconvolution**
   - **Sigma window value**: Strongly influenced by deconvolution resolution. A higher value (0.7–1.0) will lower peak top resolution, reducing the number of resolved peaks.

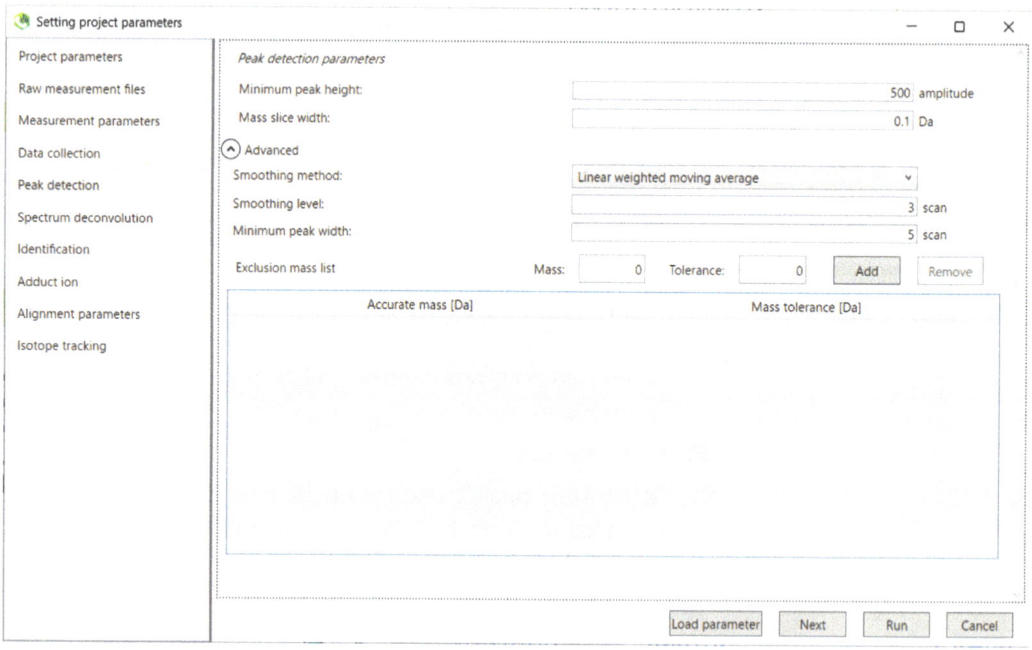

Fig. 10 Peak detection parameter settings in MS-DIAL 5

Conversely, a lower value (0.1–0.3) may result in the detection of many noisy chromatographic peaks.

- **MS/MS abundance cutoff**: The minimum intensity level of the MS/MS (tandem mass spectrometry) signals determines whether a fragment ion is considered significant and included in the analysis. Unless you are working with data-independent MS/MS datasets, you can skip this setting. However, increasing the cutoff value can help reduce MS peak noise.
- **Exclude after precursor ion**: If you want to remove the product ions after the focused precursor ion (recommended for metabolomics and lipidomics), check the box.
- **Keep the isotopic ions until**: The isotopic patterns in MS1 spectra are often disturbed. However, in some cases, the isotopic patterns in MS/MS spectra are clearer than those of MS1 spectra, allowing for more accurate molecular formula annotation. If this parameter is set to 5 Da, ions until precursor + 5 Da will be retained after the MS2Dec algorithm is completed.
- **Keep the isotopic ions w/o MS2Dec**: The MS2Dec algorithm may sometimes erase the precursor's isotopic ions owing to mathematical issues. Therefore, the raw MS/MS spectra can only be retained for the isotopic ions of the precursor using this option.

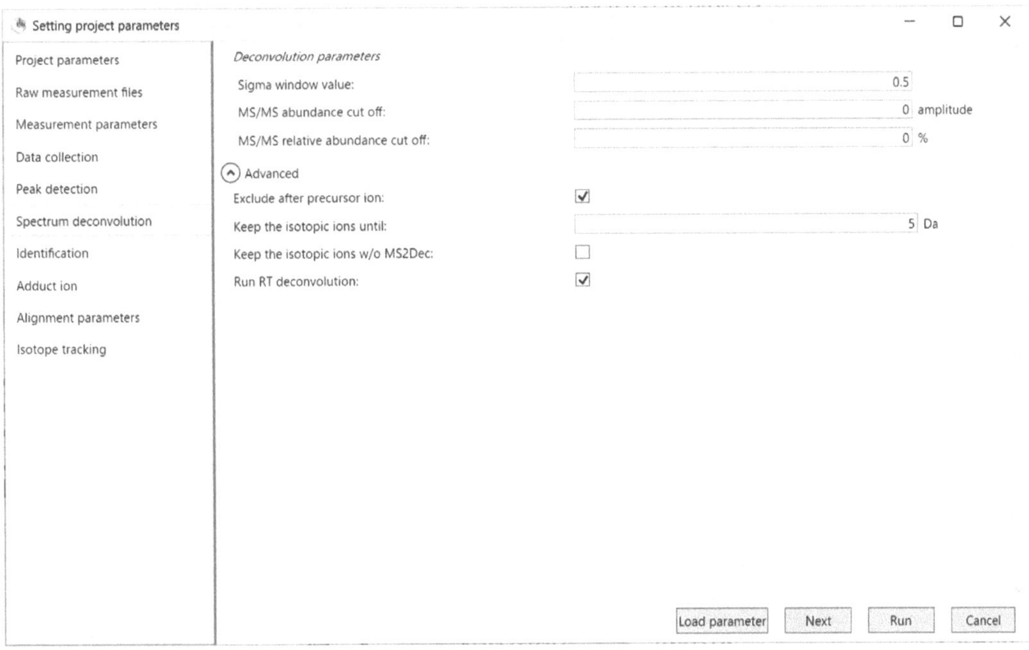

Fig. 11 Spectrum deconvolution parameter settings in MS-DIAL 5

- **Run RTdeconvolution**: By unchecking this, no deconvolution is performed; therefore, the raw data can be checked using MS-DIAL (Fig. 11).

7. **Identification** (Fig. 12a, b)

- Database **setting**:

  i. **Lipidomics project**: MS-DIAL automatically browses the LBM library using a software folder. Use "Configure lipid classes" to include or exclude lipid classes or adduct types.

  ii. **MSP file**: MS-DIAL supports the MSP file format in the ASCII text. The software can utilize "RETENTIONTIME:", "PRECURSORTYPE:", and "FORMULA:" information for metabolite identification (case-insensitive). Retention time information must be specified in minutes (min) whenever possible. The adduct ion information, i.e., here "Precursor type," will be used for the adduct ion search algorithm (Fig. 13).

  iii. **Text format library**: MS-DIAL supports a tab-delimited text format library for peak identification using retention time and MS1 accurate mass information. This option is useful for internal standard identification and can be used even without MS/MS libraries. The minimum requirement for this text library is two columns: "metabolite name" and "*m/z*." Retention time and adduct ion fields can be added to provide additional information for peak identification and adduct ion assignment in MS-DIAL (Fig. 14).

  iv. Add or delete a database by clicking the "+" icon. Select the database and click the upper arrow in the annotation method settings so that the database annotation is preferred.

- **Annotationmethod**: You can set the RT tolerance value. If suitable RT information is unavailable, set the tolerance to 100 (default) or larger (larger than your LC time). The two mass tolerances for MS1 and MS2 are required for the compound search and are dependent on your instrument performance.
- **MS2 spectrum cutoff**: The minimum intensity threshold for ions to be considered in the analysis. Fragment ions with intensities below this value are excluded from the spectrum to reduce background noise and improve the quality of data interpretation. Peaks below the user-defined threshold are not used for the MS/MS similarity calculations.

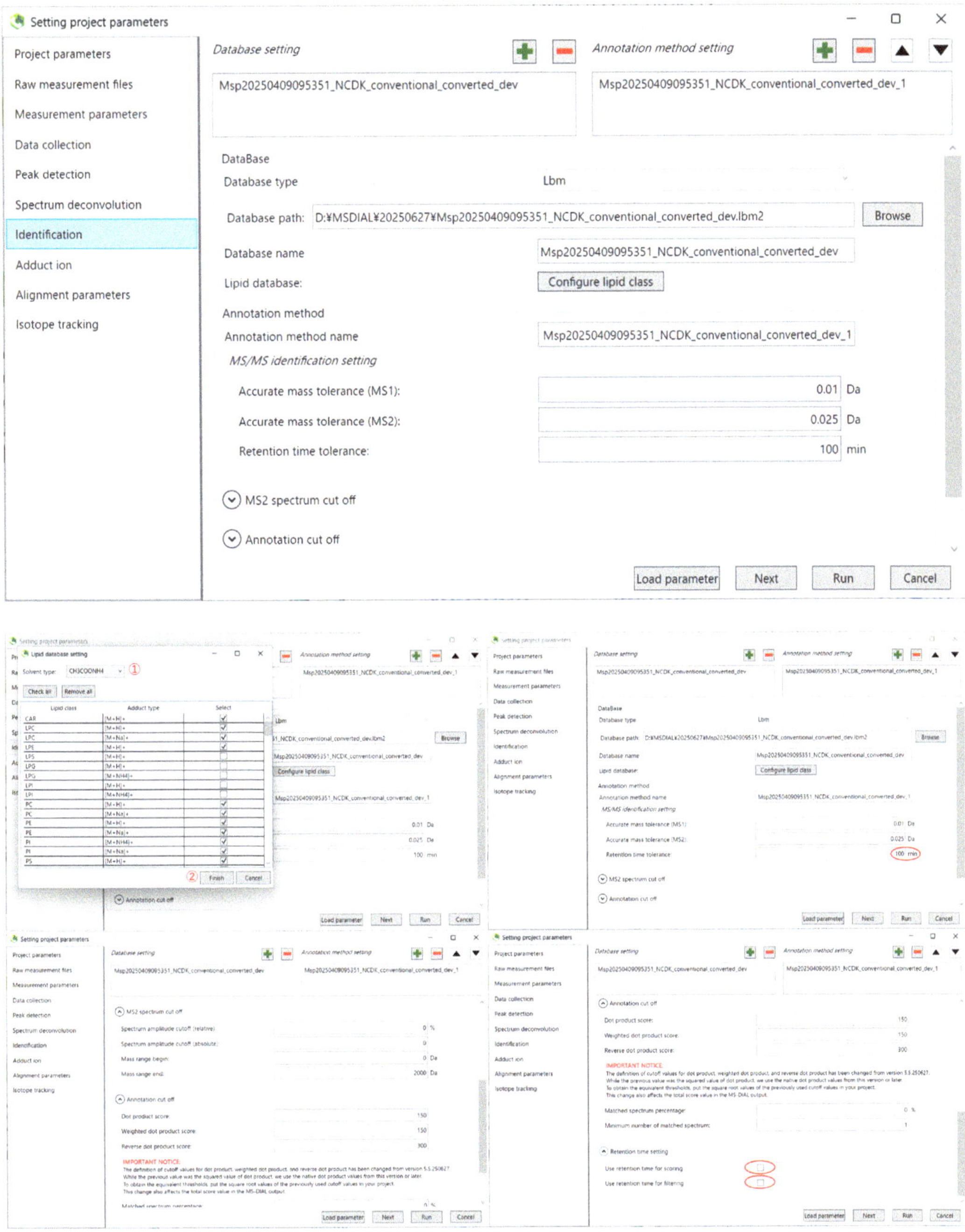

Fig. 12 Identification parameter settings in MS-DIAL 5

Fig. 13 MSP library format in MS-DIAL 5

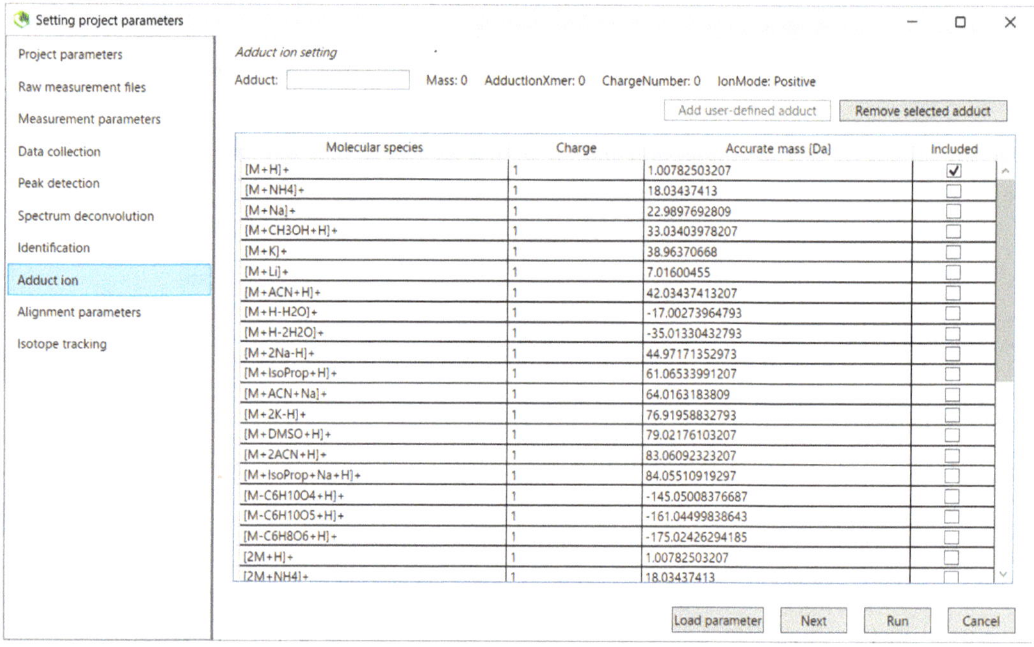

Fig. 14 Text library format in MS-DIAL 5

- **Annotationcutoff:** In MS-DIAL, these parameters are crucial for determining how closely the experimental spectra match the reference spectra from the databases during the identification process. They help score the similarity of fragmentation patterns compared to records in the database.

  i. **Dot product score**: Cosine similarity of experiment spectrum.

ii. **Weighted dot product**: Cosine similarity of the experiment spectrum when considering the intensity.

iii. **Reverse dot product**: Cosine similarity of reference spectrum.

iv. **Matched spectrum percentage**: Percentage of how many reference spectra that are matched.

v. **Minimum number of matched spectra**: Minimum number of matched spectra for annotation

Retention timesetting: If the tolerance is set and filtering is selected, peaks with different RT will not be detected. If you do not tick this, the processing will not be affected, even if the RT tolerance or range value is changed. If RT filtering is not selected, the total score will be calculated by also considering the RT score, which compares the observed RT of a lipid or metabolite in your data with the RT of the corresponding peak in the reference library (Fig. 15).

8. **Adduct ion**

- It is possible to select suitable adduct ions for analysis or to manually enter the adduct type (Fig. 16).

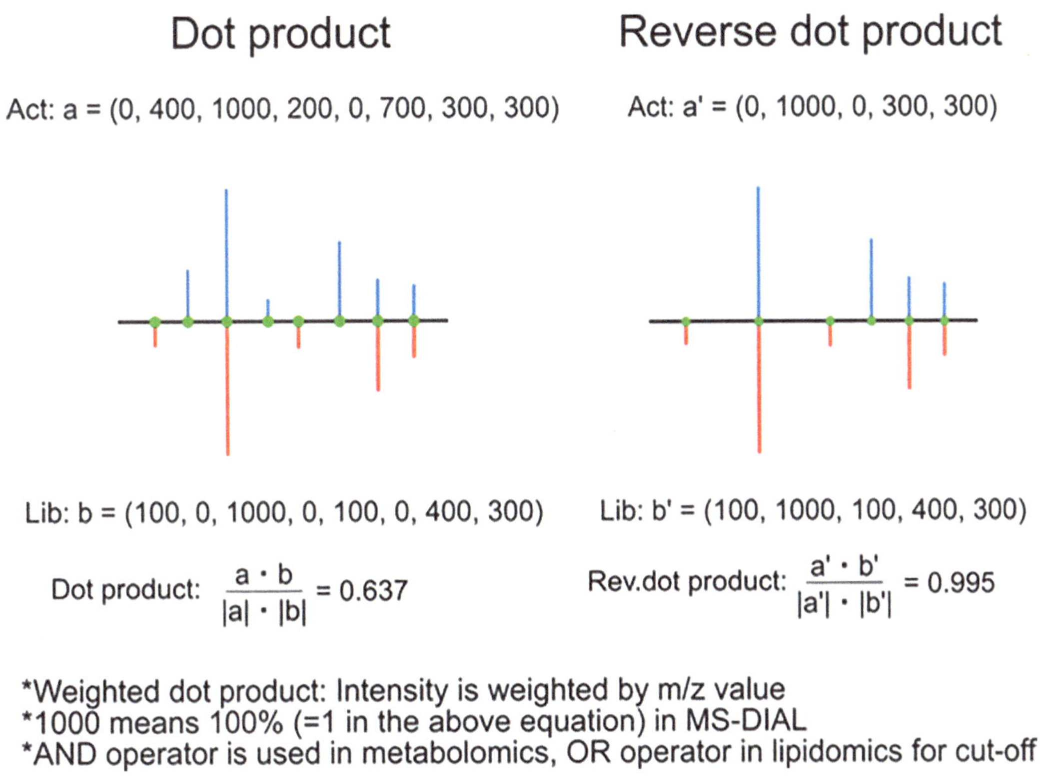

Fig. 15 Score calculations in MS-DIAL 5

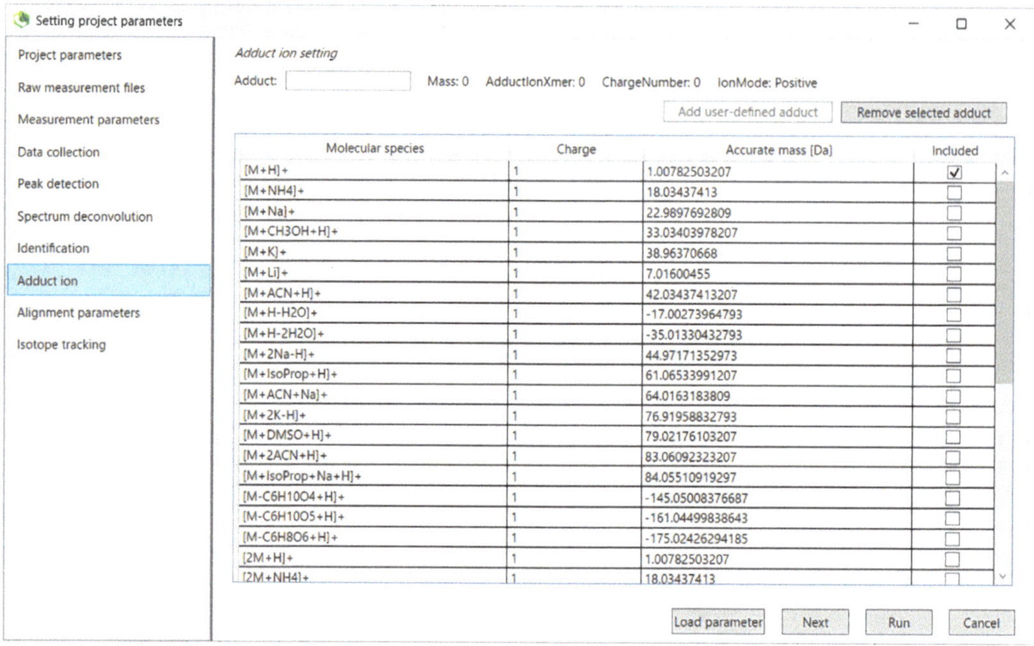

Fig. 16 Adduct ion settings in MS-DIAL 5

9. **Alignment parameters**

- **Result name**: The name of each alignment is shown in the tab "Alignment navigator" in the main window of MS-DIAL.
- **Reference file**: If you already have suitable quality control (QC) data, typically mixed sample data, specify the QC file in the reference file. All the sample data will be aligned to this QC file.
- **RT and MS1 tolerance**: The RT and MS1 values of the peaks will be compared with the library RT and MS spectra.
- **Retention timeand MS1 factor**: These values indicate the importance of either RT or MS1 for comparing and evaluating the similarity of spectra among samples based on RT and MS1 tolerances.

$$\text{RT factor} = \exp\left(-\frac{1}{2}\left(\frac{\text{RT}_1 - \text{RT}_2}{\text{RT}_{\text{tol}}}\right)\right)$$

$$m/z \text{ factor} = \exp\left(-\frac{1}{2}\left(\frac{m/z_1 - m/z_2}{mz_{\text{tol}}}\right)^2\right)$$

- **Peak count filter**: Use this setting to remove peaks that are not fully detected in the alignment. For example, if three biological replicates had the same peak information, and a total of 24 datasets. Then you may set the peak count filter as (3/24) × 100. This means peaks will be removed if they contain missing values for more than the specified percentage.

- **N% detected in at least one group**: Filtering is applied within each sample group. If set to 100%, peaks must be detected in all samples of that group.
- **Remove features based on blank information**: If you include blank sample information in your analysis and want to incorporate it into the results, enabling this option is suggested. This allows MS-DIAL to evaluate and filter peaks based on the blank sample data. Once this option is checked, you can also adjust the settings below.
- **Keep "reference matched" metabolite features**: This option ensures that any metabolites matched to a reference library (such as MSP or text libraries) are retained, even if they appear in the blank sample.
- **Keep "suggested (without MS2)" metabolite features**: This retains metabolites that are annotated based on their MS1 feature alone, without MS2 data, even if they exceed the blank feature threshold.
- **Keep removable features and assign the tag**: Any features that do not pass the blank feature threshold are still included in the results; however, they will be marked as removable. You can review these features by using the "Blank Filter" checkbox in the Peak Spot Navigator pane. If this option is not checked,

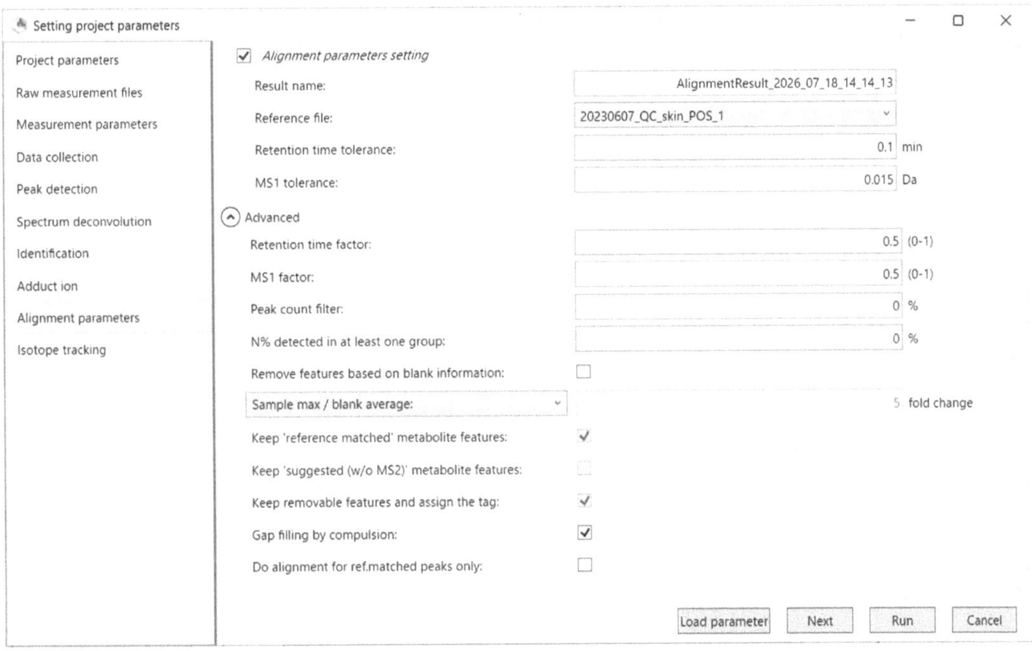

Fig. 17 Alignment parameter settings in MS-DIAL 5

blank-related features will be fully removed from the analysis, and you will no longer be able to view them after alignment.

- **Gap filling by compulsion**: This enables the recognition of peaks based on the average peak width of the detected metabolites across samples, even if a local maximum is not observed in the chromatogram.
- **Do alignmentfor ref.matched peaks only**: Only reference-matched peaks will be aligned so that the alignment process can be performed more quickly (Fig. 17).

Click Run and wait for the results. The initial window will show you "Peak spot viewer," so if you want to see the alignment result ("Alignment spot viewer"), double-click on the result from the alignment navigator (Fig. 18).

4 Curation in Detail

Materials for demonstration data

- **LC instrument**: Reverse-phase (RP) columns
- **MS instrument**: Soft ionization (CID)
- **Sample preparation**: Extract lipids using the MTBE method.
- **Software**: MS-DIAL 5.5.250627

Please use the MS-DIAL 5 tutorial webpage for further information (https://systemsomicslab.github.io/msdial5tutorial/lipidomics.html).

Tips for curation

1. **Noise cut-off**: Adjusts the peak picking process by setting a cutoff threshold for confident annotation. This can be defined as a relative value (where the highest peak is set to 100%) or as an absolute value (based on absolute intensity).

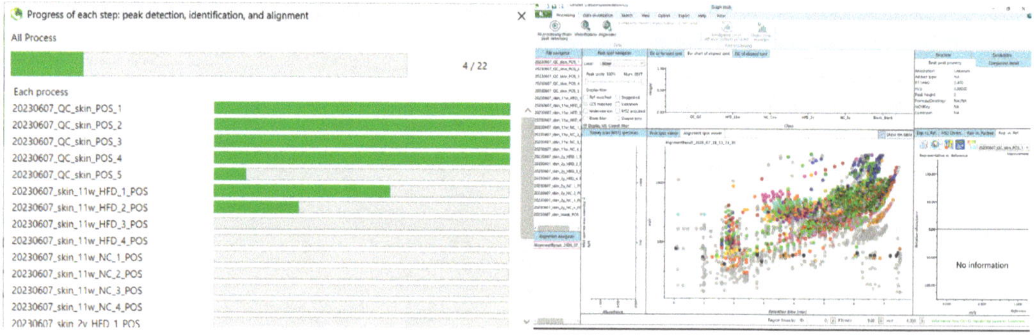

Fig. 18 Analysis running view and user interface after analysis in MS-DIAL 5

2. **Normalization**: Normalize your data to reduce batch effects and ion suppression (available in the "Data visualization tab) (Fig. 19).

3. **Use RT or RT order**: Using the RT or RT order can help prevent false positives. For example, in RP-LC, if FA 16:0 elutes at 7.53 min, FA 17:0 should elute after 7.53 min (Fig. 20).

4. **Chromatogram**: In some cases, the chromatogram shape indicates whether a peak is co-eluted. In MS-DIAL, selected

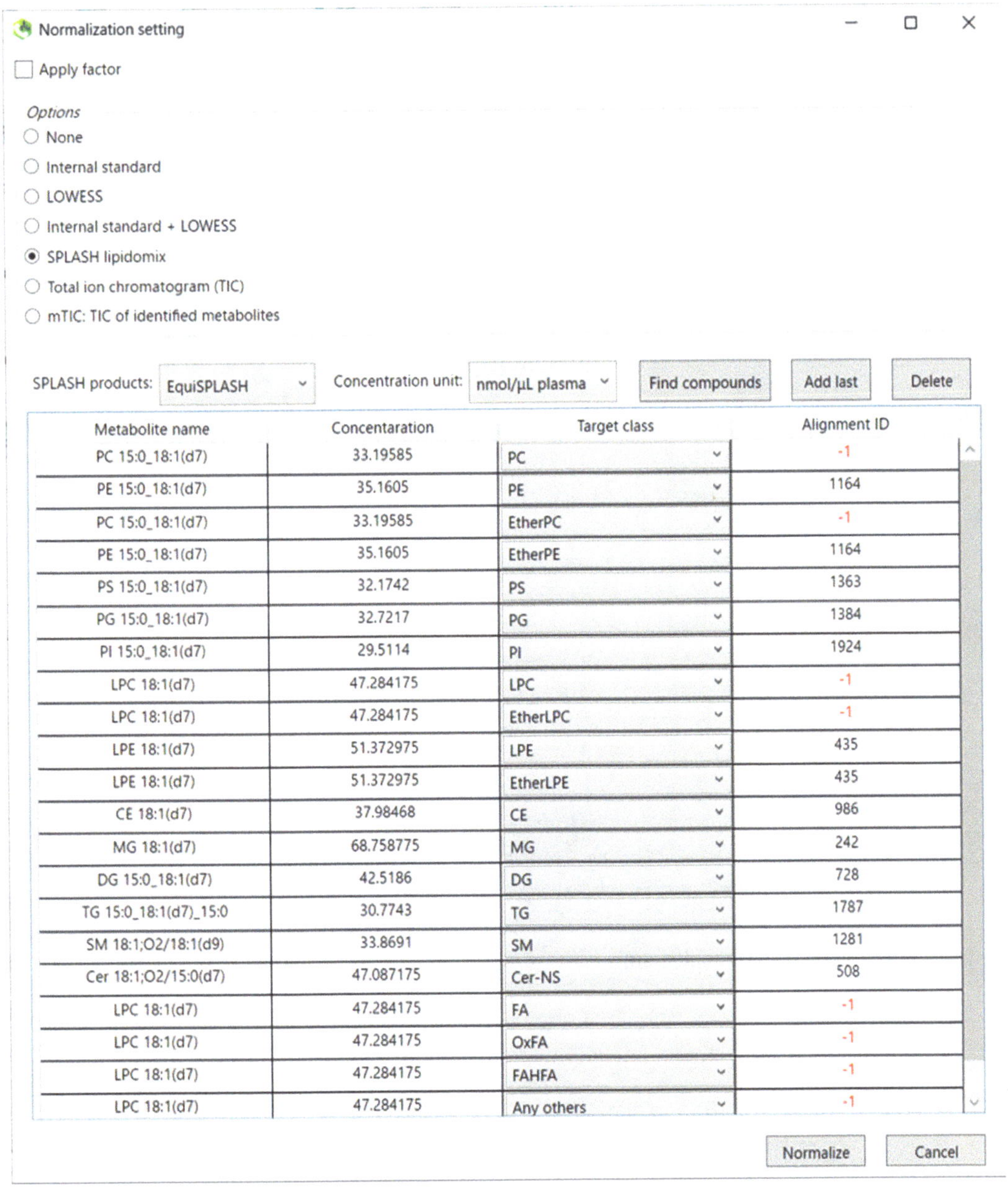

Fig. 19 Normalization settings in MS-DIAL 5

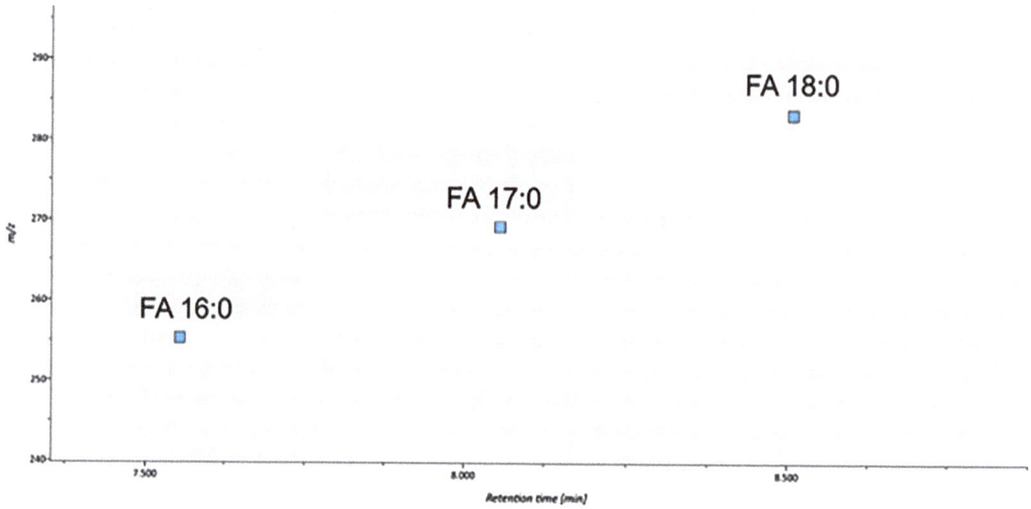

Fig. 20 Example of fatty acid elution

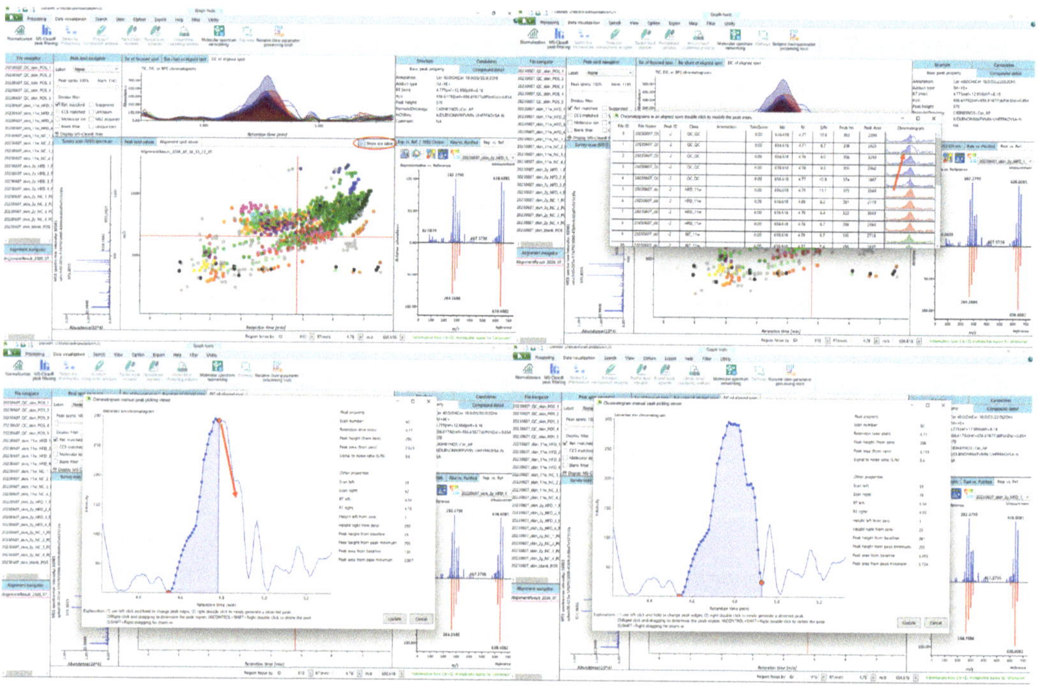

Fig. 21 Manual editing of a chromatogram

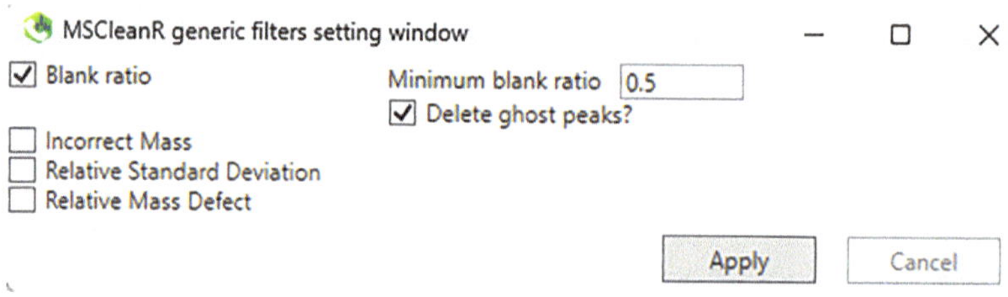

Fig. 22 MS-CleanR settings in MS-DIAL 5

chromatograms of the samples can be modified manually. For manual modification, double-click on the chromatogram that you want to edit and edit it on "Chromatogram manual peak picking viewer" by dragging the selected data point and clicking "Update" (Fig. 21).

5. **Blank filter**: Using blank samples can help if the peak originated from the sample or background (from the tube or tip used in the experiment). Click MS-CleanR peak filtering from the "Data visualization" tab, and for using a blank ratio, tick the box (to delete ghost peaks, insert a minimum blank ratio)
   - For further information, see MS-CleanR in MS-DIAL 5 (https://www.youtube.com/watch?v=c-_mP8A9I04) (Fig. 22).

5 Troubleshooting

- **Retention timeshifts**: Use QC samples to correct alignment issues.

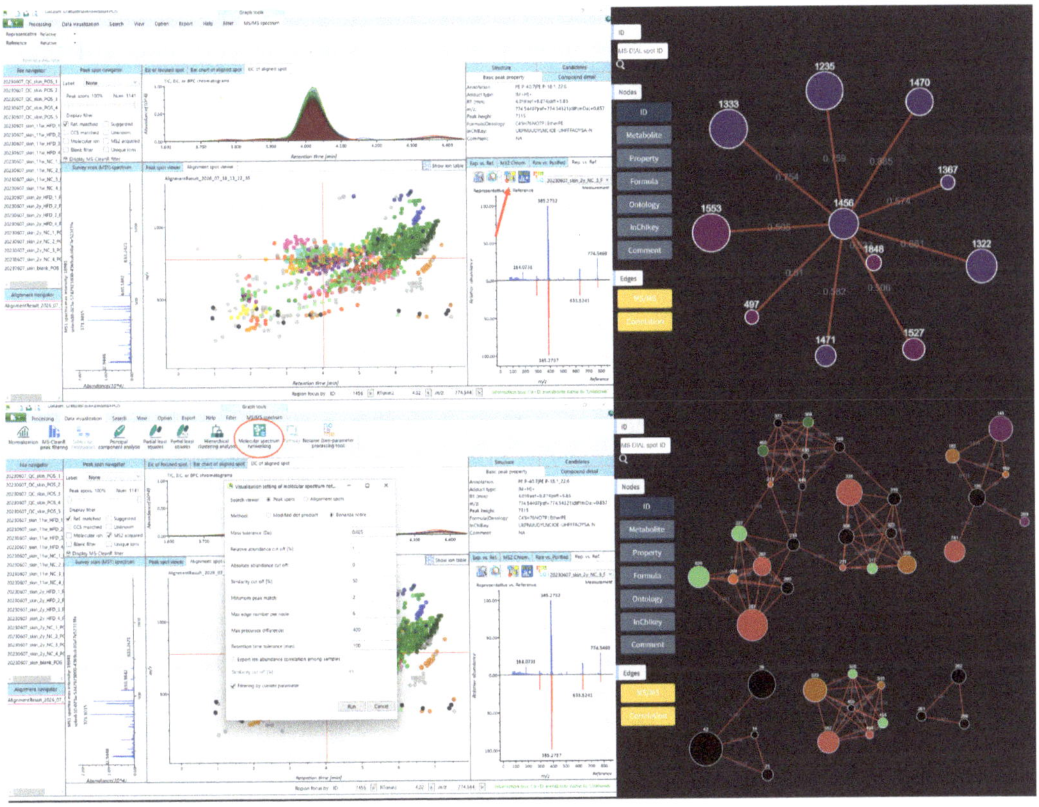

Fig. 23 Molecular networking of single spot and peak spots in MS-DIAL 5

Notame preprocessing report

By Hanhineva Lab

July 19, 2026

1 About notame

Notame (NOn-TArgeted MEtabolomics) is an R package designed to preprocess LC-MS metabolomics data and provide quality metrics, statistics, and results visualizations. The package was originally developed by Anton Kl?vus at the Kati Hanhineva lab, University of Eastern Finland, and it utilizes several other R packages, such as MUVR, to perform the data processing. The notame publication, including a description of a complete LC-MS workflow, is freely available. [1]

2 Drift correction

Signal drift is a typical issue in LC-MS metabolomics, affecting sequences with 30 to 50 injections or more. It causes a gradual change in the signal intensity of the molecular features during the LC-MS analysis, which should be corrected for optimal data quality. Notame utilizes information from the QC samples to correct the drift. The QC samples are pooled from the biological samples and injected before and after the samples as well as after every 10 to 12 samples in the sequence. Notame performs drift correction by fitting a smoothed cubic spline to the signal intensities of the QC samples, separately for each molecular feature. The smoothing function prevents overfitting the curve in case there are single deviating QC samples.

Please note that at least four QC samples are needed in the dataset to perform this step. If your data has less, it is advisable to check the PCA instead, where the QC samples should be tightly clustered.

Fig. 24 Notame report (result) exported directly from MS-DIAL 5

- **Low annotationrates**: Ensure that the library matches the experimental conditions (e.g., polarity and column type). Or check your analysis parameter of MS-DIAL.
- **Error showed up**: Let us know about the GitHub issue, and we will help you solve it, except for disk storage issue.

Features of MS-DIAL

1. **Molecular spectrum networking** (Fig. 23):
2. **Notamewith MUVR** [2]: Notame (NOn-TArgeted MEtabolomics) is an R package designed to preprocess LC–MS metabolomics data and provide quality metrics, statistics, and result visualizations. The package was originally developed by Anton Klavus at the Kati Hanhineva lab, University of Eastern Finland, and it utilizes several other R packages, such as MUVR, to perform the data processing. Notame publications, including descriptions of the complete LC–MSworkflow, are freely available. The Notame package was used to preprocess big data projects. MUVR is a processing tool that selects the top 20 features with significant changes. MUVR is highly recommended for

datasets with two parameters (time, age, diet, or sex) and more than five QCs to normalize drift correction (Fig. 24).

For more information, please proceed to our tutorial page [3] or our YouTube channel [4].

References

1. Takeda H, Yuki M, Takeuchi M et al (2024) MS-DIAL 5 multimodal mass spectrometry data mining unveils lipidome complexities. Nat Commun 15:9903. https://doi.org/10.1038/s41467-024-54137-w
2. Klåvus A, Kokla M, Noerman S et al (2020) "Notame": workflow for non-targeted LC–MS metabolic profiling. Metabolites 10:135. https://doi.org/10.3390/metabo10040135
3. MS-DIAL 5 tutorial webpage. https://systemsomicslab.github.io/msdial5tutorial/
4. MS-DIAL project YouTube channel. https://www.youtube.com/@msdialproject

Chapter 7

LC–MS/MS-Based Untargeted Hydrophilic Metabolomics Using MS-DIAL

Yoshimasa Todoroki, Bujinlkham Buyantogtokh and Hiroshi Tsugawa

Abstract

Liquid chromatography–tandem mass spectrometry (LC–MS/MS) is widely used in the field of metabolomics because it enables the simultaneous detection of hundreds to thousands of metabolites in a single analysis. MS-DIAL 5 is a software platform that facilitates the analysis of mass spectrometry data, including those obtained from LC–MS/MS, and provides a function for creating molecular spectral networks. In the analysis of hydrophilic metabolites, the availability of standard compounds is limited compared with that of lipids, and the utilization of MS/MS libraries is often restricted. To address this challenge, molecular spectral networking serves as an effective strategy for extending annotations to unknown ions. In this chapter, we introduce an analytical strategy for hydrophilic metabolite analysis using MS-DIAL 5, which utilizes retention time prediction and molecular spectral networking to enhance the reliability and comprehensiveness of metabolite annotation.

Keywords LC–MS/MS, Hydrophilic metabolomics, MS-DIAL, Retention time prediction, Molecular spectral network

1 Introduction

Liquid chromatography–tandem mass spectrometry (LC–MS/MS) is one of the most widely used analytical techniques in metabolomics because it enables simultaneous detection of hundreds to thousands of metabolites [1]. The detected ions were annotated by comparing their mass-to-charge ratio (*m/z*), retention time, and MS/MS spectra with those of a standard compound library obtained under the same analytical conditions [2]. However, in the case of hydrophilic compounds, including plant-specific metabolites, the availability of standard compounds is limited compared to that of lipids, restricting the information that can be used for annotation.

To address this limitation, predictive approaches are sometimes applied to supplement the retention time information in mass

Michael Witting (ed.), *Computational Methods and Data Analysis for Metabolomics*, Methods in Molecular Biology, vol. 3063, https://doi.org/10.1007/978-1-0716-5452-1_7,

spectrometry data [3]. Reversed-phase (RP) liquid chromatography is widely used because of its high stability and reproducibility in terms of retention time and peak shape. The separation mechanism of RP-LC is based on hydrophobic interactions, and the availability of abundant training data makes retention time prediction using machine learning relatively straightforward.

Although tools have been developed to predict MS/MS spectra based on information such as bond stability, accurately predicting the relative intensity of fragment ions remains a challenge [4, 5]. Moreover, experimental libraries based on actual measurements are limited, making it difficult to rely solely on conventional annotation methods using MS/MS spectral matching

Since conventional library-based annotation is often insufficient for metabolites without reference spectra, molecular spectral networking is one of the approaches used to extend annotation beyond existing libraries [6–8]. This approach calculates the similarity between the MS/MS spectra of the detected ions and connects ions when their similarity exceeds a defined threshold, thereby visualizing a network of compounds with similar MS/MS spectra. This method is based on the theory that structurally similar compounds exhibit similar tandem mass spectrometry patterns.

The ions connected within the network are likely to share common fragmentation patterns, suggesting the presence of shared substructures. By identifying the characteristic fragmentation patterns of the annotated ions, this approach is expected to facilitate the structural elucidation of unknown compounds.

In this chapter, we introduced a protocol for hydrophilic metabolite analysis using MS-DIAL 5, which utilizes retention time prediction and molecular spectral networking to enhance annotation reliability and comprehensiveness [9]. This documentation describes the analysis performed using the release version "MS-DIAL v5.5.250627." Please note that these features or settings may be subject to change in future versions.

2 Material

MS-DIAL is available from the GitHub releases page (https://github.com/systemsomicslab/MsdialWorkbench/releases). All data processing in this chapter was performed using "MS-DIAL v5.5.250627." We recommend using the latest available release version whenever possible and ensuring that the computing environment satisfies the minimum system requirements (i.e., Windows 11 or later, ≥32 GB RAM).

3 Methods

3.1 Preparation of Annotation Libraries

3.1.1 Construction of a Retention Time Prediction Model

This section describes the construction of a machine-learning model that uses compound descriptors generated from SMILES as the input and predicts the retention time as the output.

First, the retention time data of the standards were obtained under the LC conditions used for sample measurement. The SMILES of the standards for which the retention times were successfully measured were extracted and saved in .smi format, and a dedicated folder containing only these files was prepared. Next, compound descriptors were generated from the SMILES using PaDEL [10]. The following four types of descriptors were selected and extracted from PaDEL: 1D, 2D, Fingerprinter, and PubChem Fingerprinter.

To output the four types of molecular descriptors using PaDEL, we began by preparing a .smi file containing only the SMILES of the compounds and placing it in a designated folder. It is recommended that this folder contain only a single .smi file to avoid processing issues. After launching PaDEL, the first step is to configure the "General" tab. This directory containing the .smi file is specified in the Molecules directory/file field, and the output file name and location are set in the Descriptor output file field. In the Descriptors section, both 1D & 2D and Fingerprints options were selected. Next, in the Standardize section, the following options are enabled: Remove salt, Detect aromaticity, and Standardize nitro groups. In the Advanced section, it is confirmed that both Log and Retain molecule order are selected. To prevent the program from running indefinitely when the descriptor calculation fails for certain compounds, a sufficiently large number (100,000 in this case) is entered into the Max. running time per molecule field. The tab is then switched to "1D & 2D," where all descriptor checkboxes are selected. Finally, in the "Fingerprint" tab, Fingerprinter and PubchemFingerprinter are selected (Fig. 1). Once all settings are confirmed, clicking Start begins the descriptor calculation process.

Similarly, compound descriptors were generated for the compounds included in the annotation library. Before the descriptor calculation, compounds with duplicate ShortInChIKeys were removed from the list of SMILES extracted from the annotation library, leaving only one representative per ShortInChIKey. This step was necessary because the InChIKey, a 25-character hashed representation of the molecular structure, encodes stereochemical information. The first 14 characters (ShortInChIKey) represent the molecular skeleton, and the remaining 10 characters encode stereochemical and isotopic

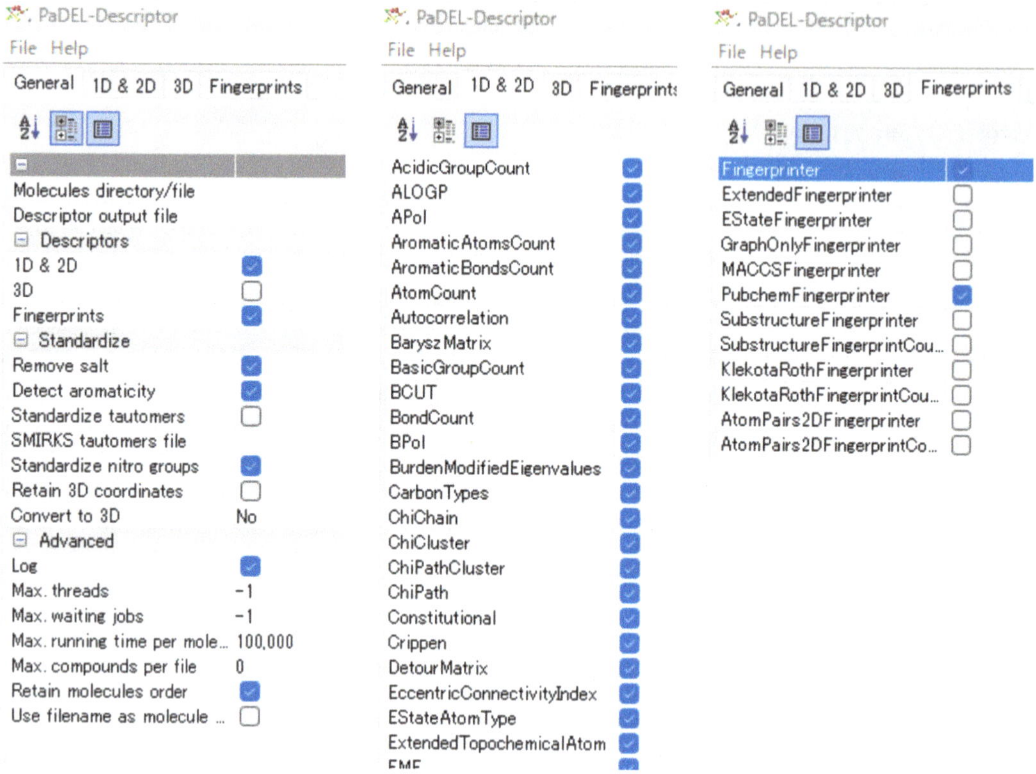

Fig. 1 PaDEL-Descriptor interface. Settings in the "General," "1D & 2D," and "Fingerprints" tabs are shown for exporting 1D descriptors, 2D descriptors, Fingerprinter, and PubChemFingerprinter

information. Since retention time in RP-LC does not distinguish between stereoisomers, compounds with different full InChIKeys but identical ShortInChIKeys are expected to have the same retention time. To reduce computational complexity, only one compound per ShortInChIKey was retained in the SMILES list before descriptor calculation. Among the extracted descriptors, those with identical values across all compounds and those containing nonnumeric characters were removed. The remaining descriptors were then used to construct a retention time prediction model using Retip [3].

The standards for which the retention times were successfully measured were divided into training data (80%) and test data (20%) in a 4:1 ratio. The training data, including retention time information and compound descriptors, were used to train the XGBoost and Random Forest models. After training, the performance of the model was evaluated using test data, and the optimal model was selected. Four evaluation metrics were used to assess the model performance: root mean square error (RMSE), R-squared (R^2) , mean absolute error (MAE), and 95% confidence interval. The final trained model was used to predict the retention times of the compounds in the annotation library based on their compound descriptors (Fig. 2).

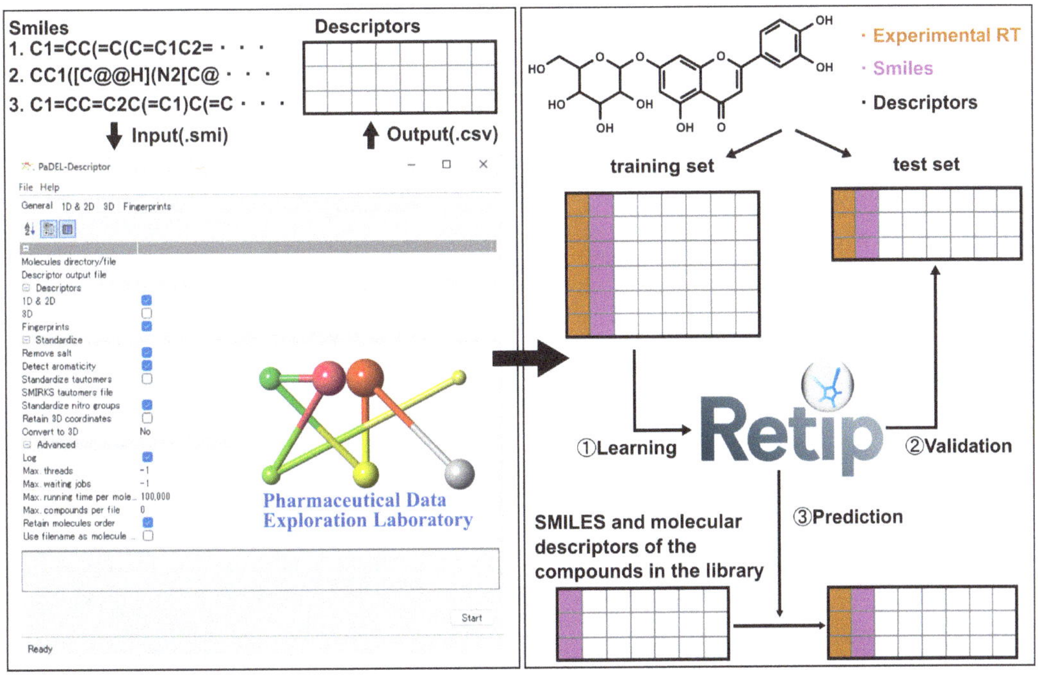

Fig. 2 Workflow for retention time prediction using PaDEL and Retip. Retention time prediction is performed by training a model with the Retip package using molecular descriptors exported from PaDEL based on SMILES and the experimentally measured retention times. The constructed model is then used to predict the retention times of compounds in the library

3.1.2 Retention Time Correction by Linear Transformation

Theoretically, under identical analytical conditions, the retention times should remain constant. However, variations in the column conditions or daily fluctuations in the system performance can lead to slight changes in the retention time. Therefore, it is necessary to adjust the retention time appropriately. In this section, we describe a method for correcting retention time by adding internal standards during actual sample measurements and using their retention times as a reference. First, internal standards were added during measurement of the actual samples, and their retention times were recorded. A correction can be applied by comparing these retention times with those obtained during the acquisition of training or test data. Specifically, a linear regression model was constructed by plotting the retention times of the internal standards used in the training or test data on the x-axis and the retention times of the internal standards measured in the actual sample analysis on the y-axis. This regression equation was then used as a correction function, allowing the adjustment of both the experimentally obtained retention times and those predicted by the model (Fig. 3).

3.1.3 Storage of Retention Time

The predicted or experimentally measured retention times were stored in a library based on the ShortInChIKey, ensuring that all

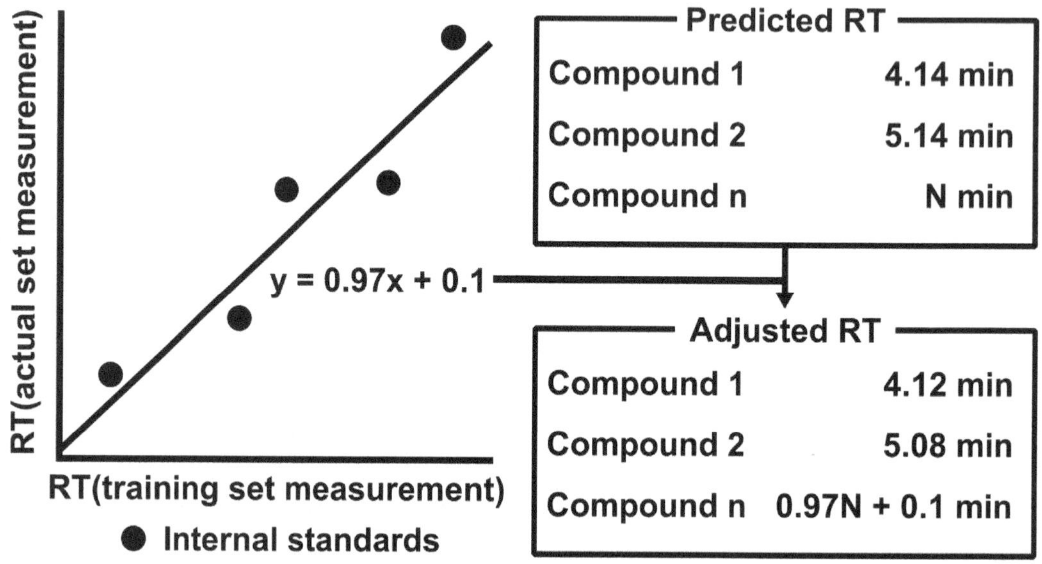

Fig. 3 Schematic of linear transformation. Retention times assigned to library compounds are adjusted based on the retention times of internal standards

compounds with the same ShortInChIKey were assigned the same retention time. If an experimentally measured retention time was available, it was assigned to all compounds that shared the same ShortInChIKey. If no experimental data were available, the predicted retention time from the retention time prediction model was applied to all compounds with the same ShortInChIKey. This resulted in the creation of three types of libraries. MSP1 contains the experimentally measured retention times, MSP2 includes the predicted retention times obtained from the retention time prediction model, and MSP3 corresponds to the original library and does not contain any retention time information. Among these, annotations derived from MSP1, where both the measured retention time and MS/MS spectra matched, were the most reliable. MSP2 was reliable because it used the predicted retention times for annotation, and MSP3 had the lowest confidence because it relied solely on MS/MS spectral matching without retention time information (Fig. 4).

3.2 Analysis in MS-DIAL 5

3.2.1 Setting in MS-DIAL 5

The basic parameter settings are described in the lipid analysis section. This section highlights modifications specific to the analysis of hydrophilic metabolites. For hydrophilic metabolite analysis, "Metabolomics" is selected in the Measurement parameters setting. Three previously created libraries (MSP1, MSP2, and MSP3) were registered in the MS/MS identification setting. Libraries were added

Annotation and Manual Curation

MS/MS library	Retention Time	MS/MS	RT Tolerance	Priority
MSP1	Experimental	✓	± 0.2min	1
MSP2	Predicted	✓	± 95% confidence interval of the model	2
MSP3	×	✓	×	3

Fig. 4 Handling and prioritization of the three constructed MSP libraries. MSP1 contains MS/MS spectra of compounds with experimentally measured retention times. Although it includes the smallest number of compounds, it is considered the most reliable for annotation. MSP2 contains spectra with predicted retention times, enabling the narrowing of candidate structures based on retention time prediction. MSP3 does not incorporate retention time information, and annotations are assigned solely based on MS/MS spectral similarity. Therefore, while MSP3 provides the lowest annotation reliability among the three, it covers the widest range of compounds

by clicking the green "+" button in the Database setting. From the Database type menu, "msp" is selected, and the appropriate MSP file is chosen using the "Browse" button in the Database path. The Retention time tolerance settings differed depending on the library used. For MSP1, which contained experimentally measured retention times and minimal deviations, a retention time tolerance of 0.2 min is applied. For MSP2, a 95% confidence interval of the retention time prediction model was used to determine the retention time tolerance. As MSP3 does not incorporate retention time information, the default settings remain unchanged. In Retention time settings, "Use retention time for scoring" and "Use retention time for filtering" must be manually enabled for MSP1 and MSP2 to ensure that retention time information is applied only to these libraries. In the Annotation method settings, annotations were assigned from top to bottom on the list. Thus, it is essential to arrange MSP1, MSP2, and MSP3 to ensure that the annotations are assigned to the correct sequences (Fig. 5).

The three scores configurable under the Annotation cut off settings—Dot product score, Weighted dot product score, and Reverse dot product score— represented the degree of similarity between the MS/MS spectrum of the library and that of the measured data. How these scores are evaluated depends on the mode selected in the Measurements parameters setting, specifically, whether "Lipidomics" or "Metabolomics" is used. When "Lipidomics" was selected, an annotation was assigned if at least one of the three scores exceeded the specified threshold. In contrast, when "Metabolomics" is selected, all three scores must exceed the threshold for annotation to be considered valid. As a result, the criteria for assigning annotations in "Metabolomics" are more stringent than those in "Lipidomics." If a more lenient matching condition is desired in "Metabolomics," it is

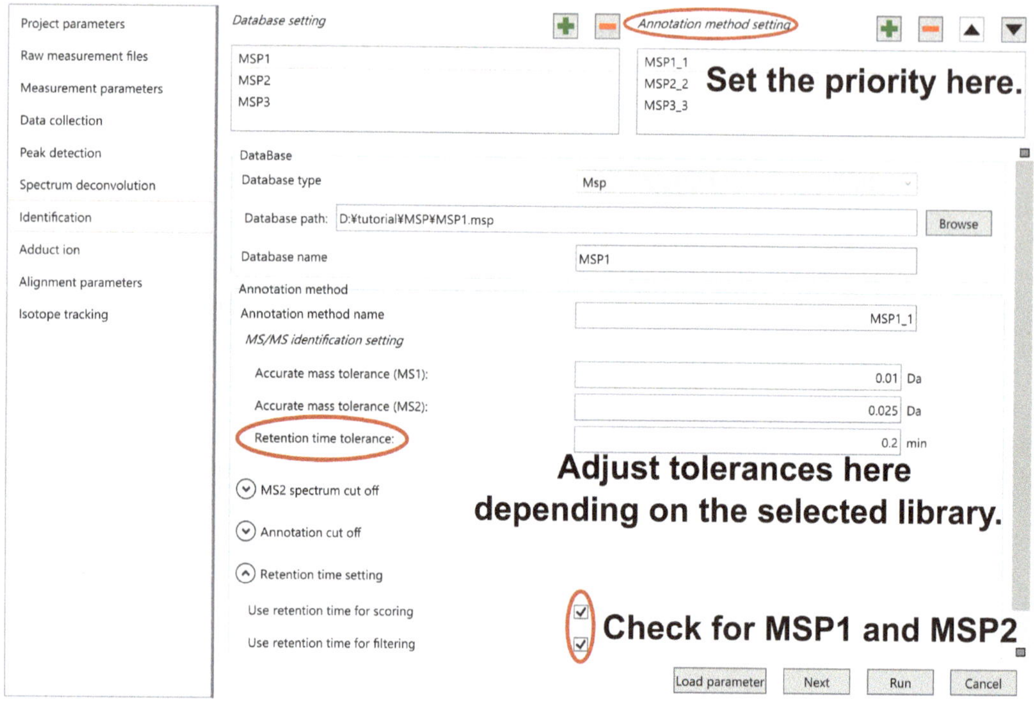

Fig. 5 Identification settings in MS-DIAL 5 for using the three MSP files. To utilize the three imported MSP files properly, users must confirm the annotation priority order, retention time tolerance, and whether the retention time is used for scoring and filtering

recommended to uniformly lower the thresholds for all three scores. However, lowering the thresholds significantly increases the risk of misannotation. Therefore, adjustments should be made carefully depending on the analytical objective.

3.2.2 Manual Validation

This section describes the manual validation process that involves manually verifying the accuracy of the automatically assigned annotations.

In MS-DIAL 5, the reliability of the annotations was evaluated based on three scoring metrics: Dot product, Weighted dot product, and Reverse dot product. Although the assigned annotations met the predefined cut-off thresholds for these scores, misannotations may still occur. In the analysis of hydrophilic metabolites, particular attention should be paid to the following three cases.

(1) The fragment ion *m/z* matches, but intensity ratios differ.

This phenomenon occurs when the fundamental substructures match; however, differences in the structural composition affect the fragmentation patterns. Additionally, MS/MS spectra obtained using different instruments or under different analytical conditions may exhibit variations in intensity ra-

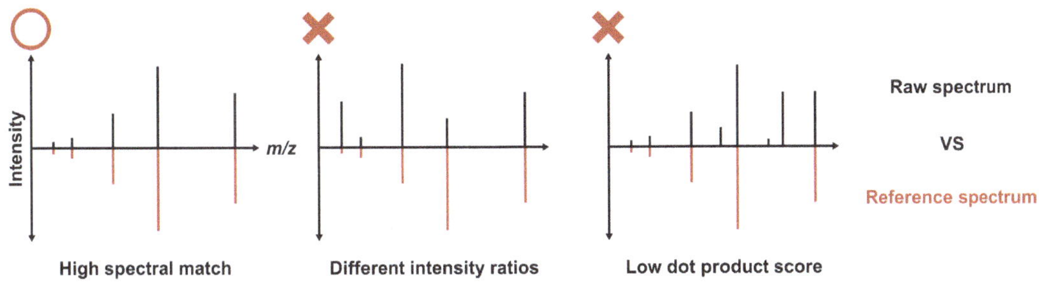

Fig. 6 Examples of spectra that require caution during manual validation. When there are fragment ions with significantly different intensity ratios or clearly structure-specific fragments that do not match the annotation, it may indicate that the compound has a different structure than the annotated one

tios. However, if the intensity ratios were drastically reversed, they are likely to be misannotated (Fig. 6).

(2) High Reverse dot product score, but low Dot product score.

This issue arises when the fragment ions in the library spectrum are present in the acquired MS/MS spectrum, but additional fragment ions are also detected.

In some cases, these may result from contamination in the MS/MS spectrum due to low precursor ion intensity, which reduces the Dot product score. However, the structural differences may also contribute to the presence of additional product ions. Thus, consideration of the precursor ion intensity can help determine whether the discrepancy is due to contamination or misannotation (Fig. 6).

(3) Multiple peaks were assigned the same annotations.

This phenomenon often occurs when an in-source fragment ion is misidentified as a precursor ion. This phenomenon is particularly common in plant metabolite analysis, where modifications are retained in LC but cleaved in MS, leading to the detection of the base skeletons (Fig. 7). Typically, the retention time can be used to differentiate in-source fragment ions; however, if this is insufficient, identifying the original structure can help resolve ambiguities.

Before conducting manual validation, the annotations were filtered to display only the peaks with assigned annotations. This can be achieved by checking "Ref. matched" in the Peak spot navigator, which restricts the displayed peaks to those with library-based annotations.

During manual validation, the retention time does not need to be strictly matched to the raw data because tolerance values are already applied based on experimentally measured or predicted retention times. Instead, MS/MS spectral similarity should be carefully examined to determine whether an annotation falls into one of the three misannotation categories described above.

To facilitate the identification of in-source fragment ions, MS-DIAL 5 provides two key functionalities (Fig. 8): (1) It can

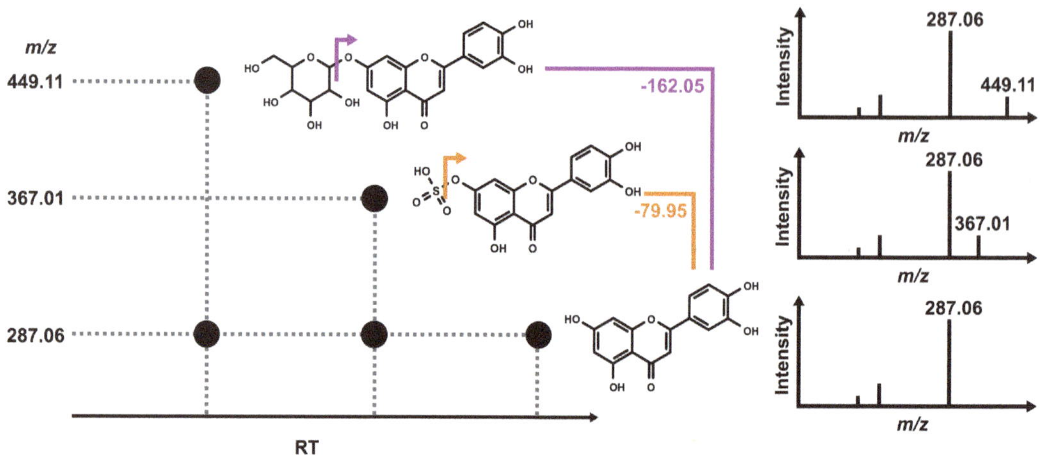

Fig. 7 Example of identical annotations caused by in-source fragment ions. Compounds that share a common aglycone and possess easily cleavable substructures may lead to the same aglycone being detected at different retention times, resulting in identical annotations

display ions with similar chromatographic profiles, which can help identify structurally related compounds that may result from in-source fragmentation. (2) It can detect cases in which the selected precursor ion appears as a fragment ion in the MS/MS spectra of other ions and highlight the corresponding peaks. By utilizing these features, it is easier to determine whether a detected ion represents a true precursor or an in-source fragment.

After manual validation, verified annotations should be recorded by opening the Ion Table via "Show ion table" and checking the "Confirmed" box.

For peaks identified as misannotations, the "Misannotation" checkbox should be selected, and the reason for misannotation should be documented in the "Comment" field to aid further analysis.

3.2.3 Construction of Molecular Spectral Networks Using MS-DIAL 5

Molecular spectral networking (MN) is a powerful tool for visualizing relationships between detected ions based on various analytical criteria. Before constructing a network, it is essential to understand the two fundamental concepts of nodes and edges. In a molecular spectral network, a node represents a detected ion, whereas an edge indicates the connection between nodes when a relationship is identified.

MS-DIAL 5 provides two approaches for constructing molecular spectral networks.

One approach is direct visualization within MS-DIAL 5, while the other involves exporting data for visualization using external tools such as Cytoscape [11].

To provide a practical explanation for the molecular spectrum networking, we used raw data acquired in positive ion mode from 12 different × 12C plant species contained in the DM0027 dataset

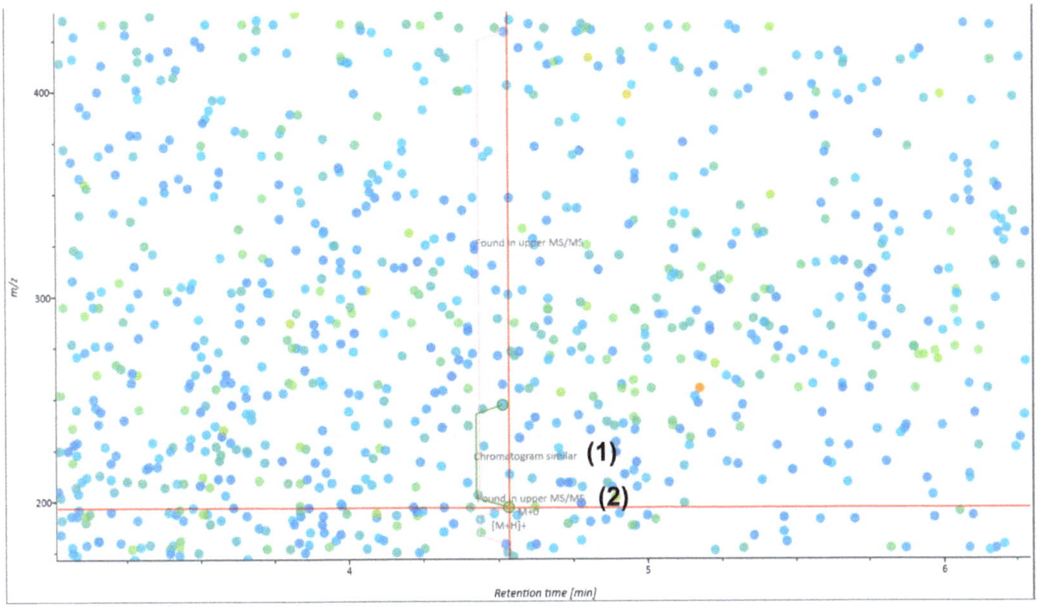

Fig. 8 MS-DIAL 5 functions that support the identification of in-source fragment ions. (1) Lines indicating ions with similar chromatographic profiles. (2) Lines indicating that a precursor ion appears in the MS/MS spectrum of another ion

available from RIKEN PRIME (https://prime.psc.riken.jp/menta.cgi/prime/drop_index) [12]. The analysis was performed using the released version "MS-DIAL v5.5.250627." Default parameters were generally used; however, this section describes key parameter settings that require attention or were specifically modified. In the Raw measurement files setting, labels were assigned in the "Class ID" column to identify each plant species regardless of tissue type (Fig. 9). In the Measurement parameters setting, "Positive ion mode" was selected under Ion mode, and "Metabolomics" was selected under Target omics. In the Identification setting, the MSP file "MSMS-Public-Pos-V14. msp" included in "Hydrophilic metabolome profiling of Wine," available on the CompMS website (https://systemsomicslab.github.io/compms/msdial/main.html#Demonstration), was used as the database. In this analysis, retention time information was not used for annotation (Fig. 10).

To visualize a molecular spectral network in MS-DIAL 5, select "Molecular spectrum networking" from the "Data visualization" tab. This opens up a new window in which various network settings can be configured (Fig. 11).

First, select "Search viewer" and specify the target dataset for network construction. If a network is to be generated based on individual peak data, select "Peak spots," whereas if it is to be generated based on alignment results, select "Alignment spots."

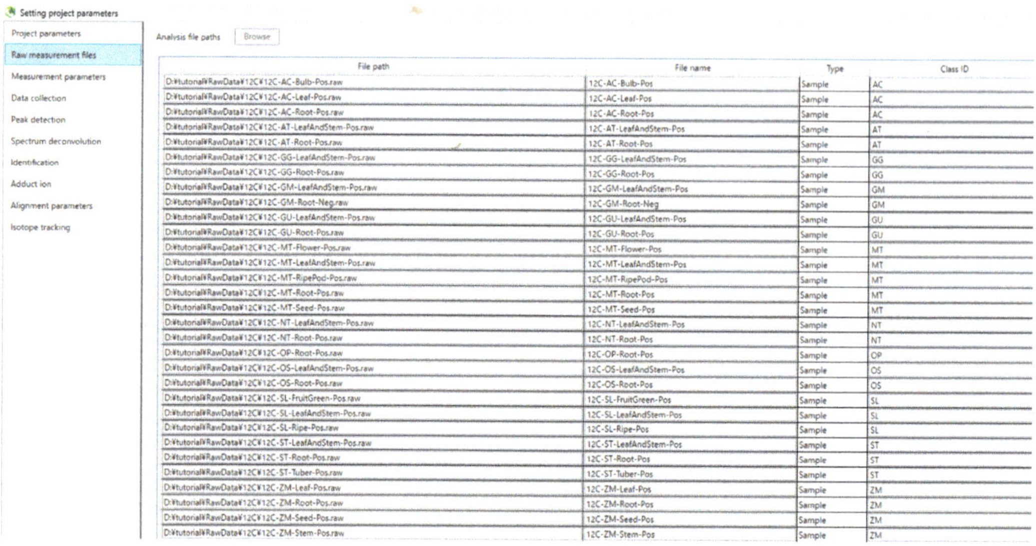

Fig. 9 Configuration of the Raw measurement files tab in this analysis. All positive ion mode raw data files from 12C plants were loaded, and the Class ID was assigned according to plant species

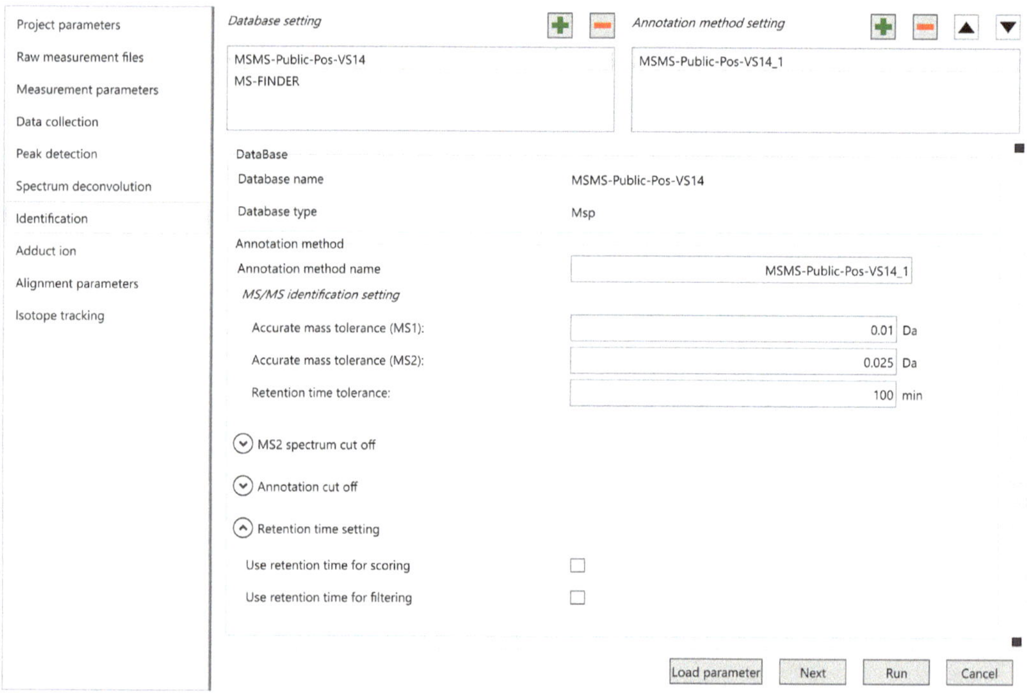

Fig. 10 Configuration of the Identification settings in this analysis. Retention time information was not used for either scoring or filtering

Fig. 11 Procedure for directly visualizing a molecular spectrum network in MS-DIAL 5. The network configuration window can be accessed from the "Molecular spectrum networking" option under the "Data visualization" tab

Next, we chose a method for calculating the spectral similarity. MS-DIAL 5 offers two calculation methods: "Modified dot product" and "Bonanza score." These methods differ in their similarity calculation approaches, and users can select either method based on their analytical requirements.

After selecting the similarity calculation method, we configured the parameters necessary for network construction. Mass tolerance was defined as the acceptable *m/z* range for considering fragment ions to be identical. The relative and absolute abundance cut-offs determine which fragment ions are retained based on the intensity thresholds during standardization. The similarity cutoff sets the minimum similarity score required to form an edge between two nodes, whereas the minimum peak match specifies the minimum number of matching fragment ions required for the similarity calculation. The maximum number of edges per node limits the number of edges that a single node can possess. The maximum precursor difference sets the maximum allowable *m/z* difference between the precursor ions for similarity calculations. The retention time tolerance restricts similarity calculations to ions with retention time differences below a defined threshold.

Two other options are also available. "Export ion abundance correlation among samples" becomes available when "Alignment spots" is selected, enabling the formation of additional edges based on ion abundance similarity across samples, in addition to MS/MS similarity. "Filtering by current parameter" applies any filters set in

"Peak spot navigator" and "Show ion table," ensuring that only displayed ions are included in the network.

In this analysis, the following parameters were applied: mass tolerance, 0.025 Da; relative abundance cutoff, 1%; absolute abundance cut-off, 0; similarity cut-off, 75; minimum peak match, 4; maximum edge number per node set to 10; maximum precursor difference, 400; and retention time tolerance, 100 min. The options "Export ion abundance correlation among samples" and "Filtering by current parameter" were left unchecked. These parameters are provided as examples, and it is recommended to adjust them based on the data, with the goal of increasing or decreasing the number of connections as needed.

Once the parameters are set, clicking "Run" initiates the network construction, and the generated molecular spectral network is displayed (Fig. 12).

The generated molecular spectral network contains four types of edges, each distinguished by a color. The red edges indicate *ms_similarity*, representing the MS/MS spectral similarity between ions. Green edges correspond to *Chromatogram-based annotation*, reflecting similarities in chromatographic peak shapes. The blue edges represent *MS2-based annotation*, where one ion appears as a fragment ion in the MS/MS spectrum of another. Orange edges indicate *Adduct annotation*, which links ions that have the same molecular structure but differ in adduct type. These edge colors were automatically assigned based on the connection type. A molecular

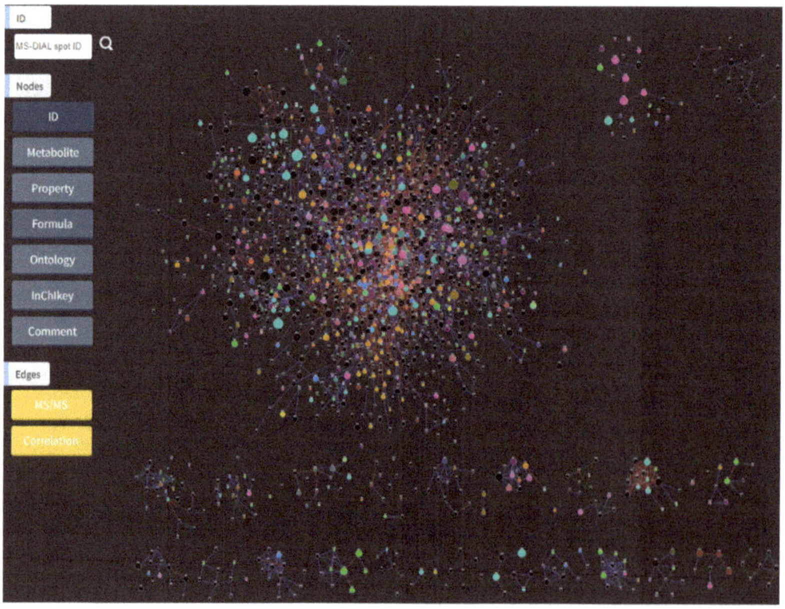

Fig. 12 Molecular spectrum network generated in this analysis. Each node represents a detected ion, and the ions are connected by four types of edges

spectral network based solely on ms_similarity can be visualized using the method introduced in the next section, which involves removing all non-*ms_similarity* entries from the edge file before importing them into Cytoscape.

The "Nodes" panel allows customization of the displayed node information. Similarly, the "Edges" panel provides options to toggle between edges based on MS/MS similarity or ion abundance correlation.

If a specific ion needs to be highlighted, enter its MS-DIAL 5 Peak ID in the "ID" field at the top left of the window. If a corresponding node exists in the network, it will be located and highlighted. Clicking on a node displays information about the corresponding metabolite, including the Peak ID, retention time, *m/z*, MS/MS spectrum, and class-based ion abundance. If the ion has been annotated using a library, additional information such as molecular formula, ontology, InChIKey, SMILES, and structural data is also displayed (Fig. 13).

Furthermore, it was possible to display only the selected ion and its directly connected nodes based on MS/MS similarity (*ms_similarity* edges only).

To do this, in the MS/MS visualization window within MS-DIAL 5, ensure that "Rep. vs. Ref." is selected and then click the central icon among the five displayed icons. This action generates a molecular spectral network consisting only of the selected ion and its related nodes, based on the settings in "Molecular spectrum networking" under the "Data visualization" tab (Fig. 14).

In addition to visualizing molecular spectral networks within MS-DIAL 5, Cytoscape can also be used for flexible network visualization and in-depth structural analyses. Cytoscape allows for greater customization and improved graphical representation of molecular

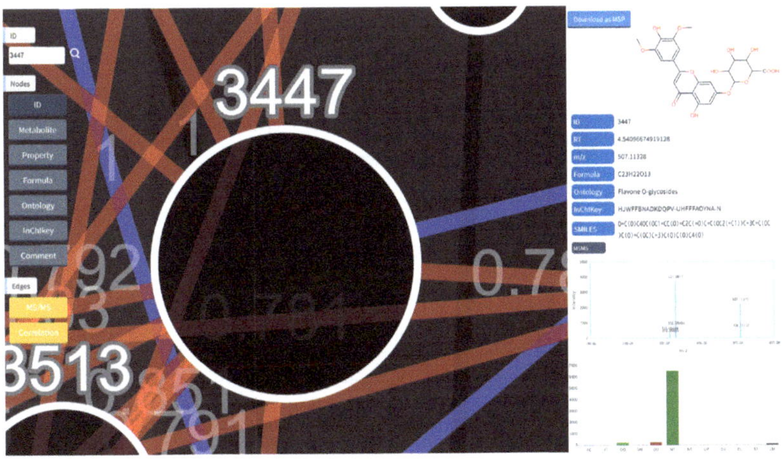

Fig. 13 Specific node and its associated mass spectrometry information. In this analysis, the ion assigned as Alignment ID 3447 was searched and clicked, displaying its corresponding mass spectrometry data

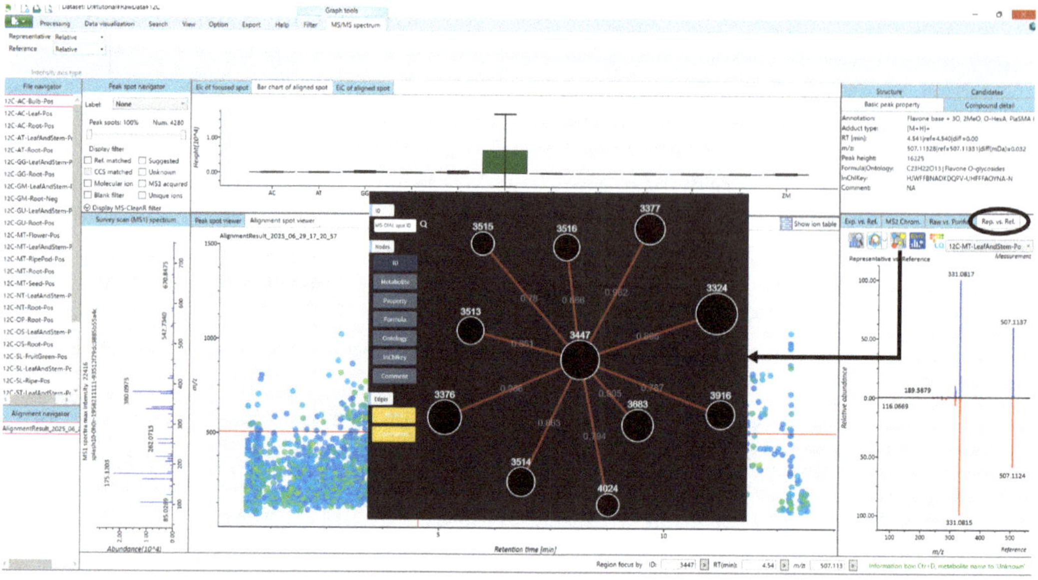

Fig. 14 Method for visualizing a molecular spectral network limited to ions connected to a specific ion. In MS-DIAL 5, this network can be visualized by clicking the icon located within the area where the MS/MS spectrum is displayed

Fig. 15 Method for visualizing a molecular spectral network limited to ions connected to a specific ion. In MS-DIAL 5, this network can be visualized by clicking the icon located within the area where the MS/MS spectrum is displayed

networks. To generate files for Cytoscape, select "Molecular spectrum networking export" from the "Export" tab in MS-DIAL 5. The fundamental settings for molecular spectral networking remained the same as those for visualization within MS-DIAL 5 (Fig. 15).

However, when exporting from Cytoscape, an additional option is available to specify the folders in which the files will be saved. This section describes how to export a molecular spectral network based on the alignment results and generate files for Cytoscape.

After configuring the required parameters in MS-DIAL 5, clicking the "Run" button generates two files in the specified folder: an edge file and a node file. The edge file contains information describing the relationships between the nodes. Edges are created based on Source ID and Target ID, which correspond to two connected nodes. Each edge is assigned a score that varies depending on the Type of edge. If the Type is *ms_similarity*, the value represents the MS/MS similarity score, whereas for the other edge types, the score is set to 1.

The Type column defines the nature of each edge in the molecular spectral network and can take one of four values. The value *ms_similarity* represents the edges based on MS/MS spectral similarity. *Chromatogram-based annotation* indicated that the connected ions exhibited similar chromatographic peak shapes. *MS2-based annotation* was used when an ion appeared as a fragment in the MS/MS spectrum of another ion. Finally, *Adduct annotation* connects ions that share the same molecular structure but differ in their adduct types. The column Color was automatically assigned based on the edge type. The Comment column provides supplementary information for all edge types except *ms_similarity*. For *Chromatogram-based annotation*, the comment field notes a "Similar chromatogram." In the case of *MS2-based annotation*, the comment includes information specifying which precursor ion contained the other ion as a fragment in its MS/MS spectrum. For *Adductannotation*, this comment describes the adduct types of the connected ions and the mass differences between them. If the molecular spectral network is intended to reflect only MS/MS spectral similarity, it is recommended to remove all edges with types other than *ms_similarity* from the edge file.

To import the exported files into Cytoscape, launch the software and drag the edge file into the area labeled "Drag network files here." During the import process, the Source ID is displayed as a green circular icon and the Target ID as an orange circular icon. If they are not assigned correctly, the triangular icon is clicked to adjust the settings. Once verified, clicking "OK" initiates the network visualization (Fig. 16).

At this stage, the network is displayed, but only the nodes are assigned ID information. To enhance visualization, the node file must be imported separately to include additional node attributes. The node file exported from MS-DIAL 5 contains various attributes for each node, including ID, Metabolite name, Retention time (Rt), mass-to-charge ratio (m/z), Formula, Ontology, InChIKey, SMILES, Size, BorderColor, BackgroundColor, and MS2. Rt and m/z values were assigned to all nodes, whereas attributes such as Metabolite name, Formula, Ontology, InChIKey, and SMILES were only provided for ions annotated using a library. Additionally, MS2 data were only available for ions with recorded MS/MS spectra.

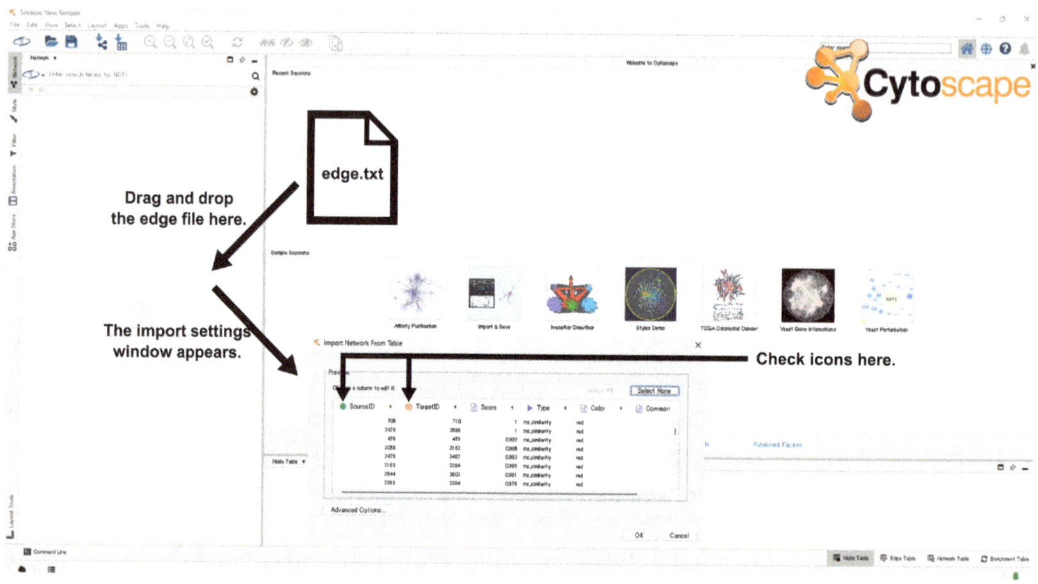

Fig. 16 Procedure for importing the edge file into Cytoscape. After launching Cytoscape, drag and drop the edge file into the blank area on the left to open the import settings window. If the Source ID column is marked with a green circle icon and the Target ID column with an orange circle icon, click "OK" to generate the network

The exported node file can be modified to include additional information. For example, adding the average intensity of each sample class allows for the visualization of class intensity ratios as pie charts in Cytoscape. To incorporate relevant data, it is recommended to use results exported from MS-DIAL 5.

To obtain additional information, select "Alignment result" from the "Export" tab in MS-DIAL 5. Choose the directory for saving the file and specify the data to be included under "Peaks." In most cases, "Raw data (Height)" is sufficient, so the default settings can be used while ensuring that "Dictionary" is selected before exporting the file.

The exported data contains a row labeled "Average," which represents the average intensity for each sample class. Additionally, a row labeled "Comment" contains tag information and annotation library details. These details can be extracted to create new columns in the node file for improved organization. The tag and annotation library information can then be linked to each node using an ID column, allowing for expanded node attributes (Fig. 17).

To import the modified node file into Cytoscape, drag and drop it into the lower section of Cytoscape or navigate to "File" → "Import" → "Table from File" and select the appropriate node file.

During the import process, ensure that the "ID" column is marked with a key icon.

This key assignment links the ID values of the node file with the corresponding node IDs that are already present in the network

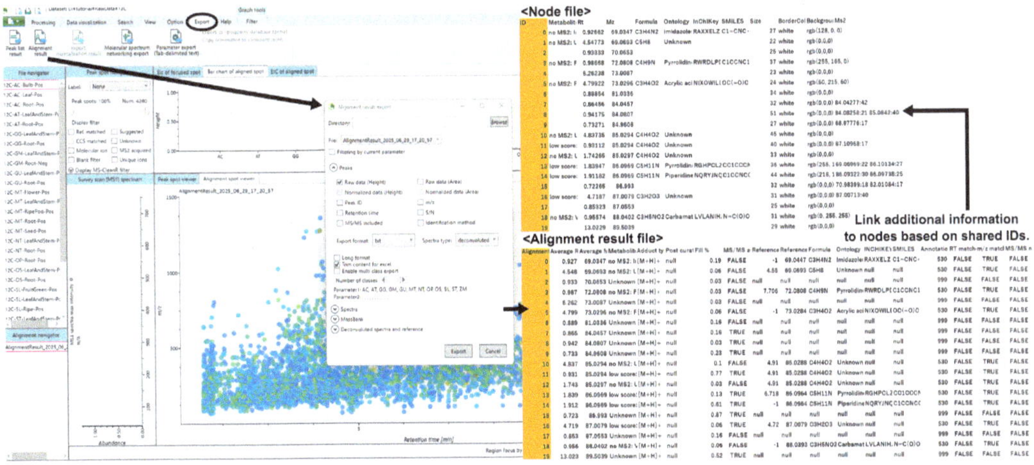

Fig. 17 Merging the node file with the Alignment result file. From the Export tab in MS-DIAL 5, the Alignment result table can be exported. By using the shared ID column, this table can be merged with the node file. The merged file, used as a new node file, enables additional information to be assigned to each node

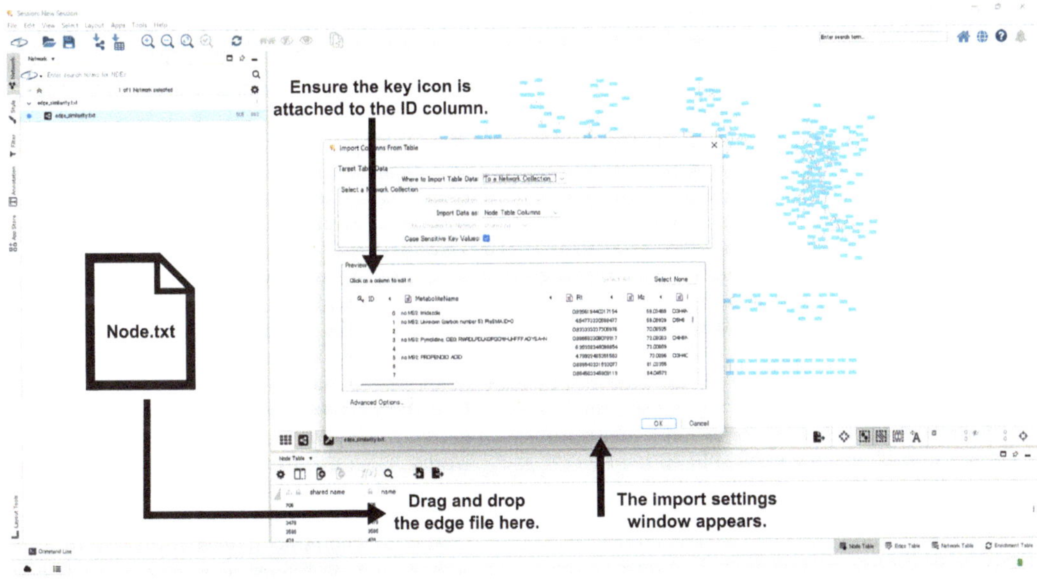

Fig. 18 Importing the node file into Cytoscape. After loading the edge file into Cytoscape, drag and drop the node file into the lower table panel to open the import settings. If the ID column displays a key icon, click OK to append the node information to the network

(Fig. 18). If the key icon is missing, the imported data will not be correctly associated with the existing nodes, thereby preventing proper integration.

The Style tab in Cytoscape allows users to customize the appearance of nodes, edges, and the overall network. The visualization settings can be configured using data imported from the edge and node files. For instance, the edge thickness can be adjusted based on the similarity scores; edges representing higher similarity can be drawn thicker, whereas those with lower similarity are thinner. Node shapes can be set to circles, and pie charts can be used to visualize the relative intensities across sample groups. It is also possible to change the stroke colors of the nodes with annotations. This makes it easier to visually compare annotated and unannotated ions, potentially aiding the structural inference of unknown compounds. Users are encouraged to customize molecular spectral networks according to their specific analytical objectives.

References

1. Dührkop K, Nothias LF, Fleischauer M et al. (2021) Systematic classification of unknown metabolites using high-resolution fragmentation mass spectra. Nat Biotechnol 39:462. https://doi.org/10.1038/s41587-020-0740-8
2. Li Y, Kind T, Folz J et al. (2021) Spectral entropy outperforms MS/MS dot product similarity for small-molecule compound identification. Nat Methods 18:1524. https://doi.org/10.1038/s41592-021-01331-z
3. Bonini P, Kind T, Tsugawa H et al. (2020) Retip: retention time prediction for compound annotation in untargeted metabolomics. Anal Chem 92:7515–7522. https://doi.org/10.1021/acs.analchem.9b05765
4. Ruttkies C, Schymanski E, Wolf S et al (2016) MetFrag relaunched: incorporating strategies beyond in silico fragmentation. J Cheminform 8:3. https://doi.org/10.1186/s13321-016-0115-9
5. Wang F, Liigand J, Tian S et al. (2021) CFM-ID 4.0: more accurate ESI-MS/MS spectral prediction and compound identification. Anal Chem 93:11692–11700. https://doi.org/10.1021/acs.analchem.1c01465
6. Li X, Zhou H, Xiao N et al. (2022) Expanding the coverage of metabolic landscape in cultivated rice with integrated computational approaches. Genomics Proteomics Bioinformatics 20:702–714. https://doi.org/10.1016/j.gpb.2020.06.018
7. Bouslimani A, Sanchez LM, Garg N et al. (2014) Mass spectrometry of natural products: current, emerging and future technologies. Nat Prod Rep 31:718–729. https://doi.org/10.1039/c4np00044g
8. Watrous J, Roach P, Alexandrov P et al. (2012) Mass spectral molecular networking of living microbial colonies. Proc Natl Acad Sci U S A 109:E1743–E1752. https://doi.org/10.1073/pnas.1203689109
9. Takeda H, Yuki M, Takeuchi M et al (2024) MS-DIAL 5 multimodal mass spectrometry data mining unveils lipidome complexities. Nat Commun 15:9903. https://doi.org/10.1038/s41467-024-54137-w
10. Yap CW (2011) PaDEL-Descriptor: an open source software to calculate molecular descriptors and fingerprints. J Comput Chem 32:1466–1474. https://doi.org/10.1002/jcc.21707
11. Shannon P, Markiel A, Ozier O et al. (2003) Cytoscape: a software environment for integrated models of biomolecular interaction networks. Genome Res 13:2498–2504. https://doi.org/10.1101/gr.1239303
12. Tsugawa H, Nakabayashi R, Mori T et al. (2019) A cheminformatics approach to characterize metabolomes in stable-isotope-labeled organisms. Nat Methods 16:295. https://doi.org/10.1038/s41592-019-0358-2

Check for updates

Chapter 8

Computational Tools for LC–IMS–MS Data Processing in Metabolomics

Dylan H. Ross, Nathalie Muñoz, Harsh Bhotika, Xueyun Zheng and Aivett Bilbao

Abstract

This chapter provides resources and step-by-step processing guidelines for analyzing liquid chromatography–ion mobility spectrometry–mass spectrometry (LC–IMS–MS) data in metabolomics. The methods described here are based on open-source software and freely available executables developed at Pacific Northwest National Laboratory (PNNL), including PNNL-PreProcessor, MZA, mzapy, LipidOz, PeakQC, and IonToolPack. Importantly, the same software ecosystem is broadly applicable to IMS–MS workflows both with and without LC and includes algorithms that support other modalities such as proteomics, making it suitable for a wide range of experimental designs. The chapter is written for scientists seeking to establish reproducible workflows to analyze multidimensional metabolomics data, regardless of prior experience with IMS. Demonstrations for both Python-based programmatic data processing and graphical user interface (GUI) workflows are provided to facilitate implementation by users with different levels of computational expertise. Following these procedures, researchers can successfully process, visualize, and interpret LC–IMS–MS data using freely available software and data resources.

Keywords Mass spectrometry, Metabolomics, Liquid chromatography, Ion mobility spectrometry, Software, MZA, mzapy, AI

1 Introduction

Liquid chromatography–ion mobility spectrometry–mass spectrometry (LC–IMS–MS) comprises an analytical approach increasingly used for the characterization of metabolites in complex biological and environmental samples. Ion mobility spectrometry (IMS or IM) separates ions in the gas phase according to their size, shape, and charge, adding a valuable orthogonal dimension to chromatographic and mass-to-charge (*m/z*) separations. This combination provides superior resolution and selectivity for distinguishing isomers, adducts, and conformers, thereby enhancing metabolite annotation confidence and reducing ambiguity in untargeted analyses [1–3].

Michael Witting (ed.), *Computational Methods and Data Analysis for Metabolomics*, Methods in Molecular Biology, vol. 3063, https://doi.org/10.1007/978-1-0716-5452-1_8,

Recent advances in IMS technology, such as drift tube IMS (DTIMS) [4], traveling wave IMS (TWIMS) [5], trapped IMS (TIMS) [6], and structures for lossless ion manipulations (SLIM) [7], have revolutionized the achievable resolving power and throughput of multidimensional measurements. The developments in computational tools have contributed to enabling large-scale LC–IMS–MS applications in metabolomics, proteomics, and lipidomics [3].

This chapter presents procedures that researchers can apply to successfully process, visualize, and interpret LC–IMS–MS data using freely available software resources. Demonstrations for both Python-based programmatic data processing and graphical user interface (GUI) workflows are provided to facilitate implementation by users with different levels of computational expertise. Although artificial intelligence (AI)-assisted coding is becoming more accessible, programming remains an essential skill for researchers processing MS data. A foundational competency in Python, machine learning, and MS algorithm development is a fundamental skill to enable researchers to critically assess AI-generated outputs, understand biases, and apply rigorous validation methods (e.g., generating step-by-step code to validate AI predictions against experimental data) to avoid pitfalls such as hallucinated results or overfitting to existing knowledge [8].

1.1 Overview of LC–IMS–MS Data

LC–IMS–MS data are inherently multidimensional, combining liquid chromatographic retention time (RT), ion mobility drift time (DT) or arrival time (AT), mass-to-charge ratio (*m/z*), and signal intensity. The analytical workflow and data structure are illustrated in Fig. 1. While IMS provides an additional dimension of separation, boosting sensitivity and supporting higher confidence annotations, this added dimension increases computational complexity. For instance, raw LC–IMS–MS data files can easily exceed several gigabytes, making efficient data storage and retrieval challenging. Similarly, collision-cross-section (CCS) values derived from IMS measurements provide molecular shape information that can be leveraged for compound identification and cross-platform comparison, but careful consideration is necessary to ensure these CCS values are determined robustly [9].

1.2 Tools and Workflows for LC–IMS–MS Data

Although IMS–MS analysis originally relied almost exclusively on proprietary vendor software, the landscape has expanded significantly, with many free and open-source tools now available to meet the growing computational needs of the community. Figure 2 shows the general steps of IMS–MS data processing workflows. At Pacific Northwest National Laboratory (PNNL), developments in both hardware and computational tools have contributed to enabling large-scale LC–IMS–MS omics applications. Among these, PNNL-PreProcessor [11, 12] provides various algorithms and functionalities for improving signal quality and transforming raw data; MZA [13]

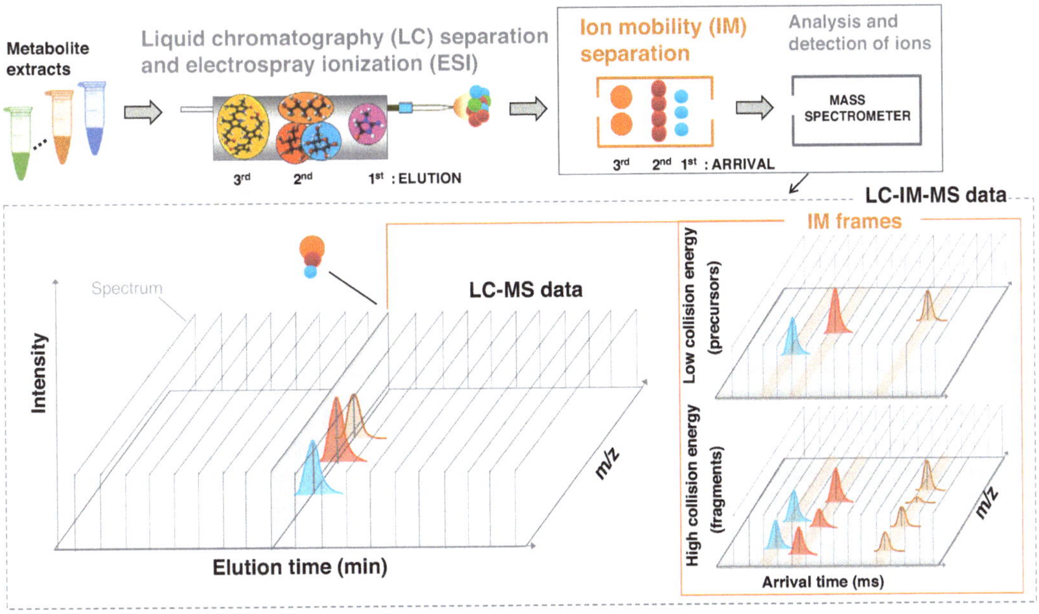

Fig. 1 Analytical workflow for multidimensional metabolite profiling by LC–IMS–MS and data structure. Metabolite extracts are separated by LC, followed by IMS, and analyzed by MS in the All-Ions DIA mode, which alternates between low and high collision energies to capture precursor and fragment ion spectra within the same run. Spectra are represented by gray dashed lines. Rather than collecting a single spectrum at every LC time point, coeluting ions (i.e., with close elution times) in this example, at the second order of elution and represented by spheres and peaks, in blue, red, and orange colors, could be further distinguished by the ion mobility separation, where multiple spectra are collected into IMS frames. Fragments are detected within the same elution and mobility time window as their precursors. (Reprinted with permission from [10])

and mzapy [14] frameworks provide efficient, open-source solutions for data conversion and extraction; PeakDecoder [10] enables machine-learning-based metabolite annotation; LipidOz [15, 16] offers deep-learning-guided lipid structure elucidation with high structural detail; and PeakQC provides user-friendly, accessible, and comprehensive quality control for raw MS data [17]. These tools constitute modular algorithms for a multidimensional data analysis pipeline. Demonstrations for both Python-based programmatic data processing and graphical user interface (GUI) workflows are provided in the following sections.

2 Materials

The following software tools and datasets are publicly available and required to complete the practical sections of this chapter:

1. A computer with the Windows operating system.
2. PNNL-PreProcessor executable, downloaded from https://pnnl-comp-mass-spec.github.io/PNNL-PreProcessor

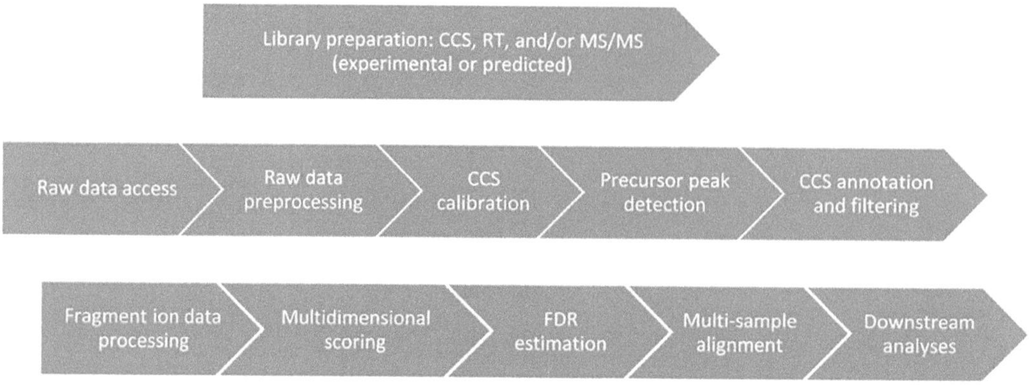

Fig. 2 General steps of IMS–MS data processing workflows. Prior to data analysis, it is often necessary to have a library of CCS, RT, and MS/MS spectra to support annotation of LC–IMS–MS features. The next set of steps corresponds to the extraction of precursor features from raw data, which includes peak detection and calibration of CCS values. The next set of steps applies to LC–IMS–MS data that include MS/MS, for which MS/MS spectra must be extracted and processed for the identified precursor features. This set of steps also includes feature annotation, assessment of annotation confidence, and other quality control steps that must occur prior to downstream analyses (Reprinted with permission from [3])

3. MZA executable, downloaded from https://github.com/PNNL-m-q/mza
4. IonToolPack executable, downloaded from https://github.com/pnnl/IonToolPack
5. Python 3.8 or later installed.
6. Example LC–IMS–MS raw data files (Agilent and Bruker .d or .mzML currently supported), downloaded from
   - Metabolomics: https://github.com/pnnl/IonToolPack/tree/master/test_data/LC-IM-MS-Agilent/RawDataAgilent
   - Lipidomics: https://doi.org/10.5281/zenodo.7181929

3 Methods

3.1 Enhancing Raw Data with the PNNL-PreProcessor

The PNNL-PreProcessor [11] is a freely available software with various algorithms generating new MS files in the same instrument format but with enhanced signal quality. It provides both command-line and graphical, workflow-style interfaces for flexible and user-friendly preprocessing of IMS–MS datasets. It supports Agilent MassHunter (.d) and UIMF mass spectrometry data files acquired on DT and SLIM IMS–MS platforms, and a wide range of acquisition modes, including LC–IMS–MS, solid-phase extraction IMS–MS (e.g., RapidFire), and direct-infusion IMS–MS, as well as All Ions data-independent acquisition files with alternating high/low collision energy fragmentation. The software enables single-click batch processing of multiple raw files and offers several data-enhancement and transformation capabilities, including frame- and mobility-based compression, retention-time filtering, interpolation of the ion mobility dimension, multidimensional smoothing, correction of saturated peaks, IMS demultiplexing, removal of spike-like noise artifacts, export of ion mobility frame metadata (e.g., field strength, pressure, temperature), and MS instrument actuals to text formats. Recent additions allow conversion of arrival time to CCS/z space within raw data, single-field CCS calibration, polygon-based extraction of user-defined regions in IMS vs. *m/z* space, and two transformations of IMS–MS datasets into LC–MS–compatible formats that facilitate interoperability with legacy LC–MS software [12]: (1) mapping the ion mobility axis to the LC axis, and (2) transforming All Ions IM/MS to pseudo-DDA format. The following steps allow preprocessing of raw IMS–MS data.

1. Double-click the "PNNL-PreProcessor.exe" file to start it (*see* **Note 1**).
2. Click the text box to search and select a single MS file (.d folder) or a directory that contains several MS files (.d folders), for example, the metabolomics files downloaded in Sect. 2.
3. Select files and click "»" to move the files to be processed to the right.
4. Adjust the parameters.
5. Click the "Process Data Files" button.
6. Inspect the output folder to find the newly generated raw files (*see* **Note 2**).
7. See an example demonstrating the conversion of IMS–MS raw data to CCS/z space and using standard Python and R packages to read the raw data from MZA files and make visualization plots: https://github.com/EMSL-Computing/CCSz-demo

3.2 Data Conversion and Storage Using the MZA Format

To ensure efficient processing, data are first converted into an open format that supports high-performance access and compatibility with modern programming languages. The HDF5-based MZA format [13] fulfills these needs and is particularly useful for the development of AI software. MZA was designed to simplify access to MS data by replacing proprietary vendor formats with a flexible, cross-platform, and simplified structure based on the HDF5 standard. It has become the base layer supporting the development of our prototypes and software tools. This section describes how to convert vendor data files to MZA.

1. Open a command-line terminal and navigate to the folder containing the mza.exe (for Unix-like operating systems, *see* **Note 3**).

2. Convert a single ".d" file using the following command (example metabolomics files downloaded in Sect. 2):
mza -file test_data\LowHigh_PC_160_180_frames1-10.d -intensityThreshold 20
3. Convert all ".d" files in a directory using the following command:
mza -file [full_path_to_raw_files] -extension.d -intensityThreshold 20
4. The converted file can now be accessed directly using mzapy or standard HDF5 packages and viewers. Example R and Python scripts using standard HDF5 packages for accessing raw data in .mza files can be found at https://github.com/PNNL-m-q/mza.

3.3 Modular Data Extraction and Processing Using mzapy

The mzapy [14] Python library enables efficient extraction, processing, and visualization of data stored in the MZA format. It provides utilities for calibration, signal processing, peak finding, and CCS calibration (Fig. 3, *see* **Note 4**). The following workflow guides the user through basic mzapy usage from Python scripts and some typical operations to reproduce the panels in Fig. 3.

1. Install mzapy via pip: pip install mzapy
2. Start a Python script or type in the Python console to implement and execute the code in the following steps.
3. Import utilities from the library:

```
from matplotlib import pyplot as plt, rcParams
import numpy as np
from mzapy import MZA
from mzapy.isotopes import MolecularFormula,
monoiso_mass
from mzapy.peaks import find_peaks_1d_gauss
from mzapy.view import (
plot_chrom, plot_atd, add_peaks_to_plot)
```

4. Load the lipidomics MZA file downloaded in Sect. 2:

```
mza = MZA('demo_data.mza', mza_version='old')
```

5. Extract the TIC chromatogram and a chromatogram for target *m/z* (this code section reproduces Fig. 3A):

```
# calculate some info for PC(18:0/18:1) [M+H]+
(C44H87NO8P)
pc_formula = MolecularFormula(C=44, H=87, N=1, O=
8, P=1)
pc_mz = monoiso_mass(pc_formula)
# extract the chromatogram for the monoiso m/z +
/- 0.02
mz_min, mz_max = pc_mz - 0.02, pc_mz + 0.02
```

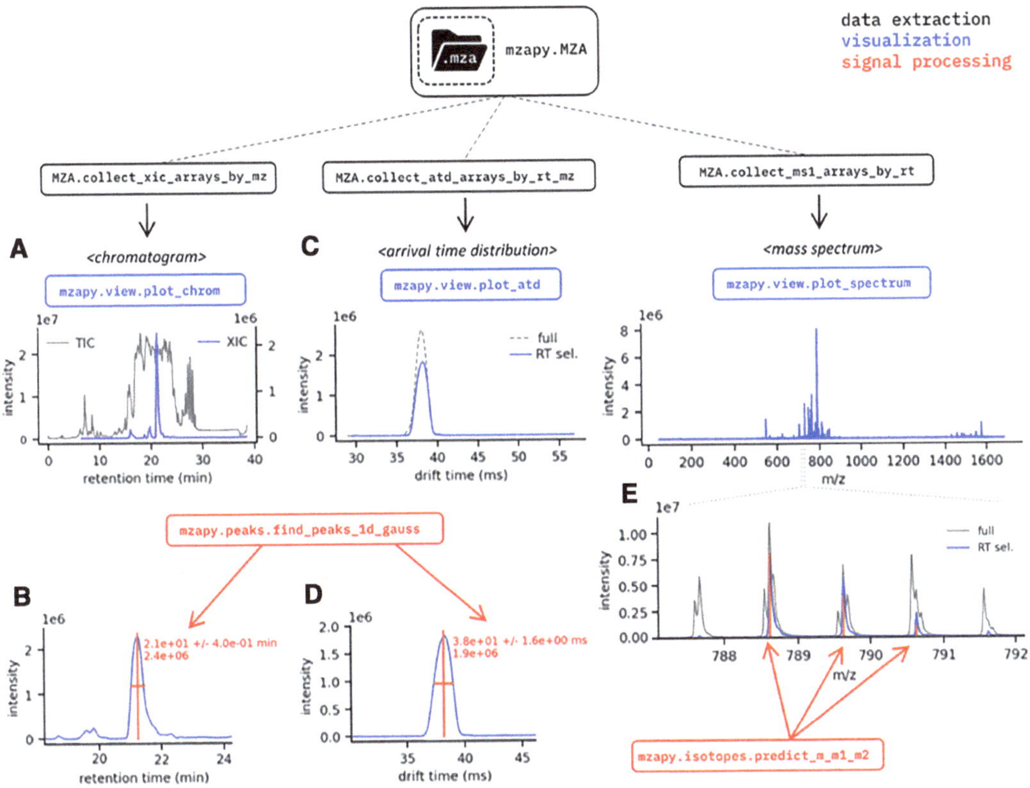

Fig. 3 Functional overview of the *mzapy* library. *mzapy* includes utilities that support data extraction, signal processing (e.g., peak fitting, molecular mass, and isotope calculations), and visualization of data from the LC, IMS, and MS dimensions. (A) Chromatogram with total ion current in gray and extracted ion current in blue. (B) Chromatogram of peak found with red lines marking peak position and width. (C) Arrival time distribution of full data (dashed gray) and a selected retention time subset (solid blue). (D) Arrival time of peak found with red lines marking peak position and width. (E) Mass spectrum of full data and selected retention time. Red arrows point to peak detection and isotope prediction. (Reprinted with permission from [14])

```
xic_rt, xic_i = mza.collect_xic_arrays_by_mz(mz_
min, mz_max)
# plot TIC and XIC, without fit
print('TIC and XIC (no fitting)')
ax = plot_chrom(*mza.collect_xic_arrays_by_mz(mza.
min_mz, mza.max_mz),
c='grey', label='TIC', figsize=(5.5, 2.5))
ax2 = ax.twinx()
for d in ['top', 'left', 'bottom']:
ax2.spines[d].set_visible(False)
ax.legend(frameon=False, loc='upper left')
plot_chrom(xic_rt, xic_i,
ax=ax2, figname='show', label='XIC')
```

6. Perform data extraction and peak fitting in LC and IMS dimensions (this code section reproduces Fig. 3C, D):

```
# extract RT and m/z selected ATD
atd_dt, atd_i = mza.collect_atd_arrays_by_rt_mz(mz_
min, mz_max, rt_min, rt_max)
# It takes a very long time to get the non-RT-
selected ATD, so don't run this every time
# plot ATD
ax = plot_atd(*mza.collect_atd_arrays_by_rt_mz(mz_
min, mz_max, mza.min_rt, mza.max_rt),
c='grey', ls='--', label='full', figsize=(5.5, 2.5))
plot_atd(atd_dt, atd_i, label='RT sel.', ax=ax)
ax.legend(frameon=False)
# show the plot and close
print('ATD with and without RT selection')
plt.show()
plt.close()
# fit ATD to get observed drift time
atd_means, atd_heights, atd_widths = find_peaks_1d_
gauss(atd_dt, atd_i,
0.05, 1e3, 1,
20, 1, True)
atd_mean, atd_height, atd_width = atd_means[0],
atd_heights[0], atd_widths[0]
# plot the ATD with fit
ax = plot_atd(atd_dt, atd_i,
dt_range=[atd_mean - 8, atd_mean + 8], figsize=
(5.25, 2.5))
# annotate with the fitted peak
add_peaks_to_plot(ax, atd_means, atd_heights,
atd_widths,
add_text_lbl=True, x_units='ms', fontsize=8)
# show the plot and close
print('ATD with gaussian fit')
plt.show()
plt.close()
```

3.4 Double-Bond Lipidomics Analysis Using LipidOz

LipidOz [15, 16] integrates deep learning to automatically assign double-bond positions in lipidomics data acquired with ozone-induced dissociation (OzID) coupled to IMS–MS (Fig. 4). This section outlines its application from Python code to fungal lipid profiling experiments. For LipidOz source code and documentation, *see* **Note 5**.

1. Install LipidOz via pip, command line:

```
pip install lipidoz
```

2. Start a Python script or type in the Python console to implement and execute the code in the following steps.
3. Import utilities from the library:

```
from lipidoz.workflows import (
hybrid_deep_learning_and_isotope_scoring,
save_isotope_scoring_results,
write_isotope_scoring_report_xlsx)
```

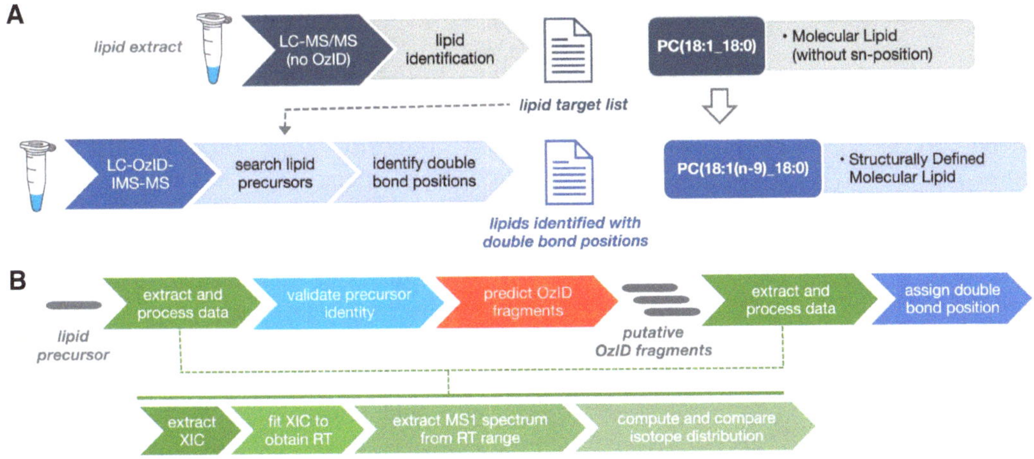

Fig. 4 Data processing workflow in *LipidOz*. (**A**) Experimental protocol: a non-OzID analysis is performed to obtain initial lipid annotations, then OzID data are analyzed to update lipid annotations with double-bond positions. (**B**) OzID data extraction and processing protocol: data are extracted for precursor ions, and identity is validated from isotope distribution; then data are extracted for predicted OzID fragments, and scoring is performed to determine whether sufficient evidence exists to assign double-bond positions. (Adapted with permission from [15, 16])

4. Perform data analysis

```
results = hybrid_deep_learning_and_isotope_scoring(
"example_data.mza",
"targets.csv",
0.1,
1.5,
0.02,
"example_dl_params.pt",
debug_flag="text")
```

5. Save and export results.

```
save_isotope_scoring_results(
results,
"results.loz")
write_isotope_scoring_report_xlsx(
results,
"results.xlsx")
```

3.5 Quality Control of Raw Data and Omics-Agnostic Tools

Omics-agnostic tools able to process various types of MS-based omics datasets, such as proteomics, metabolomics, lipidomics, and glycomics, and from different MS instrumentation platforms, could advance multiomics applications and contribute to improving standardization, harmonization, and reproducibility of results. A recent omics-agnostic tool developed at PNNL moving in this direction is PeakQC [17], which provides user-friendly, accessible, and comprehensive QC for raw MS data from various instrumentations, including LC–MS, DDA, DIA, ion mobility–MS (IMS–MS), and infusion-MS data. Several PNNL software tools are being extended into the IonToolPack suite (https://github.com/pnnl/IonToolPack), which houses PeakQC and various algorithms implemented in Python to provide raw data QC and visualization, and prototype tools for annotation and quantitation that are in progress.

IonToolPack provides a simplified and intuitive graphical user interface with each tool organized into its own tab with a single text box for parameter input, which avoids a complicated list of controls (i.e., dropdowns and checkboxes). In a way that is automatic and transparent to the user, raw MS data from various instrument vendors are converted to the MZA format. Other tools in progress incorporate modern AI approaches, such as PeakDecoder. While originally demonstrated for metabolomics [10], the new version of PeakDecoder enables sophisticated molecular annotation and quantitation of multidimensional MS datasets by adapting computer vision AI. It can annotate protein, metabolite, and lipid signatures in an automated fashion, which will enable holistic and iterative multiomics analyses [8] and will be available in IonToolPack. The following steps demonstrate how to perform QC analysis, generating plots to identify a sample with large mass errors (Fig. 5).

1. Double-click the "IonToolPack.exe" file to start it.
2. Go to the PeakQC tab.
3. In the parameters text box, update the path to the file "LC-IM-MS-Agilent/MS-runs_relative-path_LC-IM-MS.txt," which contains the list of the example metabolomics files downloaded in Sect. 2, along with sample group labels.
4. Adjust other parameters as needed.
5. Click the "Process" button.
6. Inspect the output folder to find the results generated, including PCA and raw data visualization plots (*see* **Note 6**).

4 Notes

1. PNNL-PreProcessor installer: Although the PNNL-PreProcessor can run without installation, some systems may block execution due to security policies or antivirus software. If this occurs, the installer should be used.
2. PNNL-PreProcessor temporary files: temporary files with suffixes such as "_temp1.d" will be generated with intermediate results. Do not delete the temp files while any preprocessing is

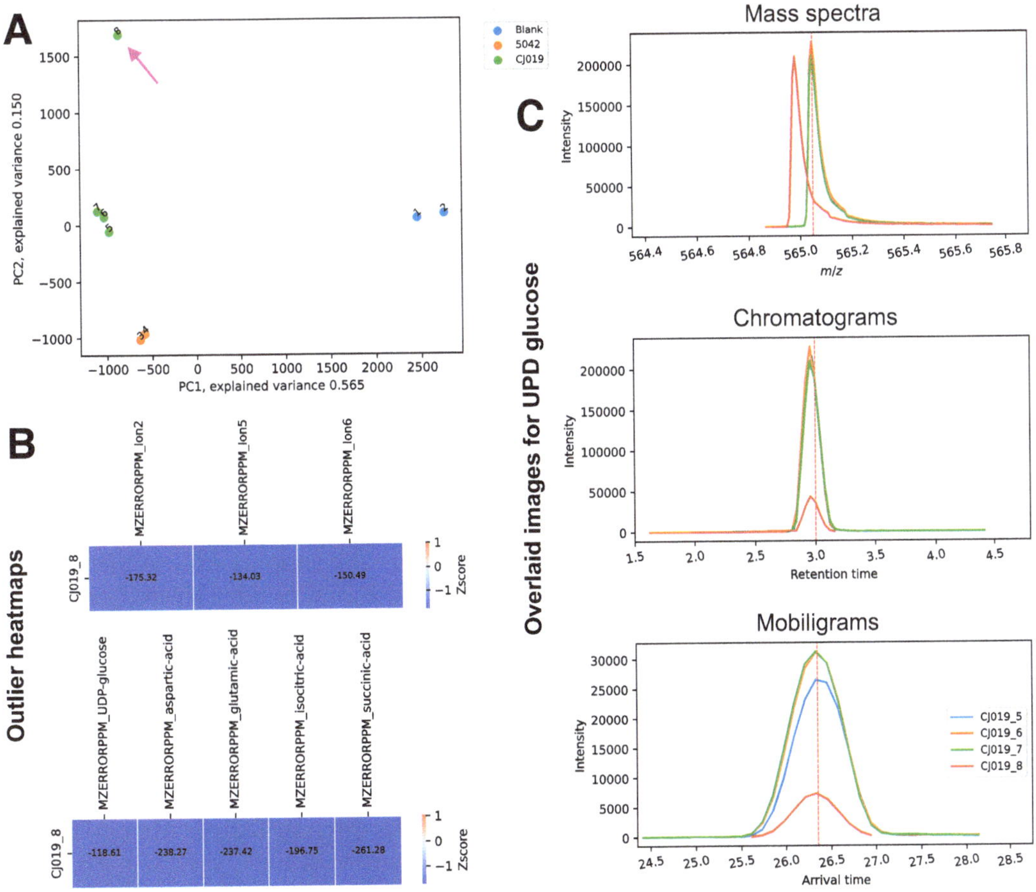

Fig. 5 Metabolomics LC–IMS–MS analysis showing detection of an outlier MS run with mass error issues. (**A**) PCA indicates the MS run CJ019_8 with a pink arrow. (**B**) Outlier heatmaps of auto-tracked ions (top) and user-specified ions (bottom) reporting MS run CJ019_8 with large *m/z* errors. (**C**) Overlaid mass spectra, chromatograms, and mobiligrams for the user-specified ion UDP glucose (*m/z* 565.05, RT 3.0 min, AT 26.35 ms) in sample group CJ019. PeakQC generates this type of figure for each user-specified ion. Dashed red lines show the reference values, which are indicated in the input file for the user-specified ions. The mass shift can be observed in the top panel for the red trace (CJ019_8). Mobiligrams are extracted if the AT is specified in the input, and the MS data contain an ion mobility spectrometry separation

in progress, and verify that all temp files were deleted after all preprocessing is finished.

3. MZA in Unix-like operating systems: The MZA executable has no requirements (i.e., it is self-contained) and can run without installation on Windows. For Unix-like operating systems, a container (see example in https://github.com/PNNL-m-q/mza/blob/main/docker/Dockerfile) or the Wine compatibility layer (www.winehq.org) can be used.

4. Documentation for mzapy: Comprehensive and up-to-date documentation is available at https://mzapy.readthedocs.io/en/latest/index.html.
5. LipidOz: the source code is freely available at https://github.com/PNNL-m-q/lipidoz, with up-to-date documentation available at https://lipidoz.readthedocs.io/en/latest.
6. IonToolPack/PeakQC: The initial implementation of PeakQC performed a detection of ions for auto-tracking by default; the current version does not use auto-tracked ions if user ions are specified. For an example of user-defined ions, see the file "Ion-Targets.csv" in the example metabolomics data.

Acknowledgments

The authors would like to thank the support from Agilent Technologies (to A. B.) and the Agile BioFoundry (http://agilebiofoundry.org), funded by the U.S. Department of Energy (DOE), Energy Efficiency and Renewable Energy, Bioenergy Technologies Office. Part of this work was supported by grant R35GM156826 from the National Institute of General Medical Sciences (to X. Z.) and the Pacific Northwest National Laboratory (PNNL) Laboratory Directed Research and Development Program, and is a contribution of the m/q Initiative. PNNL is operated by Battelle Memorial Institute for the United States Department of Energy (DOE). This work was performed at the Environmental Molecular Sciences Laboratory (EMSL), a DOE Office of Science User Facility sponsored by the Biological and Environmental Research program under Contract No. DE-AC05-76RL01830.

References

1. Dodds JN, Baker ES (2019) Ion mobility spectrometry: fundamental concepts, instrumentation, applications, and the road ahead. J Am Soc Mass Spectrom 30(11):2185–2195
2. Naylor CN, Nagy G (2025) Recent advances in high-resolution traveling wave-based ion mobility separations coupled to mass spectrometry. Mass Spectrom Rev 44(4):581–598
3. Ross DH et al. (2024) Computational tools and algorithms for ion mobility spectrometry-mass spectrometry. Proteomics 24(12-13):e2200436
4. May JC et al. (2014) Conformational ordering of biomolecules in the gas phase: nitrogen collision cross sections measured on a prototype high resolution drift tube ion mobility-mass spectrometer. Anal Chem 86(4):2107–2116
5. Shvartsburg AA, Smith RD (2008) Fundamentals of traveling wave ion mobility spectrometry. Anal Chem 80(24):9689–9699
6. Michelmann K et al. (2015) Fundamentals of trapped ion mobility spectrometry. J Am Soc Mass Spectrom 26(1):14–24
7. Ibrahim YM et al. (2017) New frontiers for mass spectrometry based upon structures for lossless ion manipulations. Analyst 142(7):1010–1021
8. Bilbao A (2025) The future of a myriad of accelerated biodiscoveries lies in AI-powered mass spectrometry and multiomics integration. J Mass Spectrom 60(8):e5157
9. Gabelica V et al. (2019) Recommendations for reporting ion mobility mass spectrometry measurements. Mass Spectrom Rev 38(3):291–320

10. Bilbao A et al. (2023) PeakDecoder enables machine learning-based metabolite annotation and accurate profiling in multidimensional mass spectrometry measurements. Nat Commun 14(1):2461
11. Bilbao A et al. (2022) A preprocessing tool for enhanced ion mobility-mass spectrometry-based omics workflows. J Proteome Res 21(3):798–807
12. Stow SM et al. (2024) Exploring ion mobility mass spectrometry data file conversions to leverage existing tools and enable new workflows. J Am Soc Mass Spectrom 35(8):1991–2001
13. Bilbao A et al. (2023) MZA: a data conversion tool to facilitate software development and artificial intelligence research in multidimensional mass spectrometry. J Proteome Res 22(2):508–513
14. Ross D et al. (2023) mzapy: an open-source Python library enabling efficient extraction and processing of ion mobility spectrometry-mass spectrometry data in the MZA file format. Anal Chem 95(25):9428–9431
15. Ross DH et al. (2023) LipidOz enables automated elucidation of lipid carbon-carbon double bond positions from ozone-induced dissociation mass spectrometry data. Commun Chem 6(1):74
16. Ross DH et al. (2024) A deep learning-guided automated workflow in LipidOz for detailed characterization of fungal fatty acid unsaturation by ozonolysis. J Mass Spectrom 59(9):e5078
17. Harrison A et al. (2024) PeakQC: a software tool for omics-agnostic automated quality control of mass spectrometry data. J Am Soc Mass Spectrom 35(11):2680–2689

Chapter 9

Tandem Mass Spectral Databases and Their Use in Non-Targeted Metabolomics

Valentina Ramundi and Michael Witting

Abstract

Liquid chromatography–mass spectrometry (LC–MS)-based non-targeted metabolomics produces intricate datasets that need advanced tools for identifying metabolites (MetID). Metabolite annotation and identification in non-targeted metabolomics require high-quality fragmentation data from biological samples and reference libraries. Public and commercial databases, such as NIST, MassBank, MoNA, GNPS, HMDB, mzCloud, and METLIN, are vital resources for both spectral matching and providing training data for machine-learning-driven MetID tools. These libraries differ in their coverage, curation, and access models, and they are often supplemented by in-house databases that cater to specific laboratory conditions, ensuring the highest level of confidence in identifications. To maximize the benefits of MS^2 libraries and ensure they work seamlessly together, we rely heavily on standardized file formats. Standard formats, such as mzML, MGF, MSP, JSON, and the MassBank format, each come with different levels of metadata richness and compatibility with various software tools. This chapter gives an overview of key MS^2 libraries, discusses the strengths and weaknesses of standard data formats, and introduces R-based solutions for better integration.

Keywords Metabolomics, MS^2 libraries, Metabolite identification, Mass spectrometry, Data formats, Spectra package, R

1 Introduction

Liquid chromatography-mass spectrometry (LC–MS)-based non-targeted metabolomicsdata analysis incorporates multiple steps, which include conversion of obtained data to a human-readable data table (e.g., mass (re)calibration, peak picking,alignment, grouping, etc.). To interpret the results in a biological context, the features obtained need to be converted into metabolites, a task known as metabolite identification (MetID). MetID is a crucial yet inherently challenging aspect of non-targeted metabolomics. The complexity of this task arises from the vast chemical diversity of metabolites, their intricate structures, and the variability in their fragmentation patterns. High-resolution tandem mass spectrometry (HRMS/MS) has proven

Michael Witting (ed.), *Computational Methods and Data Analysis for Metabolomics*, Methods in Molecular Biology, vol. 3063,
https://doi.org/10.1007/978-1-0716-5452-1_9,

to be a powerful tool for MetID, capable of generating detailed fragmentation patterns unique to specific molecular structures. However, these fragmentation spectra can vary significantly due to differences in instrumentation, ionization sources, or fragmentation techniques, such as collision-induced dissociation (CID), higher-energy collisional dissociation (HCD), and electron-induced dissociation (EID). These variations can complicate the interpretation and comparison of spectra between instruments and laboratories. Robust references, resources, and computational annotation strategies are essential to accurately identify unknown compounds. HRMS/MSdatabases are indispensable in addressing these challenges, serving as a cornerstone for advancing MetID. They offer a dual utility: directly facilitating compound identification through spectral matching and supporting the development of computational tools to expand predictive capabilities for compounds that have not yet been cataloged.

HRMS/MS databases are invaluable for identifying and annotating metabolites through spectral matching. By comparing experimental MS^2 spectra against reference spectra in a curated database, researchers can determine the identity of compounds with high confidence. This approach is efficient for well-characterized metabolites that are already present in various databases. For example, libraries such as METLIN [1], MassBank [2], and GNPS [3] have extensive collections of experimental MS^2 spectra for known compounds, enabling accurate annotations across diverse sample types. However, spectral matching is limited to compounds already present in the database, leaving many unknowns unresolved, especially in untargeted metabolomics studies.

Beyond the direct annotation of experimental data, MS^2 databases are essential as training datasets for developing and refining in silico tools for metabolite annotation, such as CSI:FingerID [4] or CFM-ID [5]. These computational models, often powered by machine learning, aim to predict or simulate MS^2 fragmentation patterns or molecular fingerprints for compounds not yet included in the mass spectral reference databases. By leveraging the wealth of existing spectra, such tools learn to recognize fragmentation rules, interpret substructure patterns, calculate molecular fingerprints, and extrapolate large structural databases to compounds, e.g., PubChem [6]. This predictive capability significantly broadens the scope of MetID, enabling insights into metabolites that do not have (publicly) available reference spectra deposited.

As the field advances, integrating MS^2 databases with computational tools continues to enhance their functionality. Initiatives such as hybrid approaches, where spectral matching is combined with predictive modeling, push the boundaries of metabolite annotation. Furthermore, the expansion of open-access databases and the incorporation of diverse datasets, including synthetic and bioactive compounds, make MS^2 databases an indispensable resource for

MetID. This chapter will cover some of the most well-known MS^2 databases used in the metabolomics field and provide an overview of the exchange formats currently used for MS^2 data.

MS^2 libraries are of utmost importance for metabolite annotation and identification. They serve a twofold purpose. First, for direct matching against experimental spectra, and second, for the development of computational tools. With several closed and open-source libraries available, one can select from a multitude of suitable libraries. While the large majority of libraries are based on CID and HCD, new fragmentation technologies, such as EID, Ultraviolet Photodissociation (UVPD), Oxygen Attachment Dissociation (OAD), are emerging.

2 MS^2 Libraries Used in MetID

Different public or closed libraries storing MS^2 data are available, each with a specific focus and metabolite coverage. The number of available spectra increases rapidly, while the number of covered compounds rises only slowly. The number and conditions covered in the different libraries can differ vastly. This starts with the collision energies used: spectra are either recorded at fixed collision energies (e.g., 10, 20, and 40 eV), merged from different energies, or collected with ramping collision energies. It is worth noting that, besides collision energies, the exact type of mass spectrometer used is crucial, as these energies can vary even with the same settings. This is due to the geometry of specific fragmentation devices and voltage settings. Furthermore, different adducts might be covered to varying extents in various databases.

2.1 In-House Libraries

In-house libraries constructed from chemical reference standards are the most valuable libraries, as they are typically created using the same analytical conditions for analyzing biological samples. Different identification schemes, such as those published by the Metabolomics Standard Initiative or by Schymanski et al., require additional orthogonal parameters for accurate identification, for example, retention time [7, 8]. This is only achieved by measuring a chemical reference standard at the moment. If analytical conditions remain constant, an in-house database enables the annotation and identification of metabolites with the highest possible levels of confidence. Therefore, every metabolomics laboratory should be advised to create its own in-house library and, if possible, share this library publicly to allow others to reproduce annotations or use spectra for further processing.

2.2 External Libraries

Besides in-house libraries and metabolites not covered by them, external libraries are essential for metabolite annotation. Not every laboratory can hold all possible chemicals and include them in its in-house library. Therefore, sharing spectral libraries within the community has become increasingly important to advance the field further and improve annotation rates. Multiple libraries have evolved, and some of them are outlined below. It is essential to note that some libraries may contain identical spectra due to the reuse of spectra in the evaluation of tools, copying, and curation. For example, the SPectraL hASH (SPLASH), a hashed identifier, can help to identify duplicated spectra and potentially remove redundancies between libraries [9]. Ideally, matching of spectra from biological samples is performed against public spectra from the same or very similar instruments to achieve the highest possible confidence. However, this is not always possible. The multiple available MS2 reference libraries discussed below are used in MetID.

2.2.1 National Institute of Standards and Technology (NIST) Mass Spectral Library

The NIST Mass Spectral Library is considered a "golden standard" in mass spectrometry for several key reasons. It offers comprehensive and extensive data collection, covering a wide range of small molecules and metabolites. The NIST library comprises mass spectra for thousands of compounds, ranging from common to rare, across a broad spectrum of chemical classes, including organic, inorganic, and biological compounds. It also covers diverse ionization techniques, such as ESI (Electrospray Ionization Source), Atmospheric Pressure Chemical Ionization (APCI), and Electron Ionization (EI), making the database highly applicable to different types of analyses. The NIST Tandem Mass Spectral Library (NIST23), which includes spectra obtained from ESI and APCI, comprises a vast collection of 2.4 million spectra.

In contrast, the EI Library within NIST23 contains 394,054 spectra. Another critical factor is the quality and accuracy of the data. The mass spectra in the NIST library are curated by experts in the field of mass spectrometry, ensuring that the published data are reliable, precise, and consistent. Additionally, the database's data are processed using standardized methods, which ensures consistency across the entire library, helping researchers compare their experimental results with high confidence. The reproducibility of results obtained by comparing experimental spectra with those in the NIST library is another factor that contributes to its high regard.

Furthermore, NIST continuously updates and improves its databases by adding new compounds and refining existing entries. This commitment to staying current makes the NIST library a dynamic and up-to-date resource. The validation of data, through comparison with experimental results, ensures that the spectra in the database remain accurate and trustworthy [10]. The database also includes extensive metadata for each mass spectrum, providing critical information such as chemical structure, molecular weight, ionization methods, and experimental conditions. This added context helps researchers interpret their results more effectively.

Additionally, the NIST Mass Spectral Database is designed to be easily searchable, making it user-friendly and accessible to both novice and experienced users of mass spectrometry. NIST is also capable of integrating with other analytical techniques. For example, NIST also provides complementary data, such as infrared (IR) spectra and nuclear magnetic resonance (NMR) data, which are helpful for structural elucidation of complex compounds. In regulatory and industrial applications, the NIST Mass Spectral Database is widely utilized due to its reliability and accuracy. It plays a key role in areas such as environmental testing, forensic analysis, and pharmaceutical development, where high-quality, standardized data are crucial. Many mass spectrometers and software systems also integrate the NIST library into their platforms, making it an indispensable tool for both research and industry. However, the NIST database is not freely accessible. Full access requires purchasing a license from authorized distributors. The only freely accessible part is the NIST Chemistry WebBook, which offers access to basic spectral data for selected compounds.

2.2.2 MassBank

MassBank was one of the first publicly available repositories for mass spectra of small chemical compounds (less than 3000 Da) [2]. Its unique approach involves contributors managing their own data on independent servers while sharing it through a unified interface. One of its most notable features is the creation of merged spectral data, which combines spectra obtained under varying CID energies to improve the precision of compound identification by 21%–23% without sacrificing recall. MassBank also offers advanced search tools, such as spectral, quick, substructure, peak, and peak difference searches, all designed to help users identify compounds with high accuracy. The repository standardizes data records, including chemical information, analytical conditions, and spectral details, ensuring easy sharing and integration with other databases. To support contributors, MassBank provides tools such as the Record Editor and Administration Tool to simplify data preparation.

Additionally, the system is open source, offering free access to its software and data formats. Data are currently hosted on GitHub for version-controlled maintenance. The repository's broad data coverage encompasses a diverse range of metabolites, synthetic drugs, and lipids, making it a comprehensive resource for life sciences research. By fostering international collaboration, MassBank serves as the first globally allied mass spectral database, promoting data sharing and advancing metabolomics research.

2.2.3 MassBank of North America (MoNA)

The MassBank of North America (MoNA), which is a comprehensive repository for mass spectrometry data of natural products and metabolites, was launched in 2016. MoNA is a comprehensive, metadata-driven repository designed to store, share, and query mass spectrometry data, with a primary focus on metabolite identification and annotation. It hosts over 200,000 mass spectral records, derived from both experimental libraries and in silico predictions, covering a wide array of compounds, including metabolites, lipids, and small molecules. MoNA integrates detailed metadata, including compound identities, chemical structures, and biological data, ensuring that users can access not only the raw spectra but also relevant contextual information. The database features auto-curation mechanisms to maintain data quality, and it offers advanced spectral visualization and analysis tools for comparing experimental spectra to reference data. MoNA supports collaborative data sharing, allowing global contributions and ensuring that it reflects the latest advancements in mass spectrometry research. With powerful search capabilities and the inclusion of predictive spectral data, MoNA facilitates accurate metabolite identification, making it a valuable resource for metabolomics, drug discovery, and environmental analysis. Its open-access model promotes transparency and collaboration, enhancing its utility across academic and industry research.

2.2.4 Human Metabolome Database (HMDB)

The Human Metabolome Database (HMDB) was originally launched in 2005 as a metabolite structure database and has since become a highly comprehensive and valuable resource for humanmetabolomics. The HMDB version 5.0 offers an expanded and enriched collection of data, providing in-depth information about human metabolites [11]. This version significantly enhances its coverage and functionality, now including over 217,000 annotated metabolites and over 1.5 million unannotated derivatized metabolites, primarily intended for gas chromatography-mass spectrometry (GC-MS) analysis. Although initially developed as a metabolite structure database, enriched with detailed chemical structures, biological roles, physiological concentrations, tissue and biofluid locations, metabolic pathways, and links to associated diseases and genetic information, this database also includes reference MS^2 spectra. One of the most notable improvements in HMDB 5.0 is the inclusion of new, more accurate spectral data, including predicted NMR, MS spectra, retention indices, and (predicted) collision cross-section (CCS) values. These data, generated using advanced computational models, significantly enhance the ability to identify metabolites through mass spectrometry and nuclear magnetic resonance techniques.

Additionally, new interactive tools for structure visualization, spectral viewing, and pathway exploration provide a more user-friendly interface for researchers. The search capabilities have also been upgraded, supporting more refined compound identification using these expanded and predicted datasets. For example, the new Chemical Functional Ontology (ChemFOnt) has enhanced machine-readability, enabling a more automated and systematic extraction of metabolite functions and origins, which complements the more traditional text-based descriptions. Moreover, extensive work has been invested in enhancing the quality and depth of metabolite descriptions, making them more accurate and informative, thereby aiding both novice and advanced users in metabolomic studies. Overall, HMDB 5.0 represents a significant upgrade, providing a crucial tool for metabolomics research, exposomics, lipidomics, and related scientific fields. These enhancements make it an essential resource for a deeper understanding of human metabolism, facilitating the identification and annotation of metabolites, and enabling more accurate analyses in clinical, environmental, and nutritional sciences.

2.2.5 Global Natural Products Social Molecular Networking (GNPS)

The Global Natural Products Social Molecular Networking (GNPS) platform, launched in 2016, is a freely accessible resource for analyzing and sharing mass spectrometry (MS) data, with a strong emphasis on natural products [3]. The platform uses molecular networking to organize and identify metabolites, allowing researchers to annotate complex biological data easily. GNPS supports both positive and negative ionization modes, making it versatile for analyzing a wide range of compound classes. The most recent version, GNPS2, builds upon the capabilities of its predecessor, enabling more efficient data-driven discovery and identification of metabolites, particularly natural products. As a community-driven platform, GNPS2 is open-access, allowing researchers to freely access and contribute data, thereby helping the platform grow continuously through global scientific collaboration. The platform hosts a comprehensive spectral library that includes MS/MS data from various sources, such as natural products (including secondary metabolites from plants, microbes, and other natural sources), synthetic compounds such as drugs and pharmaceuticals, environmental metabolites, and food and plant-derived compounds. One of GNPS2's most powerful features is its automated spectral comparison tool. When users upload their MS/MS data, the platform automatically matches it against the vast library of reference spectra. It uses cosine similarity scoring to align peaks in the uploaded spectrum with those in the reference spectra. A higher cosine similarity score, typically set at a default value of 0.7, indicates a stronger match, suggesting that the unknown molecule is closely related to a known compound in the library. Even if an exact match is not found, GNPS2 can provide partially matching spectra, offering insights into possible structural similarities and unknown analogs. This is particularly useful for dereplication, helping to identify known compounds early in the analysis to avoid redundant isolation and characterization.

2.2.6 mzCloud

mzCloud was launched in 2018 as a comprehensive database for high-resolution mass spectrometry (HRMS) data, offering tools for the identification of chemical compounds [12]. As of recent reports, mzCloud contains data on over one million compounds, making it a valuable resource for researchers in areas such as metabolomics,proteomics, and environmental analysis. It includes high-quality, experimentally acquired MS/MS spectra and supports both positive and negative ionization modes for various compound classes, across multiple collision energies, and utilizing different fragmentation methods. It is a commercial database offered by Thermo Fisher Scientific, and access to its full range of features typically requires a subscription or is available through licensing for users of Thermo Fisher mass spectrometers. Unlike many other databases that rely on community contributions, mzCloud ensures high-quality, accurate data, including high-resolution MS/MS and multistage MS^n spectra. These spectra are collected across a range of collision energies and using various fragmentation techniques, providing deeper insights into compound structures.

In comparison, databases such as MassBank or GNPS are open-access, community-driven, and may lack the same level of curation and consistency. mzCloud also integrates advanced search algorithms, such as tree-based matching and similarity scoring, which allow for more detailed and accurate compound identification. Other databases, such as NIST or MassBank, typically use simpler spectral matching techniques. Additionally, mzCloud is optimized for use with Thermo Fisher Scientific instruments, such as Orbitrap-based mass spectrometers, and generally is part of a subscription-based service. In contrast, many other databases, such as MassBank and GNPS, are publicly available at no cost. Lastly, mzCloud focuses heavily on small-molecule identification and applications in metabolomics,lipidomics, and forensic studies, while other databases may cover broader or different scientific areas, such as proteomics or human metabolomics.

2.2.7 METLIN

The METLIN tandem mass spectrometry (MS/MS)database, created in 2003 and made publicly available in 2005, was the first of its kind for identifying metabolites and chemical entities [1]. Initially, it contained MS/MS spectra for only 100 metabolites, but by 2012, it had expanded to over 10,000 metabolites, with an additional 12,000 metabolites and compounds added in the following years. METLIN's data are highly versatile, supporting various types of tandem mass spectrometry instruments, and include high-resolution spectra obtained under different ionization modes and collision energies. METLIN provides MS/MS data under varying collision energies, which generates detailed fragmentation patterns for various compounds. This variability allows for a more comprehensive analysis of metabolites and other small molecules by observing a broader range of fragment ions. The database includes a vast range of compound classes, such as lipids, amino acids, nucleotides, organic acids, steroids, hormones, vitamins, and xenobiotics (e.g., drugs, pesticides). For example, lipids such as fatty acids often fragment better in negative ionization mode, while lipid metabolites such as ceramides or sphingolipids are detected in positive mode. Similarly, amino acids and peptides are commonly analyzed in positive ion mode, while organic acids, such as citric acid, are analyzed in negative ion mode. For compounds that lack experimental MS/MS data, METLIN generates in silico MS/MS spectra using advanced machine learning algorithms. This helps to extend its coverage and enables the identification of previously unknown metabolites.

3 Data Format for MS^2 Libraries

Several of the mentioned databases have a web interface, but one may also wish to integrate them into data-processingsoftware, such as XCMS [13], MZmine [14], or MS-DIAL [15], or even into their own workflows. In most databases, data can be downloaded from the web pages in different formats. The following paragraphs give a short overview of various commonly used data formats.

3.1 mzML

mzML is a widely recognized community standard for the storage and exchange of mass spectrometry (MS) data [13]. It was developed under the guidance of the HUPO Proteomics Standards Initiative (PSI) to ensure compatibility, reproducibility, and efficient data sharing in proteomics and related disciplines. This format is essential for harmonizing data representation across different instruments and software platforms, reducing conversion errors, and ensuring data integrity. It provides a rich framework for capturing detailed metadata about instrument settings, acquisition parameters, and experimental conditions. Spectral data can be efficiently stored using binary encoding, significantly reducing file size without compromising accessibility. The format is designed with extensibility in mind, enabling it to accommodate advancements in mass spectrometry technology. The development of mzML unified two earlier standards, mzData and mzXML, into a single robust framework. The first official release occurred in 2008, and subsequent updates have continued to enhance its features and address emerging needs in the field. The mzML file structure includes a header containing metadata about the software, instrument, and experiment, as well as a run section that holds the actual spectral and chromatographic data. The use of controlled vocabularies ensures consistent terminology, improving data interpretability and interoperability. In practice, mzML has become an essential format in mass spectrometry-based omics such as proteomics,metabolomics, and lipidomics. Researchers use it for data analysis, sharing datasets in public repositories such as PRIDE, and supporting downstream workflows. A wide range of software tools, including ProteoWizard for data conversion from vendor-specific formats, support the open .mzML files. All major metabolomics processing tools (e.g., XCMS,MZmine, MS-DIAL) have adopted mzML as one of the accepted input formats, making it a cornerstone of modern MS-based research. Though mzML is mostly used for storing entire LC–MS runs, single spectra can also be stored. However, the human-readability is limited as this format was mainly developed for data storage.

3.2 Mascot Generic Format (MGF)

The Mascot Generic Format (MGF) is a widely used file format in mass spectrometry for representing peak list data [14]. It is particularly well-suited for workflows involving peptide and protein identification, as it serves as the input format for many database search engines, such as Mascot, X!Tandem, and MS-GF+. The MGF format is designed to describe peak lists derived from raw mass spectrometry data in a simple and structured manner. While it lacks the comprehensive metadata capabilities of formats such as mzML, its simplicity and compatibility have made it a staple in proteomicsworkflows. MGF is a plain-text format that organizes spectral information, including the m/z (mass-to-charge ratio) and the intensity of detected ions. Specific keywords and data blocks define each spectrum in an MGF file. A typical entry includes details such as the precursor m/z, charge state, and a list of fragment ion peaks. For example, a spectrum begins with the marker "BEGIN IONS" and ends with "END IONS." Within this block, fields such as "TITLE" provide a descriptive identifier, "PEPMASS" records the precursor mass-to-charge ratio, and "CHARGE" specifies the ion's charge state. Fragment ion peaks are listed as pairs of m/z and intensity values. This format is particularly useful for submitting peak list data to database search engines for protein and peptide identification. It is also used for sharing spectral data in collaborative research and supporting workflows in proteomics and metabolomics. Despite its advantages, the MGF format has limitations. It does not include detailed experimental or instrument metadata, which are often essential for reproducibility and advanced analyses. Additionally, it only stores processed peak lists and does not handle raw data. MGF files are supported by a wide range of tools, including ProteoWizard for data conversion and search engines such as Mascot, X!Tandem, and MS-GF+. Visualization tools such as PeptideShaker and Skyline also work seamlessly with this format. Although more advanced standards such as mzML offer greater flexibility and metadata support, MGF remains a practical and efficient choice for database searches and peptide identification workflows. Its simplicity, ease of use, and widespread compatibility ensure its continued relevance in mass spectrometry research. In metabolomics, MGF is mostly used as an input file format for search engines and tools, but also for sharing MS2 libraries. However, different versions and flavors of MGF exist because, apart from some mandatory fields, user-defined fields can be used. No harmonization of fields required for metabolomics exists so far, such as adducts or MS and chromatographic metadata, making it only partially usable for library sharing.

3.3 MSP

The MSP file format is a plain-text standard used to store mass spectral libraries, widely utilized in metabolomics, proteomics, and analytical chemistry [14]. It acts as a repository for reference spectra and is essential for workflows involving compound identification through spectral matching. MSP files are supported by tools such as NIST MS Search, MS-DIAL, and SpectraST. Each entry in an MSP file corresponds to a compound or analyte and contains both metadata and spectral information, including a list of fragment ion peaks. The MSP format is flexible regarding the types of metadata it can include and the number of fragment ions it can store. However, it generally focuses on the core spectral data that support identification tasks. The format does not store raw data, chromatograms, or other detailed experimental conditions, which limits its use for more advanced analyses that require full context. NIST (National Institute of Standards and Technology) is closely associated with the MSP (Mass Spectral Peak) file format due to its role in developing and maintaining one of the most widely used mass spectral libraries in the world. The NIST mass spectral library is a comprehensive database containing thousands of mass spectra for chemical compounds, and it is a crucial resource for compound identification through mass spectrometry. The NIST library uses the MSP format to store and organize mass spectral data, making it accessible for researchers to search and compare spectra. NIST MS Search is a widely used software tool that enables researchers to match experimental mass spectra with the reference spectra in the NIST database. The NIST library, stored in the MSP format, enables the MS Search tool to facilitate compound identification by comparing experimental spectral data with the library's data. The MSP format is used to structure information about each compound, such as the compound name, molecular weight, precursor m/z, and a list of fragment ion peaks with their corresponding m/z and intensity values. When researchers perform spectral matching, NIST MS Search compares the experimental spectra to the data stored in the NIST library, all formatted in MSP. This process is vital for identifying unknown compounds, making the MSP format central to the way the NIST library is used for spectral matching in mass spectrometry. Thus, the MSP format is fundamentally tied to NIST's contributions to mass spectrometry by providing a standardized way of storing and sharing spectral data.

3.4 JavaScript Object Notation (JSON)

The JSON (JavaScript Object Notation) format is widely used for storing and exchanging data, especially in web applications [15]. It is lightweight, human-readable, and easy for machines to parse and generate, making it a versatile format. JSON is commonly used in

APIs, configuration files, and data serialization because of its simplicity. A JSON file consists of objects and arrays. An object is an unordered set of key–value pairs enclosed in curly braces {}. Each key is a string, and the corresponding value can be a string, a number, an array, another object, or a boolean. An array is an ordered collection of values enclosed in square brackets [], and the values can be strings, numbers, booleans, objects, or other arrays. Data in a JSON file are stored in a hierarchical structure, which makes it easy to represent complex datasets. The JSON format is advantageous because it is easy to read and write, both for humans and machines. It is efficient for data exchange over networks, and its lightweight nature makes it ideal for use in many applications.

Additionally, JSON is language-independent and can be easily parsed and generated in most programming languages, such as JavaScript, Python, and Java. In mass spectrometry, JSON is a valuable format for storing and sharing complex spectral data and metadata. It is particularly suitable for use in web-based tools, APIs, and other systems that require efficient data interchange, making it an essential format for modern mass spectrometry data-sharing platforms, including GNPS, MassBank, mzSpecML, MS-DIAL, MetaboLights, PRIDE (Proteomics Identifications Database), METLIN, mzCloud, ReSpect, LipidBlast, PeptideAtlas, MassIVE, and MoNA.

3.5 MassBank Format

The MassBank format is a widely used standard for representing mass spectrometry data, particularly in the context of mass spectral databases [2]. It is designed for storing the results of mass spectrometry experiments, which include both molecular ions and fragment ions, along with related metadata. The MassBank format is used by the MassBank database, which is one of the largest and most comprehensive public repositories of high-quality mass spectra, providing spectral data for various compounds, including pharmaceuticals, metabolites, and natural products. A MassBank entry begins with a set of tags that provide essential metadata about the compound and the experiment. These tags describe the compound's name, the instrument used, ionization mode, and other experimental details. Following the metadata, the entry contains a list of fragment ions, represented by their mass-to-charge ratio (m/z) and intensity. The format is structured as a plain-text file where key–value pairs describe various aspects of the spectrum, making it easy to read and share. The MassBank format allows for flexible storage of spectral data, making it suitable for a wide range of mass spectrometry experiments. It is designed to be both human-readable and machine-readable, ensuring that spectral data can be easily shared and analyzed.

The format can include a variety of optional and required fields, depending on the specific details of the analysis. This format is especially useful for storing rich metadata about experimental conditions, which can help interpret the spectrum in different contexts. One of the significant features of the MassBank format is its ability to store detailed metadata about experimental conditions. This includes information on ionization methods, instrumentation, experimental parameters, and compound identifiers such as InChI and SMILES. The MassBank format supports both tandem mass spectrometry (MS/MS) and single-stage mass spectrometry (MS) data. The MassBank format is beneficial because it allows efficient sharing of spectral data between research groups and institutions. Its text-based structure makes it easy to access and manipulate the data, and the inclusion of rich metadata ensures that spectral data can be interpreted accurately. Moreover, the format is compatible with various mass spectrometry-software tools that support spectral library searching and comparison. Here is an example of a MassBank format entry (Fig. 1):

3.6 Conversion of Formats Using R

Different metabolomics software tools support various file formats to differing extents. Among these, MassBank stands out as the format with the most comprehensive metadata fields, making it the most suitable for data exchange. However, it cannot be losslessly converted to other formats. While most formats provide a comment field for storing arbitrary information, the lack of standardized definitions for these fields often leads to compatibility issues, as other software tools frequently ignore such data. Additionally, variations, or "flavors," exist within certain file types, such as .mgf and .msp, further complicating seamless interoperability. Conversion between the different formats can be achieved by using the infrastructure of the R for Mass Spectrometry landscape. Various packages, called backends,

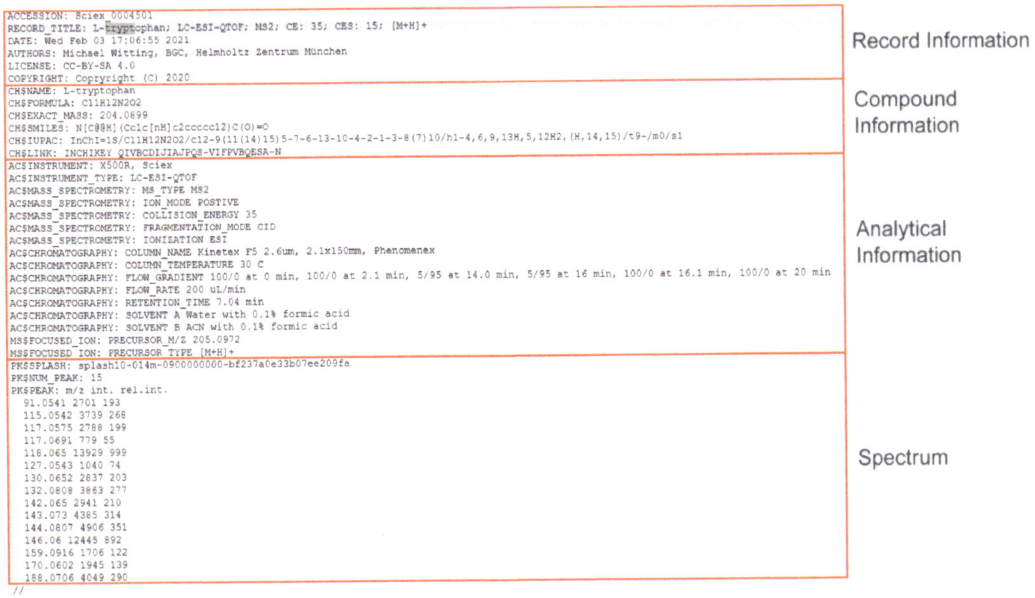

```
ACCESSION: Sciex_0004501
RECORD_TITLE: L-tryptophan; LC-ESI-QTOF; MS2; CE: 35; CES: 15; [M+H]+
DATE: Wed Feb 03 17:06:55 2021
AUTHORS: Michael Witting, BGC, Helmholtz Zentrum München
LICENSE: CC-BY-SA 4.0
COPYRIGHT: Copryright (C) 2020
CH$NAME: L-tryptophan
CH$FORMULA: C11H12N2O2
CH$EXACT_MASS: 204.0899
CH$SMILES: N[C@@H](Cc1c[nH]c2ccccc12)C(O)=O
CH$IUPAC: InChI=1S/C11H12N2O2/c12-9(11(14)15)5-7-6-13-10-4-2-1-3-8(7)10/h1-4,6,9,13H,5,12H2,(H,14,15)/t9-/m0/s1
CH$LINK: INCHIKEY QIVBCDIJIAJPQS-VIFPVBQESA-N
AC$INSTRUMENT: X500R, Sciex
AC$INSTRUMENT_TYPE: LC-ESI-QTOF
AC$MASS_SPECTROMETRY: MS_TYPE MS2
AC$MASS_SPECTROMETRY: ION_MODE POSTIVE
AC$MASS_SPECTROMETRY: COLLISION_ENERGY 35
AC$MASS_SPECTROMETRY: FRAGMENTATION_MODE CID
AC$MASS_SPECTROMETRY: IONIZATION ESI
AC$CHROMATOGRAPHY: COLUMN_NAME Kinetex F5 2.6um, 2.1x150mm, Phenomenex
AC$CHROMATOGRAPHY: COLUMN_TEMPERATURE 30 C
AC$CHROMATOGRAPHY: FLOW_GRADIENT 100/0 at 0 min, 100/0 at 2.1 min, 5/95 at 14.0 min, 5/95 at 16 min, 100/0 at 16.1 min, 100/0 at 20 min
AC$CHROMATOGRAPHY: FLOW_RATE 200 uL/min
AC$CHROMATOGRAPHY: RETENTION_TIME 7.04 min
AC$CHROMATOGRAPHY: SOLVENT A Water with 0.1% formic acid
AC$CHROMATOGRAPHY: SOLVENT B ACN with 0.1% formic acid
MS$FOCUSED_ION: PRECURSOR_M/Z 205.0972
MS$FOCUSED_ION: PRECURSOR_TYPE [M+H]+
PK$SPLASH: splash10-014m-0900000000-bf237a0e33b07ee209fa
PK$NUM_PEAK: 15
PK$PEAK: m/z int. rel.int.
  91.0541 2701 193
  115.0542 3739 268
  117.0575 2788 199
  117.0691 779 55
  118.065 13929 999
  127.0543 1040 74
  130.0652 2837 203
  132.0808 3863 277
  142.065 2941 210
  143.073 4385 314
  144.0807 4906 351
  146.06 12445 892
  159.0916 1706 122
  170.0602 1945 139
  188.0706 4049 290
//
```


Fig. 1 Example MassBank record of Tryptophan. Different sections cover different information associated with the spectrum, forming a comprehensive record

are available for reading and writing data from and to specific source files. The following code snippets can help make the conversion to different formats using this infrastructure more straightforward.

The core of this is the Spectra package, which handles mass spectral data within R [16]. The Spectra object has several core variables summarized in Table 1. If core variables are not set, the

Table 1
Core variables of a Spectra object

Variable	Type	Description
acquisitionNum	Integer	The index of acquisition of a spectrum during an MS run
centroided	Logical	Whether the spectrum is in profile or centroid mode
collisionEnergy	Numeric	Collision energy used to create an MSn spectrum
dataOrigin	Character	The origin of the spectrum's data, e.g., the mzML file from which it was read
dataStorage	Character	The (current) storage location of the spectrum data. This value depends on the backend used to handle and provide the data. For an in-memory backend such as the MsBackendDataFrame, this will be "<memory>"; for an on-disk backend such as the MsBackendHdf5Peaks, it will be the name of the HDF5 file where the spectrum's peak data are stored
intensity	Numeric	Intensity values for the spectrum's peaks
isolationWindowLowerMz	Numeric	Lower m/z for the isolation window in which the (MSn) spectrum was measured
isolationWindowTargetMz	Numeric	The target m/z for the isolation window in which the (MSn) spectrum was measured
isolationWindowUpperMz	Numeric	Upper m/z for the isolation window in which the (MSn) spectrum was measured
msLevel	Integer	The MS level of the spectrum
mz	Numeric	The m/z values for the spectrum's peaks
Polarity	Integer	The polarity of the spectrum (0 and 1 representing negative and positive polarity, respectively)
precScanNum	Integer	The scan (acquisition) number of the precursor for an MSn spectrum
precursorCharge	Integer	The charge of the precursor of an MSn spectrum
precursorIntensity	Numeric	The intensity of the precursor of an MSn spectrum
precursorMz	Numeric	The m/z of the precursor of an MSn spectrum
rtime	Numeric	The retention time of a spectrum
scanIndex	Integer	The index of a spectrum within a (raw) file
Smoothed	Logical	Whether the spectrum was smoothed

Spectra automatically returns NA. Handling of Spectra objects can be found in the Spectra vignette.

Different backends exist, and core variables are typically automatically mapped, but various versions might call them differently. Mapping of spectra variables can be used to provide export functions with instructions.

Additional spectra variables can be mapped to the variables used for export. Not every format supports all fields, but mapping allows for the correct formatting of available data fields. Generally, all fields can be integrated into all formats, but the support of software tools for these fields is not guaranteed.

The following code is an example of how to convert a spectrum from a MassBank record into different versions of msp. Complete code, including the generation of the MassBank record, is available from GitHub (https://github.com/michaelwitting/MiMB_TandemMS_Chapter).

1. First, all required R packages are loaded:

```
library(MsBackendMassbank)
library(MsBackendMgf)
library(MsBackendMsp)
```

2. Spectra can be read from MassBank records using the MsBackendMassBank. The MassBank format is quite extensive and contains many different fields. Which fields, or more specifically blocks of fields, are read can be defined during reading. In the following example, all fields are read.

```
meta <- metaDataBlocks()
meta$read <- TRUE
sps <- Spectra("spectraFiles/Trp.mb,"
source = MsBackendMassbank(),
backend = MsBackendDataFrame(),
metaBlocks = meta)
```

3. Mapping of spectra variables for export can be found by the function

```
spectraVariableMapping().
spectraVariableMapping(MsBackendMsp())
```

4. Specific mapping, for example, in the .msp format from the MassBank of North America (MoNA), can also be retrieved in the same way.

```
spectraVariableMapping(MsBackendMsp(), "mona")
```

5. The established mapping can then be used for export. The parameter allVariables controls which variables, defined in the

mapping, shall be exported. It is advised to keep the value at FALSE to avoid any issues with non-recognized fields in other software tools.

```
export(sps,
"spectraFiles/Trp_std.msp",
backend = MsBackendMsp(),
mapping = spectraVariableMapping(MsBackendMsp()),
allVariables = FALSE)
export(sps,
"spectraFiles/Trp_MoNA.msp",
backend = MsBackendMsp(),
mapping = spectraVariableMapping(MsBackendMsp(),
"mona"),
allVariable = FALSE)
```

6. Also, custom mappings can be created, e.g., for the .msp to be recognized by MZmine.

```
msp_mzmine <- c(
## minimal information
accession = "DB#",
name = "Name",
smiles = "SMILES",
exactmass = "ExactMass",
formula = "Formula",
inchi = "INCHI",
cas = "CAS">,
inchikey = "InChIKey",
collisionEnergy = "Collision_energy",
precursorMz = "PrecursorMZ",
precursorIntensity = "feature_ms1_height",
adduct = "Precursor_type",
rtime = "RT",
polarity = "Ion_mode",
splash = "Splash",
title = "Title",
## instrument information
instrument = "Instrument",
instrument_type = "Instrument_type",
## ms information
ms_ms_type = "Spectrum_type",
## ims information
ims_instrument_type = "ims_type",
ims_ccs = "CCS",
## ms information part II
focus_ion_type = "Ion_type",
## data processing information
data_processing_comment = "Processing_Comment",
## record informatio
date = "Date",
authors = "Authors",
comment = "Comments")
```

7. This custom mapping can also be used for exporting. In the case of MZmine, the retention time needs to be defined in minutes.

```
sps_mzmine <- sps
sps_mzmine$rtime <- sps_mzmine$rtime / 60
export(sps_mzmine,
"spectraFiles/Trp_mzmine.msp",
backend = MsBackendMsp(),
mapping = msp_mzmine,
allVariables = FALSE)
```

References

1. Xue J, Guijas C, Benton HP, Warth B, Siuzdak G (2020) METLIN MS2 molecular standards database: a broad chemical and biological resource. Na Methods 17(10):953–954. https://doi.org/10.1038/s41592-020-0942-5
2. Horai H, Arita M, Kanaya S, Nihei Y, Ikeda T, Suwa K, Ojima Y, Tanaka K, Tanaka S, Aoshima K, Oda Y, Kakazu Y, Kusano M, Tohge T, Matsuda F, Sawada Y, Hirai MY, Nakanishi H, Ikeda K, Akimoto N, Maoka T, Takahashi H, Ara T, Sakurai N, Suzuki H, Shibata D, Neumann S, Iida T, Tanaka K, Funatsu K, Matsuura F, Soga T, Taguchi R, Saito K, Nishioka T (2010) MassBank: a public repository for sharing mass spectral data for life sciences. J Mass Spectrom 45:703–714. https://doi.org/10.1002/jms.1777
3. Wang M, Carver JJ, Phelan VV, Sanchez LM, Garg N, Peng Y, Nguyen DD, Watrous J, Kapono CA, Luzzatto-Knaan T, Porto C, Bouslimani A, Melnik AV, Meehan MJ, Liu W-T, Crüsemann M, Boudreau PD, Esquenazi E, Sandoval-Calderón M, Kersten RD, Pace LA, Quinn RA, Duncan KR, Hsu C-C, Floros DJ, Gavilan RG, Kleigrewe K, Northen T, Dutton RJ, Parrot D, Carlson EE, Aigle B, Michelsen CF, Jelsbak L, Sohlenkamp C, Pevzner P, Edlund A, McLean J, Piel J, Murphy BT, Gerwick L, Liaw C-C, Yang Y-L, Humpf H-U, Maansson M, Keyzers RA, Sims AC, Johnson AR, Sidebottom AM, Sedio BE, Klitgaard A, Larson CB, Boya PCA, Torres-Mendoza D, Gonzalez DJ, Silva DB, Marques LM, Demarque DP, Pociute E, O'Neill EC, Briand E, Helfrich EJN, Granatosky EA, Glukhov E, Ryffel F, Houson H, Mohimani H, Kharbush JJ, Zeng Y, Vorholt JA, Kurita KL, Charusanti P, McPhail KL, Nielsen KF, Vuong L, Elfeki M, Traxler MF, Engene N, Koyama N, Vining OB, Baric R, Silva RR, Mascuch SJ, Tomasi S, Jenkins S, Macherla V, Hoffman T, Agarwal V, Williams PG, Dai J, Neupane R, Gurr J, Rodríguez AMC, Lamsa A, Zhang C, Dorrestein K, Duggan BM, Almaliti J, Allard P-M, Phapale P, Nothias L-F, Alexandrov T, Litaudon M, Wolfender J-L, Kyle JE, Metz TO, Peryea T, Nguyen D-T, VanLeer D, Shinn P, Jadhav A, Müller R, Waters KM, Shi W, Liu X, Zhang L, Knight R, Jensen PR, Palsson BØ, Pogliano K, Linington RG, Gutiérrez M, Lopes NP, Gerwick WH, Moore BS, Dorrestein PC, Bandeira N (2016) Sharing and community curation of mass spectrometry data with Global Natural Products Social Molecular Networking. Nat Biotechnol 34(8):828–837. https://doi.org/10.1038/nbt.3597
4. Dührkop K, Shen H, Meusel M, Rousu J, Böcker S (2015) Searching molecular structure databases with tandem mass spectra using CSI:FingerID. Proc Natl Acad Sci U S A 112:12580–12585. https://doi.org/10.1073/pnas.1509788112
5. Djoumbou-Feunang Y, Pon A, Karu N, Zheng J, Li C, Arndt D, Gautam M, Allen F, Wishart DS (2019) CFM-ID 3.0: significantly improved ESI-MS/MS prediction and compound identification. Metabolites 9(4):72
6. Kim S, Chen J, Cheng T, Gindulyte A, He J, He S, Li Q, Shoemaker BA, Thiessen PA, Yu B, Zaslavsky L, Zhang J, Bolton EE (2022) PubChem 2023 update. Nucleic Acids Res 51(D1):D1373–D1380. https://doi.org/10.1093/nar/gkac956
7. Schymanski EL, Jeon J, Gulde R, Fenner K, Ruff M, Singer HP, Hollender J (2014) Identifying small molecules via high resolution mass spectrometry: communicating confidence. Environ Sci Technol 48(4):2097–2098. https://doi.org/10.1021/es5002105
8. Sumner L, Amberg A, Barrett D, Beale M, Beger R, Daykin C, Fan TM, Fiehn O,

Goodacre R, Griffin J, Hankemeier T, Hardy N, Harnly J, Higashi R, Kopka J, Lane A, Lindon J, Marriott P, Nicholls A, Reily M, Thaden J, Viant M (2007) Proposed minimum reporting standards for chemical analysis. Metabolomics 3(3):211–221. https://doi.org/10.1007/s11306-007-0082-2

9. Wohlgemuth G, Mehta SS, Mejia RF, Neumann S, Pedrosa D, Pluskal T, Schymanski EL, Willighagen EL, Wilson M, Wishart DS, Arita M, Dorrestein PC, Bandeira N, Wang M, Schulze T, Salek RM, Steinbeck C, Nainala VC, Mistrik R, Nishioka T, Fiehn O (2016) SPLASH, a hashed identifier for mass spectra. Nat Biotechnol 34(11):1099–1101. https://doi.org/10.1038/nbt.3689
10. Yang X, Neta P, Stein SE (2014) Quality control for building libraries from electrospray ionization tandem mass spectra. Anal Chem 86(13):6393–6400. https://doi.org/10.1021/ac500711m
11. Wishart DS, Guo A, Oler E, Wang F, Anjum A, Peters H, Dizon R, Sayeeda Z, Tian S, Lee Brian L, Berjanskii M, Mah R, Yamamoto M, Jovel J, Torres-Calzada C, Hiebert-Giesbrecht M, Lui Vicki W, Varshavi D, Varshavi D, Allen D, Arndt D, Khetarpal N, Sivakumaran A, Harford K, Sanford S, Yee K, Cao X, Budinski Z, Liigand J, Zhang L, Zheng J, Mandal R, Karu N, Dambrova M, Schiöth Helgi B, Greiner R, Gautam V (2021) HMDB 5.0: the Human Metabolome Database for 2022. Nucleic Acids Res 50(D1):D622–D631. https://doi.org/10.1093/nar/gkab1062
12. https://www.mzcloud.org/. https://www.mzcloud.org/
13. Deutsch EW (2010) Mass spectrometer output file format mzML. Methods Mol Biol 604:319–331. https://doi.org/10.1007/978-1-60761-444-9_22
14. Deutsch EW (2012) File formats commonly used in mass spectrometry proteomics. Mol Cell Proteomics 11(12):1612–1621. https://doi.org/10.1074/mcp.R112.019695
15. Bourhis P, Reutter JL, Vrgoč D (2020) JSON: Data model and query languages. Inf Syst 89:101478. https://doi.org/10.1016/j.is.2019.101478
16. Rainer J, Vicini A, Salzer L, Stanstrup J, Badia JM, Neumann S, Stravs MA, Verri Hernandes V, Gatto L, Gibb S, Witting M (2022) A modular and expandable ecosystem for metabolomics data annotation in R. Metabolites 12(2):173

Check for updates

Chapter 10

Metabolite Annotation and Identification in R

Michael Witting and Johannes Rainer

Abstract

Metabolite annotation and identification represent one of the major bottlenecks in non-targeted metabolomics. Different levels of confidence have been defined to report results of this task, but tools often treat annotation results as equal. Here, we present example analyses based on the MetaboAnnotation R/Bioconductor package to perform annotation conforming to the Metabolomics Standard Initiative or the Schymanski scale.

Keywords R, MetaboAnnotation, Metabolite identification, Non-targeted metabolomics, RforMassSpectrometry

1 Introduction

Reproducible metabolite annotation and identification remain one of the major bottlenecks in non-targeted metabolomics. Multiple levels of information, e.g., from separation techniques (chromatography, electrophoresis, and/or ion mobility), mass spectrometry, and biological information, need to be integrated. Different levels of confidence of metabolite annotation/identification have been defined, e.g., by the Metabolomics Standard Initiative (MSI) or Schymanski et al. [1, 2]. The highest level of confidence is achieved by comparing information from the biological sample with chemical reference standards measured under the identical analytical conditions. This is typically accomplished by comparing an in-house reference database with the samples based on fragmentation/tandem MS spectra, precursor *m/z*, and retention times. The next lower level is achieved by comparing measured spectra against publicly available reference spectra, e.g., from MassBank, GNPS, MoNA, HMDB, and other sources [3–5].

Multiple LC–MSprocessing and annotation tools exist, e.g., MZmine, MS-DIAL, or SIRIUS [6–8]. For most, if not all, tools, an export of result peak tables and MS^2 spectra is possible at any time for further processing in another software. The programming language R allows multifaceted data analysis and has evolved into a major tool

Michael Witting (ed.), *Computational Methods and Data Analysis for Metabolomics*, Methods in Molecular Biology, vol. 3063,
https://doi.org/10.1007/978-1-0716-5452-1_10,

for the analysis of biological data, e.g., genomics, transcriptomics, proteomics, and metabolomics. As part of the RforMassSpectrometry initiative, a variety of interoperable R packages have been developed covering different aspects of mass spectrometry-based metabolomics and proteomics data analysis. Central to this environment is the Spectra package that provides the infrastructure to handle and process MS data, easily extendable to support additional file types and formats [9]. Additionally, metabolomics-specific packages have been developed [10], and the *xcms* package for LC–MSdata preprocessing [11] was adapted to integrate into the package ecosystem. The reuse of standardized data structures and objects ensures interoperability and allows the combination of the packages' functionality into reproducible analysis workflows, such as Metabonaut [12].

This chapter describes the use of the *MetaboCoreUtils* and *MetaboAnnotation* packages in conjunction with the *Spectra* package and multiple reader packages for metabolite annotation on different levels. This includes matching against in-house and external databases, as well as matching based on MS^1 data only. Lastly, the matching of formulae as well as potential additional values, such as collisional cross sections, is demonstrated. Table 1 summarizes the functions that can be used to achieve different levels of confidence according to the confidence levels defined by Schymanski et al. [2]. This chapter will follow these confidence levels, starting from the lowest to the highest level.

The R packages developed and maintained within the RforMassSpectrometry community allow creating reproducible workflows for annotation and identification of metabolites in non-targeted

Table 1

Levels of confidence as described by Schymanski et al. and the respective function from the MetaboAnnotation or SpectraQL packages

Schymanski level	Resource required	MetaboAnnotation/ SpectraQL function	Matching values
5	Database with exact masses	`matchValues()`	*m/z*
4	Database with formulae	`matchFormula()`	Formula
3	List of potential fragment *m/z*, e.g., specific for metabolite or lipid class	`query()`	Fragment *m/z*
2	External MS^2 database	`matchSpectra()`	*m/z*, MS^2
1	In-house MS^2 database with retention times	`matchSpectra()`	RT, *m/z*, MS^2 spectra
1	In-house MS^1database with retention times	`matchValues()`	RT, *m/z*

metabolomics. The code snippets presented in this chapter provide examples of how identification in compliance with existing reporting rules can be performed. Altogether, they can be combined into a comprehensive annotation script. Further developments, e.g., the *RuSirius* package, will also allow the integration of in silico annotation. Other packages such as *bridgeDb* can be used for standardizing metabolite names and identifiers for consistent reporting.

2 Material

Examples in this chapter have been run on a Lenovo X1 Yoga with an Intel i7 processor, 16 GB of RAM, running Windows 11 and R 4.5.1 within RStudio 2025.09.1. The R code for this chapter is available from GitHub (https://github.com/michaelwitting/MiMB_MetaboAnnotation_Chapter).

3 Methods

3.1 Preparation of Data for Annotation

To use the functions with your own data, it needs to be read into R and potentially further processed. Preprocessing of LC–MS or LC–MS/MS data can, for example, be performed with the *xcms* Bioconductor package (see Chap. 2), and MS^2 spectra for LC–MS features can be extracted from the respective result objects to continue the analysis in R. Alternatively, feature abundance tables, if provided as text or spreadsheet files, can be imported into R, and the related MS^2 spectra from file formats such as the Mascot Generic Format (.mgf). Reference library files can also be imported from a variety of formats, including, among others, .mgf, .msp, or MassBank records (see Chap. 10).

3.2 Matching Based on Exact m/z

When no further information, such as isotope pattern or MS^2 spectra, is available, exact *m/z* values can be matched against theoretical *m/z* values. This will, however, lead to many false positive annotations if a large chemical database is used as a reference. Results have to be treated with great care. The function `matchValues()` can be used to perform exact *m/z* searches, allowing absolute and *m/z*-dependent tolerances. The following code shows a simple example.

1. Load required R packages.

```
library(MetaboAnnotation)
library(MetaboCoreUtils)
library(tidyverse)
```

2. A simple `data.frame` containing some mock feature data is created as a query.

```
query_df <- data.frame(id = c("FT001", "FT002",
"FT003", "FT004"),
mz = c(166.088, 205.0966, 132.1012, 782.5733),
chem_formula = c("C9H11NO2","C11H12N2O2",
"C6H13NO2", "C44H80NO8P"))
```

3. Likewise, a `data.frame` containing some metabolites is created as the target database. Larger databases can be imported from text or spreadsheet files.

```
target_df <- data.frame(name = c("Phenylalanine",
"Tryptophan", "Leucine",
"Isoleucine", "Creatine", "PC 34:1",
"PC 36:4"),
chem_formula = c("C9H11NO2", "C11H12N2O2",
"C6H13NO2",
"C6H13NO2", "C4H9N3O2", "C42H82NO8P",
"C44H80NO8P"),
chebi = c("CHEBI:28044", "CHEBI:27897",
"CHEBI:25017",
"CHEBI:24898", "CHEBI:16919", "CHEBI:64517",
"CHEBI:64520"),
kegg = c("C02057", "C00806", NA,
NA, "C00300", NA,
NA),
hmdb = c(NA, NA, NA,
NA, "HMDB0000064", NA,
NA),
lipidmaps = c(NA, NA, NA,
NA, NA, NA,
NA))
```

4. For matching, the exact neutral masses of the reference metabolites are required in the targeted database. These are calculated with the `calculateMass()` function from the *MetaboCoreUtils* package and added as a new column "exactmass" to the target `data.frame`.

```
target_df$exactmass <
- calculateMass(target_df$chem_formula)
```

5. The algorithm to be used for the matching, as well as its settings, is configured using a parameter object. Since in this example the query contains *m/z* values and the target neutral exact masses, the `Mass2MzParam()` will be used. With this matching approach, theoretical *m/z* values are calculated for predefined adducts based on the exact mass of the molecules.

```
param <- Mass2MzParam(adducts = c("[M+H]+", "[M+
Na]+"),
tolerance = 0.05,
ppm = 0.0)
```

6. Matching is performed with the `matchValues()` function.

```
match_results <- matchValues(query_df,
target_df,
param)
```

7. Results are accessed by the `matchedData()` function that returns a `DataFrame` with one or more rows per query feature (depending on the number of target molecules the query can be matched against).

```
matchedData(match_results)
```

8. In some cases, the target database might contain precalculated *m/z* values, e.g., if adduct definitions are not included in the *MetaboCoreUtils* package and therefore theoretical *m/z* cannot be calculated. In this case, the `MzParam()` can be used. The example below converts the target database to contain *m/z* values. The matching is then performed using the `matchValues()` function combined with the `MzParam()` parameter class.

```
target_df <- bind_cols(target_df,
formula2mz(target_df$chem_formula,
adduct = c("[M+H]+", "[M+Na]+"))) %>%
pivot_longer(cols = c("[M+H]+", "[M+Na]+"),
names_to = "adduct",
values_to = "mz") %>%
as.data.frame()
# perform matching
param <- MzParam(tolerance = 0.05,
ppm = 0.0)
match_results <- matchValues(query_df,
target_df,
param)
matchedData(match_results)
```

In the example results, the third feature showed three hits in the target database: two isomers (leucine and isoleucine) and one isobar (creatine). Though potential false positives can be removed by lowering the mass error, this approach is only helpful to a certain extent, especially if the target database gets larger.

3.3 Matching Based on Chemical Formulae

Today's LC-MS setups are capable of delivering both high mass accuracy and isotope pattern fidelity. Searching for molecular formulae in a database is much more confident than searching only by mass, as performed above. Different tools for the calculation of sum formulae from isotopic patterns exist. Using the isotopic pattern instead of exact *m/z* alone reduces the number of potential candidates drastically. It has been shown that correct isotope pattern information is even more important than mass accuracy [13]. Therefore, functions to search for formulae have been implemented as well in the *MetaboAnnotation* package.

1. The same query and target data are used in the previous example.

```
# query formulas
query_df <- data.frame(id = c("FT001", "FT002",
"FT003", "FT004"),
mz = c(166.088, 205.0966, 132.1012, 782.5733),
chem_formula = c("C9H11NO2", "C11H12N2O2",
"C6H13NO2", "C44H80NO8P"))
# database of formulae
target_df <- data.frame(name = c("Phenylalanine",
"Tryptophan", "Leucine",
"Isoleucine", "Creatine", "PC 34:1",
"PC 36:4"),
chem_formula = c("C9H11NO2", "C11H12N2O2",
"C6H13NO2",
"C6H13NO2", "C4H9N3O2", "C42H82NO8P",
"C44H80NO8P"),
chebi = c("CHEBI:28044", "CHEBI:27897",
"CHEBI:25017",
"CHEBI:24898", "CHEBI:16919", "CHEBI:64517",
"CHEBI:64520"),
kegg = c("C02057", "C00806", NA,
NA, "C00300", NA,
NA),
hmdb = c(NA, NA, NA,
NA, "HMDB0000064", NA,
NA),
lipidmaps = c(NA, NA, NA,
NA, NA, NA,
NA))
```

2. The `matchFormula()` function is used to match data based on their chemical formulae. The parameter `formulaColname` allows specifying the name of the column in the query and the target `data.frame` containing the chemical formulae. Internally, before matching, the chemical formulae are normalized according to the Hill formula notation.

```
match_results <- matchFormula(query_df,
target_df,
formulaColname = c("chem_formula", "chem_formula"))
```

3. Matching results are retrieved using the `matchedData()` function.

```
matchedData(match_results)
```

3.4 Searching for Fragmentation Pattern

Annotation confidence increases with the use of MS^2 (fragment) spectra. Besides spectral matching, which will be described later, MS^2 spectra can also be searched for the presence of specific fragmentation patterns that can be indicative of functional groups or metabolite classes. Such searches can, for example, be performed using the Mass Spectrometry Query Language (MassQL) [14]. The *SpectraQL* Bioconductor package adds support for this query language to the *Spectra* R package and hence allows filtering and analyzing a `Spectra` object with MassQL. In the example below, we use MassQL to look for MS^2 spectra containing fragments with an *m/z* of 184.0733. Phosphocholine-containing lipids such as phosphatidylcholine (PC) and sphingomyelins (SM) produce a headgroup fragment at *m/z* 184.0733, while phosphoethanolamine-containing lipids, such as phosphatidylethanolamines (PE), show a neutral loss of the headgroup of 141.0191. The example below searches for putative lysoPCs in MS^2 data from fecal extracts.

1. Load required libraries: *Spectra* to handle MS data, *MsBackendMgf* to support import of MS data from .mgf files, and *SpectraQL* to support querying MS data in `Spectra` objects with the MassQL query language.

```
library(Spectra)
library(MsBackendMgf)
library(SpectraQL)
```

2. MS^2 data are read from an .mgf file using the *MsBackendMgf* backend. Subsequently, intensities in each fragment spectrum are normalized to the base peak signal.

```
query_specs <- Spectra("exampleData/fecal_ms2data.
mgf",
source = MsBackendMgf())
query_specs <- scalePeaks(query_specs, by = max)
```

3. Search MS^2 spectra for the presence of a particular fragment peak. This is expressed as a MassQL query, which is then applied to the `Spectra` object using the `query()` function. By using "QUERY *," the matching result is returned as a `Spectra` object.

```
query_results_spec <- query(query_specs, "QUERY *
WHERE MS2PROD=184.07:TOLERANCEMZ=0.005")
```

4. The spectra can be directly plotted using the `plotSpectra()` function.

```
plotSpectra(query_results_spec)
```

5. Repeating the query extraction by specifying "QUERY scaninfo(MS2DATA)" returns only the spectra metadata, not the individual fragment peaks. The result is returned as a `data.frame`.

```
query_results_df <- query(query_specs, "QUERY
scaninfo(MS2DATA) WHERE MS2PROD=184.07:TOLERANCEMZ=
0.005")
```

6. Plot the precursor *m/z* against the retention time of the resulting `data.frame` to identify potential homolog series.

```
plot(query_results_df$rtime,
query_results_df$precursorMz)
```

MassQL offers numerous possibilities to search MS data. The selected example shows only a relatively simple use case. Several examples exist for the use of MassQL in more complex tasks, e.g., to find novel lipid species of a specific lipid class [15].

3.5 Matching Fragment Spectra Against External Reference Libraries

Higher-confidence annotations are obtained if measured MS^2 data from biological samples are matched against measured spectra of reference substances. Multiple public spectral libraries containing such reference spectra are available (see Chap. 9). These can be easily included in annotation workflows employing the *MetaboAnnotation* package. While public reference databases can also be directly integrated and queried, we read in the present example reference spectra from an .mgf file. Subsequently, we use the `matchSpectra()` function to perform the pairwise spectra similarity calculation and matching of experimental query against reference target spectra. The results of this matching, which can be an m:n association between query and target spectra, are returned as a `MatchedSpectra` object.

1. Load required libraries. Both the query data and the target database are in .mgf format and require the *MsBackendMgf* to be read.

```
library(Spectra)
library(MsBackendMgf)
library(MetaboAnnotation)
```

2. Import the MS^2 data from a file in .mgf format. The peak intensities are subsequently normalized to the base peak signal per spectrum using the `scalePeaks()` function.

```
query_specs <- Spectra("exampleData/fecal_ms2data.
mgf",
source = MsBackendMgf())
query_specs <- scalePeaks(query_specs, by = max)
```

3. Import the reference spectra from an .mgf file and normalize the peak intensities.

```
target_external <- Spectra("exampleData/BILELIB19.
mgf",
source = MsBackendMgf(),
backend = MsBackendDataFrame())
target_external <- scalePeaks(target_external, by
= max)
```

4. Match query and reference spectra using `matchSpectra()`. The approach, as well as the settings and tolerances for spectra similarity calculation, can be configured using a specific parameter object. In the present example, the `MatchForwardReverseParam()` is used. This method performs first forward matching of all queries with all target spectra and then calculates, for the matching query–target spectra pairs, a reverse similarity score (target vs. query). By default, with parameter `requirePrecursor = TRUE`, similarity calculations are only performed if the precursor *m/z* of the query and target spectrum matches (given the provided tolerances, which can be configured with parameters `tolerance` and `ppm`). Parameters `THRESHFUN` and `THRESHFUN_REVERSE` allow for defining which pairs of spectra are considered matching. For the example below, matches between query and target spectra with a normalized dot product similarity score ≥0.7 are reported.

```
param <- MatchForwardReverseParam(tolerance =
0.005,
ppm = 0.0,
requirePrecursor = TRUE,
TRESHFUN = function(x) which(x >= 0.7),
TRESHFUN_REVERSE = function(x) which(x >= 0.7))
```

5. Performed the matching with the `matchSpectra()` function.

```
match_results <- matchSpectra(query_specs,
target_external,
param)
```

6. Extract matching results as a `DataFrame` using the `matchedData()` function. Due to the possible n:m mapping between query and target spectra, both query and target spectra can occur repeatedly in multiple rows.

```
matchedData(match_results)
```

7. Restrict the results to query spectra with at least one match in the reference library.

```
match_results_filtered <
- match_results[whichQuery(match_results)]
matchedData(match_results_filtered)
```

Visualize matching results using mirror plots. The `plotSpectraMirror()` function is used below to show the results for the second matched query spectrum. The upper panel in the mirror plot shows the query spectrum, the lower panel the target spectrum. Matching peaks are colored blue.

```
plotSpectraMirror(match_results_filtered [2])
```

For interactive browsing and verification, the `validateMatchedSpectra()` function can be used.

```
validateMatchedSpectra(match_results_filtered)
```

3.6 Matching MS² Spectra Against an Internal Reference Database

The highest level of confidence is achieved by matching measured fragment spectra against an in-house reference database, considering the precursor *m/z*, the retention time, and the MS^2 spectrum similarity. To allow matching based on retention times, the reference data must have been acquired with the same chromatography setup.

Retention indexing has been described as a suitable alternative to normalize differences in retention times arising from differences in void volumes, capillary lengths, etc. While used routinely in GC–MS, such an approach is not regularly used in LC–MS. Stoffel et al. have described an indexing approach based on *N*-Alkyl-pyridinium sulfonates to normalize retention information for reversed-phase-based metabolomics [16]. Such indexing could also be performed in R, and functions from the *MetaboAnnotation* package can use this information. The present example uses, however, retention times for the matching.

1. Load required libraries. In addition to the *MsBackendMgf* to allow import of MS data from .mgf files, the *MsBackendMassbank* R package is loaded. This package adds support for data files in MassBank format.

```
library(Spectra)
library(MsBackendMgf)
library(MsBackendMassbank)
library(MetaboAnnotation)
```

2. Import the query spectra and normalize the fragment intensities to each spectrum's base peak signal.

```
query_specs <- Spectra("exampleData/fecal_ms2data.
mgf",
source = MsBackendMgf())
query_specs <- scalePeaks(query_specs, by = max)
```

3. Import the target spectra from a subset of an in-house reference library in MassBank file format and perform the same peak intensity normalization as applied to the query spectra.

```
target_internal <- Spectra(list.files("exampleData/
in-house_MassBank/",
full.names = TRUE),
source = MsBackendMassbank())
target_internal <- scalePeaks(target_internal, by
= max)
```

4. Match the query spectra against the reference spectra.

```
param <- MatchForwardReverseParam(tolerance =
0.005,
ppm = 0.0,
requirePrecursor = TRUE,
TRESHFUN = function(x) which(x >= 0.7),
TRESHFUN_REVERSE = function(x) which(x >= 0.7),
toleranceRt = 0.5 * 60)
match_results <- matchSpectra(query_specs,
target_internal,
param)
```

5. Extract the full matching results.

```
matchedData(match_results)
```

6. Subset the result to query spectra with at least one match in the reference database.

```
match_results_filtered <
- match_results[whichQuery(match_results)]
matchedData(match_results_filtered)
```

7. Create a mirror plot for the first matched query spectrum.

```
plotSpectraMirror(match_results_filtered[1])
```

References

1. Sumner L et al (2007) Proposed minimum reporting standards for chemical analysis. Metabolomics 3(3):211–221
2. Schymanski EL et al (2014) Identifying small molecules via high resolution mass spectrometry: communicating confidence. Environ Sci Technol 48(4):2097–2098
3. Horai H et al (2010) MassBank: a public repository for sharing mass spectral data for life sciences. J Mass Spectrom 45:703–714.
4. Wang M et al (2016) Sharing and community curation of mass spectrometry data with Global Natural Products Social Molecular Networking. Nat Biotechnol 34(8):828–837
5. Wishart DS et al (2021) HMDB 5.0: the Human Metabolome Database for 2022. Nucleic Acids Res 50(D1):D622–D631
6. Heuckeroth S et al (2024) Reproducible mass spectrometry data processing and compound annotation in MZmine 3. Nat Protoc 19:2597–2641
7. Takeda H et al (2024) MS-DIAL 5 multimodal mass spectrometry data mining unveils lipidome complexities. Nat Commun 15(1):9903

8. Dührkop K et al (2019) SIRIUS 4: a rapid tool for turning tandem mass spectra into metabolite structure information. Nat Methods 16(4):299–302
9. Rainer J et al (OSF PREPRINTS) An Open Infrastructure for Mass Spectrometry Data in R. https://doi.org/10.31219/osf.io/cwt2v_v3
10. Rainer J et al (2022) A modular and expandable ecosystem for metabolomics data annotation in R. Metabolites 12(2):173
11. Louail P et al (2025) xcms in Peak Form: Now Anchoring a Complete Metabolomics Data Preprocessing and Analysis Software Ecosystem. Anal Chem 97(50):27639–27645. https://doi.org/10.1021/acs.analchem.5c04338
12. Louail P et al (2026) rformassspectrometry/Metabonaut: v1.5.0. https://doi.org/10.5281/zenodo.15062929
13. Kind T, Fiehn O (2006) Metabolomic database annotations via query of elemental compositions: mass accuracy is insufficient even at less than 1 ppm. BMC Bioinformatics 7(1):234
14. Damiani T et al (2025) A universal language for finding mass spectrometry data patterns. Nat Methods 22(6):1247–1254
15. Witting M et al (2025) Phosphorylated glycosphingolipids are commonly detected in Caenorhabditis elegans lipidomes. Metabolomics 21(2):29
16. Stoffel R et al (2022) N-Alkylpyridinium sulfonates for retention time indexing in reversed-phase-liquid chromatography-mass spectrometry–based metabolomics. Anal Bioanal Chem 414(25):7387–7398

Check for updates

Chapter 11

Large-Scale Metabolite Annotation in Untargeted Metabolomics Using MetDNA

Haosong Zhang and Zheng-Jiang Zhu

Abstract

MetDNA (http://metdna.zhulab.cn/) is a network-based computational platform for large-scale metabolite annotation in untargeted metabolomics using liquid chromatography–mass spectrometry (LC–MS). By using a metabolic reaction network (MRN) to guide recursive MS2 spectral similarity matching, MetDNA can accurately annotate both known and unknown metabolites, going beyond the limits of conventional spectral libraries. Since 2019, the platform has evolved from MetDNA to MetDNA2 and now to MetDNA3, with improvements in efficiency, coverage, and confidence in metabolite annotation for untargeted metabolomics. In this protocol, we outline best practices for data preparation, parameter configuration, and result interpretation, offering users a practical workflow to maximize the utility of MetDNA for high-confidence metabolite annotation.

Keywords Untargeted metabolomics, LC–MS, Metabolite annotation, MetDNA

1 Introduction

Untargeted metabolomics aims to capture a wide range of endogenous metabolites in biological systems, providing valuable insights into metabolic regulation, disease mechanisms, and biomarker discovery [1–3]. Advances in liquid chromatography–mass spectrometry (LC–MS) over the past decades have greatly improved coverage and accuracy for metabolite measurements in untargeted metabolomics [4–6]. However, metabolite annotation remains a major bottleneck due to the vast chemical diversity and structural heterogeneity of metabolites [7, 8]. Traditional strategies rely on MS2 spectral matching, which offers high confidence but is restricted to the availability of reference spectral libraries [9–11]. Thus, a large proportion of metabolites, especially those lacking chemical standards, remain unannotated. To overcome this limitation, alternative approaches such as network-based strategies have emerged as powerful complementary tools, enabling the propagation of annotations to metabolites without

Michael Witting (ed.), *Computational Methods and Data Analysis for Metabolomics*, Methods in Molecular Biology, vol. 3063, https://doi.org/10.1007/978-1-0716-5452-1_11,

reference to MS2 spectra and thereby expanding coverage to both known and previously unidentified metabolites [12–16].

Network-based strategies for metabolite annotation are generally divided into data-driven and knowledge-driven approaches. In data-driven networks, experimental MS features are represented as nodes and connected by edges that capture relationships such as MS2 spectral similarity [12, 13], intensity correlations [17], or characteristic mass differences [18]. This unsupervised modeling reveals latent connections and supports metabolite annotation. A prominent example is molecular networking (MN) in the GNPS ecosystem [12, 13], which organizes features by MS2 spectral similarity and leverages fragment-based substructure information to aid the discovery of novel metabolites. However, data-driven networks often become highly complex, complicating interpretation. Further developments refine network construction and improve annotation accuracy, while complementary tools such as NAP [19] offer additional structural information.

Alternatively, knowledge-driven networks represent metabolites as nodes and define edges based on established biochemical relationships, such as metabolic reactions [14, 20] or structural similarities [21]. This strategy uses supervised modeling to incorporate prior biochemical knowledge into the annotation process, enabling more targeted annotation and improving overall efficiency. A representative example is MetDNA [14], first introduced in 2019, which applies a metabolic reaction network (MRN) to guide MS2 spectral similarity–based annotation. Through recursive propagation along reaction pairs, MetDNA achieves automated, large-scale metabolite annotation from complex LC–MS data. The framework has since been progressively improved. In 2022, MetDNA2 [22] introduced a knowledge-guided multi-layer network to achieve global annotation from known to unknown metabolites. By expanding the MRN with in silico-generated unknown metabolites and integrating a global peak correlation network, MetDNA2 enhanced the annotation accuracy of known metabolites while facilitating the discovery of novel ones. In 2025, MetDNA3 [23] advanced this concept by incorporating graph neural network–based reaction prediction to build a more comprehensive MRN with improved coverage and network connectivity. It also established a two-layer interactive networking topology that integrates knowledge-driven and data-driven networks. This integration leverages cross-network interactions to enable efficient recursive propagation, markedly improving annotation coverage and confidence in untargeted metabolomics.

Here, we present a step-by-step protocol for using MetDNA to perform large-scale metabolite annotation in untargeted metabolomics, together with practical guidance on key prerequisites and common considerations for successful implementation.

2 Material

2.1 Demo Datasets

To illustrate the workflow of metabolite annotation using MetDNA, two demo metabolomics datasets are provided (*see* **Note 1**). Demo dataset 1 was acquired using an Orbitrap MS instrument from mouse liver tissues. The samples consisted of two groups of mice of different ages ("young" vs. "old"), with ten biological replicates in each group (*see* **Note 2**). Demo dataset 2 was acquired using a TOF MS instrument from fruit fly (*Drosophila melanogaster*) head tissues. Samples were collected from flies of two ages (3 days and 30 days). Head tissues from 15 flies were pooled together to form one biological replicate, with ten biological replicates in each age group (*see* **Note 3**).

3 Methods

3.1 Prepare Input Data Files Using Met4DX

Met4DX [24] (https://met4dx.zhulab.cn/) is a fast, robust, and user-friendly mass spectrometrydata processing tool for metabolomics. The versatile tool facilitates the processing of both three-dimensional LC–MS data and four-dimensional LC-IM-MS data, encompassing main functions, such as data conversion, peak detection, retention time correction, peak grouping, assignment of MS/MS spectra, and others. Importantly, Met4DX allows users to export results as input files for metabolite annotation in MetDNA

Specifically, raw LC–MS data were processed using Met4DX (version 2.4) (*see* **Note 4**). For demo dataset 1, raw LC–MS data files (.raw) from all samples were grouped and imported into Met4DX for preprocessing, including peak detection, retention time correction, and peak grouping. Experimental settings are summarized in Table 1. The parameter set "Orbitrap segment DDA (metabolomics)" was applied, with default parameters. After processing, results were exported for MetDNA in the following formats:

- MS1 peak table (data.csv)
- Sample information table (sample.info.csv)
- MS2 spectral data (spectra.msp)

For demo dataset 2, raw LC–MS data files (.wiff) from all samples were similarly processed in Met4DX. Experimental settings are summarized in Table 1. The parameter set "TOF segment DDA (metabolomics)" was applied, with default parameters. Processed results were exported in the same formats required for MetDNA (data.csv, sample.info.csv, and spectra.msp).

Table 1
Experimental information of two demo datasets

Content	Demo dataset 1	Demo dataset 2
Experiment type	Metabolomics	Metabolomics
Acquisition method	LC–MS	LC–MS
LC column	HILIC/RPLC	HILIC
MS type	Orbitrap	QTOF
Ionization mode*	Polarity switch*	Positive/Negative
Acquisition mode	MS1 + MS2	MS1 + MS2

* Polarity switching is a data acquisition method on the Orbitrap MS instrument, indicating that dataset 1 was acquired in both positive and negative ionization modes within a single analysis.

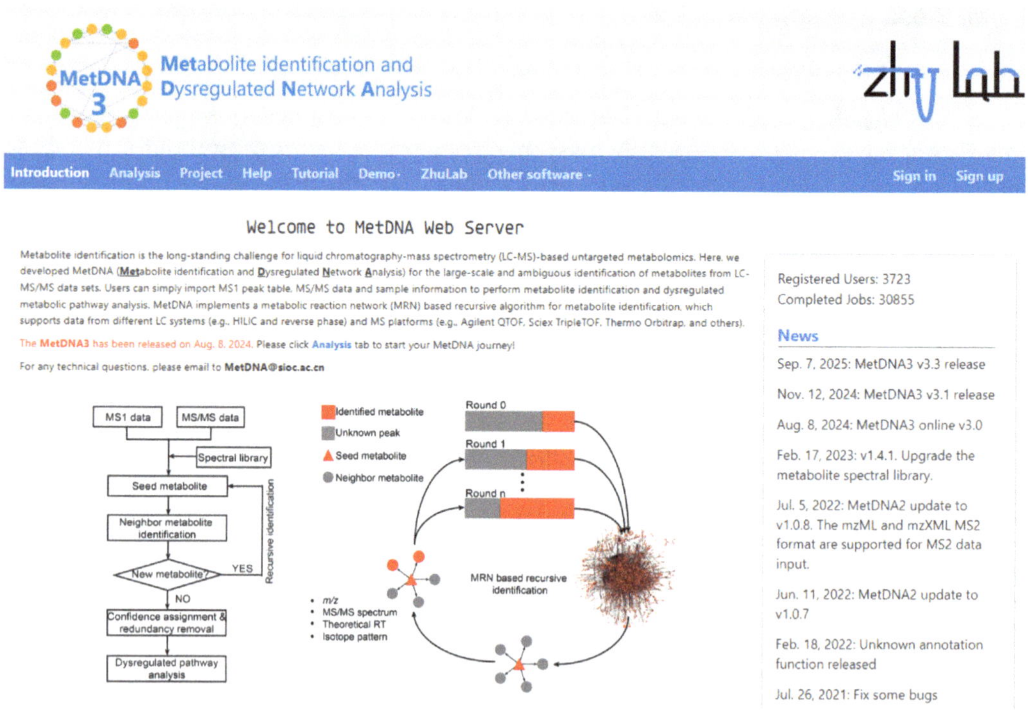

Fig. 1 Screenshot of MetDNA website interface

3.2 MetDNA Website

The MetDNA is freely accessible at http://metdna.zhulab.cn/. Upon entering the website, the user is directed to the main interface (Fig. 1). The navigation bar at the top of the interface provides access to several key modules:

- **Introduction**—This is the homepage of the website, which provides a general description of the MetDNA tool, together with important notes and the updated news.

- **Analysis**—This module enables users to perform metabolite annotation for untargeted metabolomics data.
- **Project**—This section allows users to view and manage their uploaded projects.
- **Help**—This page contains general guidance and instructions for using MetDNA.
- **Tutorial**—This section provides step-by-step tutorials to guide users through different functions of MetDNA.
- **Demo**—This page offers several demo datasets that users can download and test to familiarize themselves with the workflow.

3.3 Metabolite Annotation Using MetDNA

3.3.1 Registration and Login

Before performing metabolite annotation with MetDNA, users are required to register and log in to an account. To register, click "Sign Up" at the top-right corner of the website. Enter a valid email address and request the verification code (*see* **Note 5**). After entering the verification code, complete the registration process. Once registered, click "Sign In" at the top-right corner of the website to log in with your account credentials.

3.3.2 Preparation of Data Files

MetDNA requires several input data files for metabolite annotation (*see* **Notes 6–9**):

- MS1 peak table (.csv format, required)
- Sample information table (.csv format, required)
- MS2 spectra data file(s) (.mgf or.msp format, required)
- RT calibration table (.csv format, optional)

3.3.3 Upload and Check Data

After preparing the input files, users can upload the data and perform metabolite annotation using MetDNA (Fig. 2). The procedure is as follows:

1. Click "Analysis" in the navigation bar. Read the instructions carefully regarding file preparation. Then, scroll to the bottom of the page and click "Next" to continue.
2. Enter a project name, then select the sample information table and the corresponding data files for each ionization polarity (positive and negative should be uploaded separately) (*see* **Note 10**).
3. Click "Upload" at the bottom-right corner of the page (*see* **Note 11**).

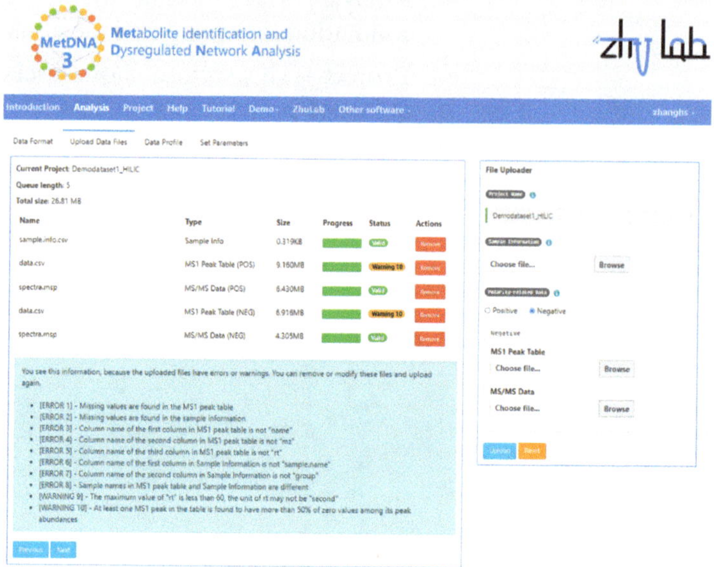

Fig. 2 Screenshot of uploaded data for MetDNA

4. The web server will automatically check the uploaded data. The status of each file is displayed in the "Status" column, with three possible outcomes (*see* **Note 12**):
   - **Valid**—the file is valid and can be processed further.
   - **Warning**—the file contains minor issues, but these will not affect processing.
   - **Error**—the file contains errors that must be corrected before processing.
5. Scroll to the bottom of the page and click "Next" to proceed to the next step.
6. A data summary and basic statistics will be displayed (Fig. 3). After confirming that there are no errors, click "Next" to continue.

3.3.4 Parameter Setting

1. Configure and verify the parameters according to the metabolomics data (Fig. 4) (*see* **Notes 13–15**). The parameters applied for the two demo datasets are summarized in Table 2.
2. *(optional)* For RT calibration and RT matching, select "Amide12min ZhuLabRT," "Amide23min ZhuLabRT," or "RP12min ZhuLabRT" under the "RT Library" option. Click the "Calibration Table" button and upload the RT calibration table generated from the corresponding metabolomics

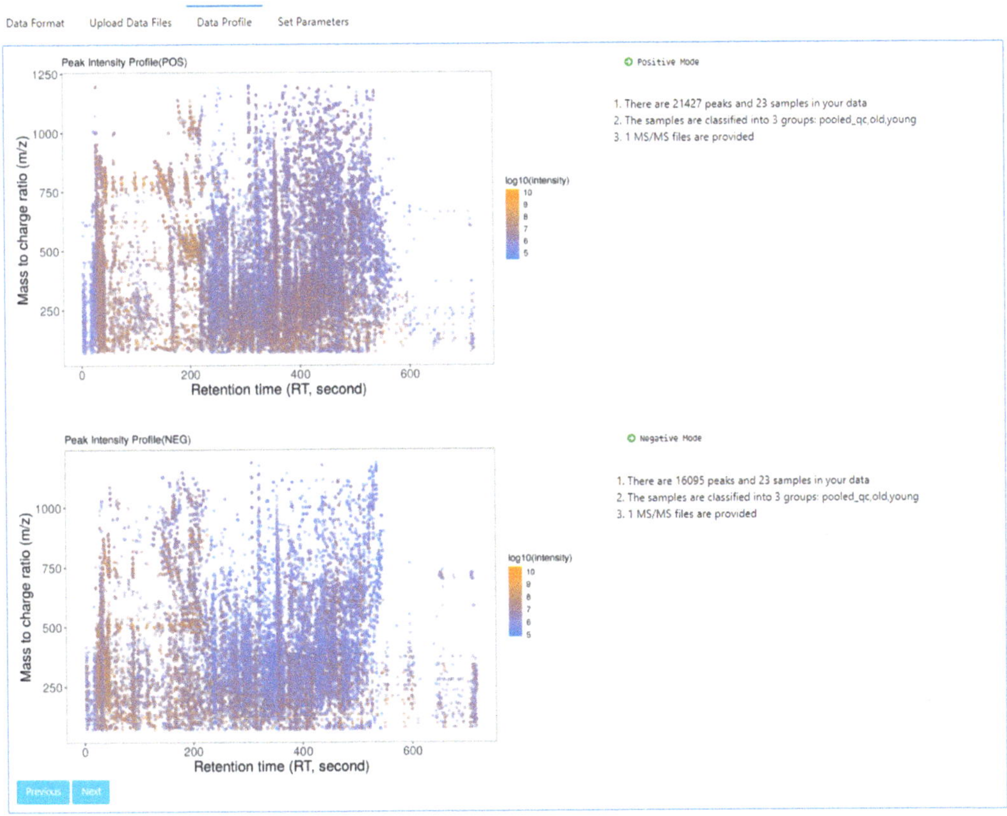

Fig. 3 Screenshot of uploaded data profile

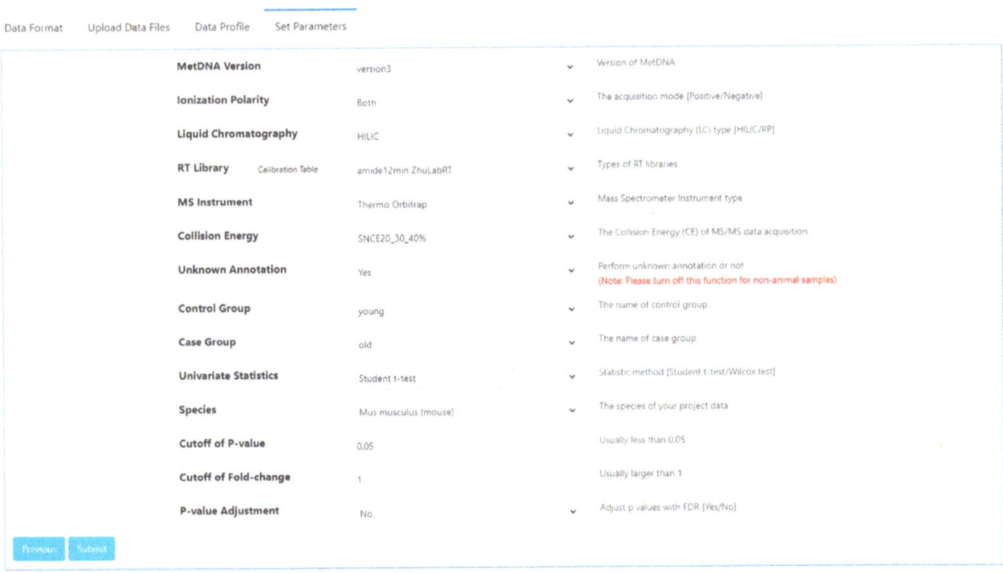

Fig. 4 Screenshot of parameter setting

Table 2
Parameter settings of two demo datasets

Content	Demo dataset 1	Demo dataset 2
MetDNA Version	version3	version3
Ionization Polarity[a]	Both	Both
Liquid Chromatography	HILIC/RPLC	HILIC
RT Library	amide12min ZhuLabRT/rp12min ZhuLabRT	amide12min ZhuLab RT
MS Instrument	Thermo Orbitrap	Sciex QTOF(TripleTOF/ZenoTOF)
Collision Energy	SNCE20_30_40%	30
Unknown Annotation[b]	Yes	No
Control Group	young	3d
Case Group	old	30d
Univariate Statistics	Student's t-test	Student's t-test
Species	*Mus musculus* (mouse)	*Drosophila melanogaster* (fruit fly)
Cutoff of P-value	0.05	0.05
Cutoff of Fold-change	1	1
p value Adjustment	No	No

[a] Process positive mode data and negative mode data sequentially. [b] The "Unknown Annotation" function is not recommended for non-animal samples, as the predicted unknown metabolites by BioTransformer [25] tend to follow mammalian metabolic patterns. Please turn off this function for non-mammalian samples

experiment (Fig. 5). If the corresponding LC method has not been used, select "No" under the "RT Library" option (*see* **Note 16**).

3. Click "Submit" to start the metabolite annotation process in MetDNA.

Users should configure the parameters according to metabolomics data (Fig. 4).

3.3.5 Download of Annotation Results

Once the uploaded project is completed, the user will receive a notification email (Fig. 6) (*see* **Note 17**). By clicking the link provided in the email, the user can review and download the results. Each project is stored on the server for 7 days, and users are advised to download the results in time.

Alternatively, projects can be accessed via the "Project" tab in the navigation bar (Fig. 7). Within each project, users can view details including processing logs, a brief analysis summary, and links to download the metabolite annotation results.

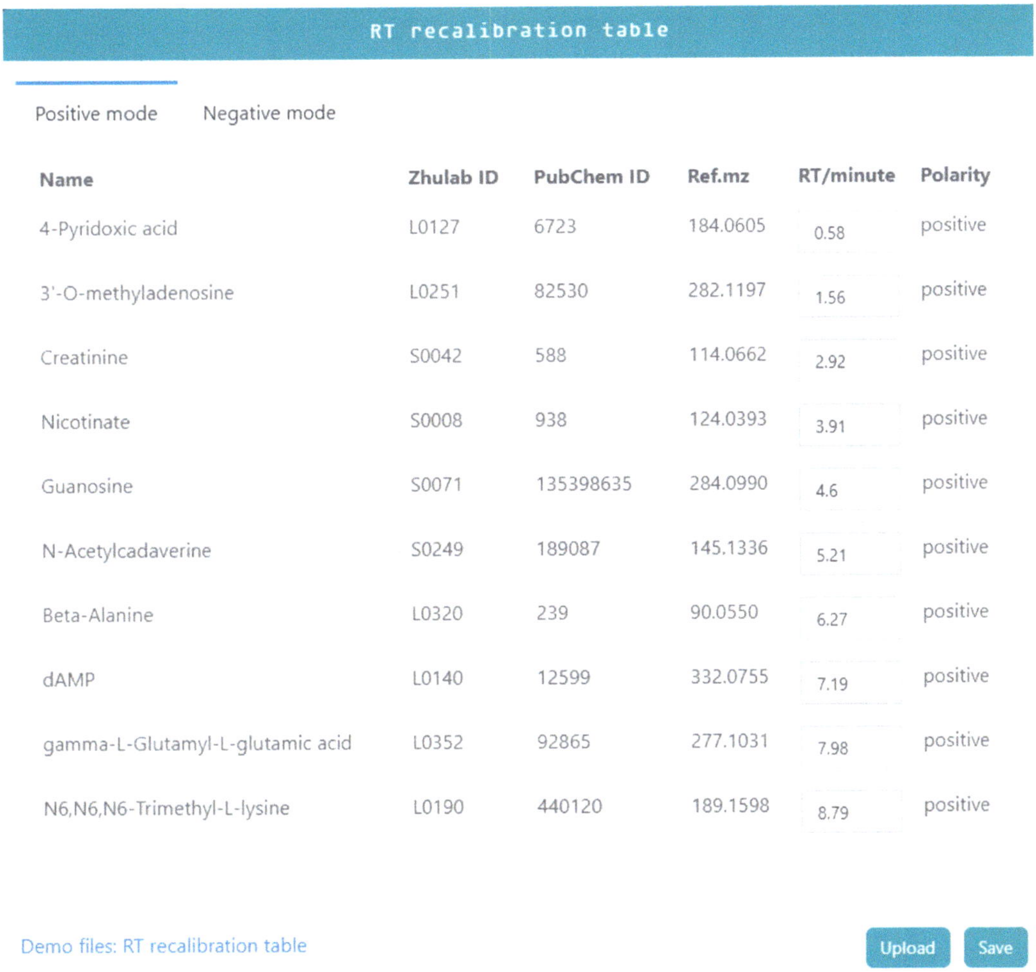

Fig. 5 Screenshot of uploading the RT calibration table

Fig. 6 Screenshot of notification email

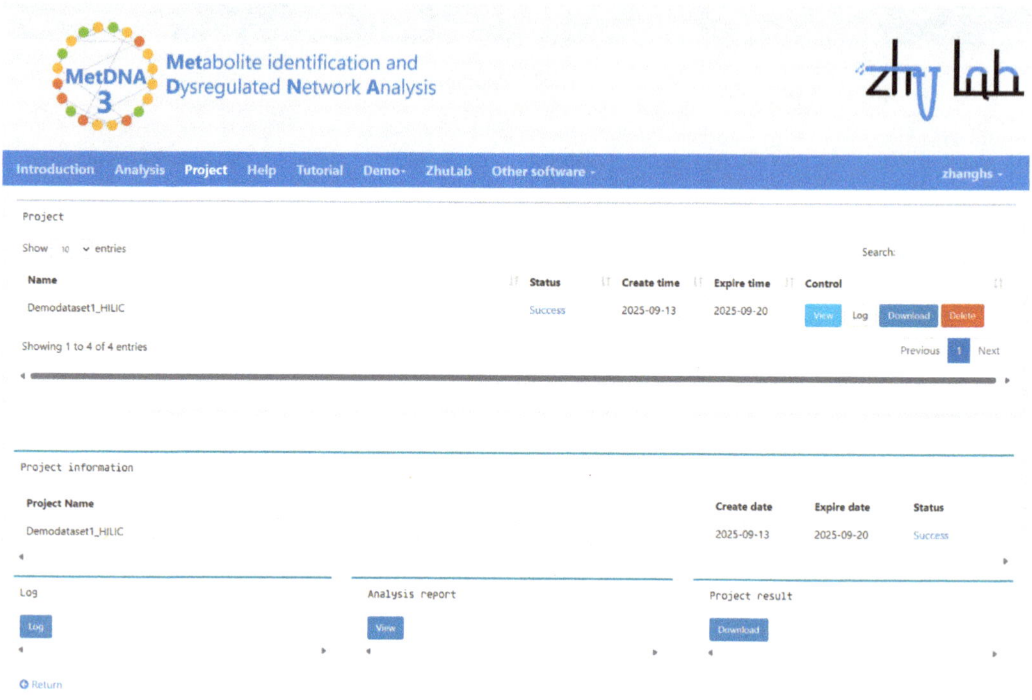

Fig. 7 Screenshot of project view of MetDNA website

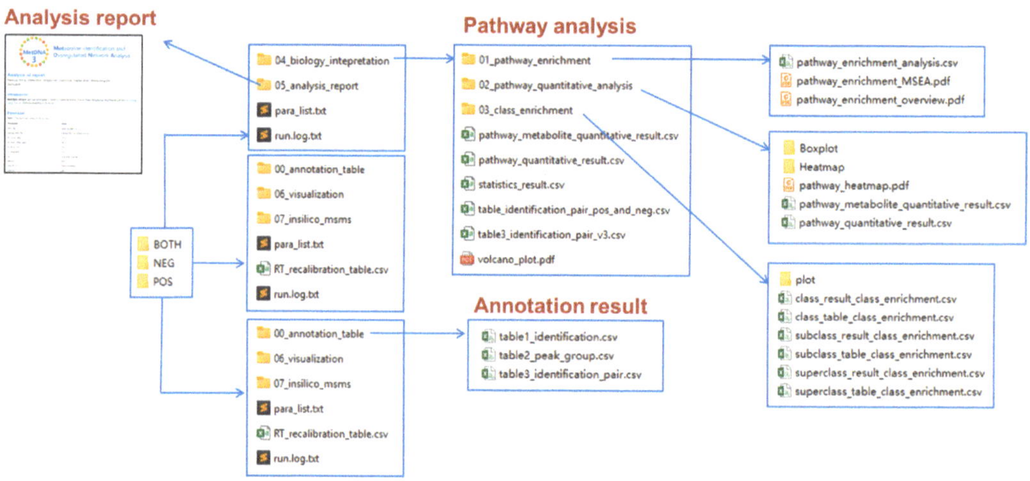

Fig. 8 Screenshot of downloaded results from MetDNA

The downloaded results are organized into folders according to the uploaded data and selected settings, typically including "POS," "NEG," and "BOTH" directories (Fig. 8). Within each polarity-specific folder (POS or NEG), annotation results are provided under the "00_annotation_table" subfolder, which contains three key tables:

- **table1_identification.csv**—annotated metabolites for each feature.
- **table2_peak_group.csv**—recognized abiotic peaks within each peak group (e.g., isotopes, adducts, neutral losses, in-source fragments).
- **table3_identification_pair.csv**—separated feature–metabolite pairs for further analysis.

Annotation confidence levels in MetDNA are reported in the "confidence_level" column of the annotation table. These levels follow the standards of the Metabolomics Standards Initiative (MSI) [26]:

- **Level 1**: m/z + experimental RT + experimental MS2 match
- **Level 2**: m/z + experimental MS2 match
- **Level 3.1**: m/z + predicted RT + surrogate MS2 match (known metabolite annotation)
- **Level 3.2**: m/z + predicted RT + surrogate MS2 match (unknown metabolite annotation)

Level 3 annotations are generated through metabolic reaction network (MRN)-based annotation propagation. The detailed algorithmic procedure has been described in the MetDNA publications.

In addition, the downloaded results include:

- **"04_biology_interpretation"**—pathway analysis results
- **"05_analysis_report"**—an HTML-formatted brief analysis report

A detailed description of each output file is provided in the documentation "Interpretation of Analysis Results from MetDNA," accessible via the "Help" page in the navigation bar.

4 Notes

1. Step-by-step instructions for the demo dataset analysis were carried out using Met4DX on the Windows platform. Users may alternatively process their raw data with tools such as XCMS [27], MZmine [28], or MS-DIAL [29], depending on their specific requirements. It should be noted that the data files used for MetDNA must strictly conform to the required format and content specifications. MetDNA does not support targeted metabolomics data, such as MRM and PRM.

MetDNA does not support metabolomics data acquired from GC–MS and NMR.

2. Untargeted metabolomics data of demo dataset 1 were collected using a UHPLC system (Vanquish, Thermo Scientific) coupled to an Orbitrap mass spectrometer (Exploris 480, Thermo Scientific). Two LC methods were applied in our laboratory: "Amide12min" (HILIC separation) and "RP12min" (reverse phase separation). For both LC methods, polarity switching was applied in the full mass range for both biological samples and pooled QC samples. In pooled QC samples, data-dependent acquisitions were additionally performed in both full and segmented mass ranges (70–300 Da, 290–600 Da, 590–1200 Da) to enhance MS2 spectral coverage. Detailed acquisition parameters are described in the "Tutorial" section of the MetDNA website (http://metdna.zhulab.cn/metdna/faq).
3. Untargeted metabolomics data of demo dataset 2 were acquired using a UHPLC system (1290 series, Agilent Technologies) coupled to a quadrupole time-of-flight mass spectrometer (TripleTOF 6600, AB Sciex). The LC method used in our laboratory was "Amide12min" (HILIC separation). Data were acquired in both positive and negative ionization modes. In all biological and pooled QC samples, data-dependent acquisition was performed over the full mass range (60–1200 Da). In addition, pooled QC samples were analyzed using both full and segmented mass ranges (60–300 Da, 290–600 Da, 590–1200 Da) to improve MS2 spectral coverage. Detailed acquisition parameters are described in the "Tutorial" section of the MetDNA website (http://metdna.zhulab.cn/metdna/faq).
4. Met4DX is freely available at https://met4dx.zhulab.cn/, and a detailed protocol can be accessed at https://met4dx.zhulab.cn/pages/19.html?name=Met4DX. The processed data required for MetDNA can be downloaded from the section "Input data files from Met4DX" on the MetDNA "Tutorial" page (http://metdna.zhulab.cn/metdna/faq).
5. If the verification code email is not received, please check the spam or junk email folder.
6. **MS1 peak table** can be generated by Met4DX, XCMS,MZmine, or MS-DIAL. The first three columns must be named "name," "mz," and "rt," with the unit of retention time (rt) specified in seconds. The remaining columns should correspond to sample names. The sample names in the MS1 peak table must exactly match those in the "sample.name" column of the sample information table.

7. **MS2 spectra data file(s)** can be derived from MS2 spectra obtained via data-dependent acquisition (DDA) or deconvolved spectra from data-independent acquisition (DIA). We recommend using pooled QC sample MS2 data instead of all individual sample MS2 files. Supported formats include *.msp* and *.mgf*. For Met4DX users, assigned MS2 spectra can be exported automatically in *.msp* format. For XCMS and MZmine users, raw MS2 data in each ionization polarity can be converted to *.mgf* format using ProteoWizard MSConvert (https://proteowizard.sourceforge.io/download.html). For MS-DIAL users, the "Representative spectra" can be exported directly in *.msp* format.
8. **Sample information table** must contain at least two columns: the first column named "sample.name," which must exactly match the sample names provided in the MS1 peak table, and the second column named "group," which should include at least two distinct group names.
9. **RT calibration table (optional)** can be used for retention time recalibration against an in-house RT library. This step enables level 1 metabolite annotation (*m/z* + experimental RT + experimental MS2 match) and ensures optimal performance of MetDNA. RT calibration requires following the same LC conditions and gradients (HILIC or RP) established in our laboratory, including the acquisition of an RTQC sample. MetDNA currently supports three LC methods: "Amide12min," "Amide23min," and "RP12min." More details are provided on the "Help" page of the MetDNA website.
10. MetDNA is capable of sequentially processing both positive and negative ionization mode data within the same project. To enable this, users must ensure that the sample information tables for the two datasets are identical. Alternatively, users may also choose to process data acquired in a single ionization mode independently.
11. The total size of all files uploaded to the server must not exceed 2048 MB. In addition, users should not upload different types of MS2 files within the same project.
12. When uploading files, the server automatically performs validation and returns specific messages if issues are detected. "Warning" indicates non-critical issues that do not prevent further analysis. For example, Warning 10 ("At least one MS1 peak in the table is found to have more than 50% zero values among its peak abundances") is commonly observed. Users may proceed with the analysis despite this warning. "Error" indicates critical issues that must be corrected before proceeding. In such cases, users should carefully review the error

message, revise the file accordingly, and re-upload the corrected version for validation.

13. Currently, MetDNA provides three versions (version 1, version 2, and version 3). The latest version is version 3, and it is strongly recommended to use this version for data processing to ensure optimal performance and updated functionalities.

14. The following options should be specified according to the experimental design:

    a) Ionization Polarity: Both, Positive, Negative

    b) Liquid Chromatography: HILIC, RP

    c) RT Library: amide12min ZhuLabRT, amide23min ZhuLabRT, rp12min ZhuLabRT, No. Users should select the appropriate method according to the actual LC gradient applied in their experiment. If the LC gradient differs from those established in our laboratory, select "No."

    d) MS Instrument: Sciex QTOF(TripleTOF/ZenoTOF), Agilent QTOF, Bruker QTOF, Waters QTOF, Shimadzu QTOF, Other QTOF, Thermo Orbitrap, Thermo Astral

    e) Collision Energy: 10, 20, 30, 40, 50, 35 ± 15, NCE10, NCE20, NCE30, NCE40, NCE50, SCE20_30_40, SNCE20_30_40

15. The following parameters are required for biological interpretation:

    a) Control Group: Name of the control group, consistent with the sample information table

    b) Case Group: Name of the case (experimental) group, consistent with the sample information table

    c) Univariate Statistics: Method for univariate testing, options include Student's *t*-test or Wilcoxon test

    d) Species: Species corresponding to the project data

    e) Cutoff of *p* value and Fold-change: Thresholds for identifying differential metabolites

    f) *p* value Adjustment: Option to apply *p* value correction in pathway analysis

16. For LC methods not supported by MetDNA, users can proceed without RT calibration and RT matching; annotation will then be restricted to *m/z* and experimental MS2 matching (level 2 annotation). When selecting "Amide12min ZhuLabRT," "Amide23min ZhuLabRT," or "RP12min ZhuLabRT" under the "RT Library" option, an RT calibration table must be provided.

17. If a project fails during processing, the user will also receive a failure notification email. In such cases, users can go to the "Project" page, click "Log," and check the detailed error information. If further assistance is needed, please refer to the "Help" page of the MetDNA website or contact the MetDNA team via email (metdna@sioc.ac.cn).

Acknowledgments

The work was supported by National Key R&D Program of China (2022YFC3400700), National Natural Science Foundation of China (22425404), Shanghai Key Laboratory of Aging Studies (19DZ2260400), Shanghai Municipal Science and Technology Major Project, and Shanghai Basic Research Pioneer Project.

References

1. Patti GJ, Yanes O, Siuzdak G (2012) Metabolomics: the apogee of the omics trilogy. Nat Rev Mol Cell Biol 13:263–269. https://doi.org/10.1038/nrm3314
2. Wishart DS (2016) Emerging applications of metabolomics in drug discovery and precision medicine. Nat Rev Drug Discov 15:473–484. https://doi.org/10.1038/nrd.2016.32
3. Johnson CH, Ivanisevic J, Siuzdak G (2016) Metabolomics: beyond biomarkers and towards mechanisms. Nat Rev Mol Cell Biol 17:451–459. https://doi.org/10.1038/nrm.2016.25
4. Alseekh S, Aharoni A, Brotman Y et al. (2021) Mass spectrometry-based metabolomics: a guide for annotation, quantification and best reporting practices. Nat Methods 18:747–756. https://doi.org/10.1038/s41592-021-01197-1
5. Perez De Souza L, Alseekh S, Scossa F, Fernie AR (2021) Ultra-high-performance liquid chromatography high-resolution mass spectrometry variants for metabolomics research. Nat Methods 18:733–746. https://doi.org/10.1038/s41592-021-01116-4
6. Luo M, Yin Y, Zhou Z et al. (2023) A mass spectrum-oriented computational method for ion mobility-resolved untargeted metabolomics. Nat Commun 14: 1813. https://doi.org/10.1038/s41467-023-37539-0
7. Cai Y, Zhou Z, Zhu Z-J (2023) Advanced analytical and informatic strategies for metabolite annotation in untargeted metabolomics. TrAC Trends Anal Chem 158:116903. https://doi.org/10.1016/j.trac.2022.116903
8. Novoa-del-Toro EM, Witting M (2024) Navigating common pitfalls in metabolite identification and metabolomics bioinformatics. Metabolomics 20: 103. https://doi.org/10.1007/s11306-024-02167-2
9. Vinaixa M, Schymanski EL, Neumann S et al. (2016) Mass spectral databases for LC/MS- and GC/MS-based metabolomics: state of the field and future prospects. TrAC Trends Anal Chem 78:23–35. https://doi.org/10.1016/j.trac.2015.09.005
10. Kind T, Tsugawa H, Cajka T et al. (2018) Identification of small molecules using accurate mass MS/MS search. Mass Spectrom Rev 37:513–532. https://doi.org/10.1002/mas.21535
11. Yi Z, Zhu Z-J (2020) Overview of tandem mass spectral and metabolite databases for metabolite identification in metabolomics. In: Li S (Ed.)Eds. Computational methods and data analysis for metabolomics. Springer, New York, NY, USA, 139–148
12. Watrous J, Roach P, Alexandrov T et al. (2012) Mass spectral molecular networking of living microbial colonies. Proc Natl Acad Sci U S A 109:E1743–52. https://doi.org/10.1073/pnas.1203689109
13. Wang M, Carver JJ, Phelan VV et al. (2016) Sharing and community curation of mass spectrometry data with global natural products social molecular networking. Nat Biotechnol 34:828–837. https://doi.org/10.1038/nbt.3597
14. Shen X, Wang R, Xiong X et al. (2019) Metabolic reaction network-based recursive metabolite annotation for untargeted metabolomics. Nat Commun 10: 1516. https://doi.org/10.1038/s41467-019-09550-x

15. Chen L, Lu W, Wang L et al. (2021) Metabolite discovery through global annotation of untargeted metabolomics data. Nat Methods 18:1377–1385. https://doi.org/10.1038/s41592-021-01303-3
16. Gao Y, Luo M, Wang H et al. (2025) Charting unknown metabolic reactions by mass spectrometry-resolved stable-isotope tracing metabolomics. Nat Commun 16: 5059. https://doi.org/10.1038/s41467-025-60258-7
17. Nothias L-F, Petras D, Schmid R et al. (2020) Feature-based molecular networking in the GNPS analysis environment. Nat Methods 17:905–908. https://doi.org/10.1038/s41592-020-0933-6
18. Schmid R, Petras D, Nothias L-F et al. (2021) Ion identity molecular networking for mass spectrometry-based metabolomics in the GNPS environment. Nat Commun 12: 3832. https://doi.org/10.1038/s41467-021-23953-9
19. Da Silva RR, Wang M, Nothias L-F et al. (2018) Propagating annotations of molecular networks using in silico fragmentation. PLOS Comput Biol 14:e1006089. https://doi.org/10.1371/journal.pcbi.1006089
20. Aguilar-Mogas A, Sales-Pardo M, Navarro M et al. (2017) iMet: a network-based computational tool to assist in the annotation of metabolites from tandem mass spectra. Anal Chem 89:3474–3482. https://doi.org/10.1021/acs.analchem.6b04512
21. Wang X, Li C, Li Z et al. (2023) A structure-guided molecular network strategy for global untargeted metabolomics data annotation. Anal Chem 95:11603–11612. https://doi.org/10.1021/acs.analchem.3c00849
22. Zhou Z, Luo M, Zhang H et al. (2022) Metabolite annotation from knowns to unknowns through knowledge-guided multi-layer metabolic networking. Nat Commun 13: 6656. https://doi.org/10.1038/s41467-022-34537-6
23. Zhang H, Zeng X, Yin Y, Zhu Z-J (2025) Knowledge and data-driven two-layer networking for accurate metabolite annotation in untargeted metabolomics. Nat Commun 16: 8118. https://doi.org/10.1038/s41467-025-63536-6
24. Yin Y, Luo M, Zhu Z-J (2024) Met4DX: a unified and versatile data processing tool for multidimensional untargeted metabolomics data. J Am Soc Mass Spectrom 35:2960–2968. https://doi.org/10.1021/jasms.4c00290
25. Wishart DS, Tian S, Allen D et al. (2022) BioTransformer 3.0—a web server for accurately predicting metabolic transformation products. Nucleic Acids Res 50:W115–W123. https://doi.org/10.1093/nar/gkac313
26. Sumner LW, Amberg A, Barrett D et al. (2007) Proposed minimum reporting standards for chemical analysis: Chemical Analysis Working Group (CAWG) Metabolomics Standards Initiative (MSI). Metabolomics 3:211–221. https://doi.org/10.1007/s11306-007-0082-2
27. Smith CA, Want EJ, O'Maille G et al. (2006) XCMS: processing mass spectrometry data for metabolite profiling using nonlinear peak alignment, matching, and identification. Anal Chem 78:779–787. https://doi.org/10.1021/ac051437y
28. Heuckeroth S, Damiani T, Smirnov A et al. (2024) Reproducible mass spectrometry data processing and compound annotation in MZmine 3. Nat Protoc 19:2597–2641. https://doi.org/10.1038/s41596-024-00996-y
29. Tsugawa H, Cajka T, Kind T et al. (2015) MS-DIAL: data-independent MS/MS deconvolution for comprehensive metabolome analysis. Nat Methods 12:523–526. https://doi.org/10.1038/nmeth.3393

Chapter 12

Lipid and Metabolite Annotation Using Lipid Data Analyzer

Leonida M. Lamp, Jürgen Hartler and Gerhard G. Thallinger

Abstract

The Lipid Data Analyzer (LDA) is a platform-independent software tool for the automated identification and quantification of lipid species and other metabolites in both untargeted and targeted mass spectrometry (MS) data. LDA mirrors the decision-making process of trained MS experts, enabling automated, high-throughput analyses across any instrumental setup through customizable, rule-based logic. These decision rule sets represent chemical and structural logic, yielding annotation results of high reliability. Furthermore, decision rules can be easily extended to accommodate new lipid classes, adduct forms, or fragmentation mechanisms. This makes LDA particularly effective for identifying novel lipid species or structural isomers that are often missed by traditional approaches, and enables users without a bioinformatics background to easily adapt to evolving analytical protocols. Moreover, LDA supports the identification of double bond positions and other modifications, such as oxidations. Here, we present a step-by-step guide and practical tips for efficient operation of LDA, and provide an introductory discussion of common pitfalls in lipidomics data analysis.

Keywords Lipidomics, LDA, Lipid identification, Metabolomics, Mass spectrometry, Untargeted, Chromatography, Oxidized lipids, double bond localization

1 Introduction

A central challenge in lipidomics is the vast structural diversity of lipids, which stems from the numerous possible combinations of different lipid building blocks [1]. Compared to metabolomics, a typical cellular lipidome contains a disproportionately higher number of isobaric (same nominal mass but different elemental composition) and isomeric (same elemental composition but different structure) molecules [2]. This substantially increases the complexity of data analysis, particularly for untargeted lipidomics, which aims to comprehensively measure lipid compounds on a large scale. Bielow et al. [3] provide a systematic overview of isobaric overlaps in lipidomics and an estimate of the mass resolution required for resolving them. In this section, we provide a brief overview of these analytical challenges,

Michael Witting (ed.), *Computational Methods and Data Analysis for Metabolomics*, Methods in Molecular Biology, vol. 3063,
https://doi.org/10.1007/978-1-0716-5452-1_12,

outline practical solutions implemented in the Lipid Data Analyzer (LDA) [4, 5], and illustrate them using representative examples.

A significant and often overlooked issue resulting from the often-minute structural differences in lipid molecules is *overannotation*, i.e., assigning more structural detail to lipid species than is supported by the available experimental evidence. A valuable resource in this context is the brief guide on common pitfalls in lipidomics by Skotland et al. [6]. Overly detailed annotations can create a false sense of certainty and ultimately lead to misinterpretation of biological results. *Overannotation* typically arises from software tools that rely on spectral library matching: when a reference spectrum is matched, the associated full structural annotation is often transferred—even if key diagnostic evidence is lacking. In contrast, LDA's decision rule-based approach [4, 5] is designed to strictly constrain annotations to the level of structural certainty (*lipid species* level, *lipid molecular species* level, etc. [5–8]) that is supported by the experimental evidence. This strategy minimizes false positive (FP) identifications and improves the reliability of downstream data analyses [5, 9].

Another key advantage of LDA's decision rule framework is the straightforward identification of novel species. Rather than requiring extensive spectral libraries containing thousands of reference spectra for all possible lipid molecular species within a lipid class, a single set of decision rules covers all molecules of a lipid class or subclass, even those that have not been reported previously.

Isobaric overlaps in lipidomics most commonly arise between the monoisotopic peak of a lipid species and the M+2 isotopic peak of a lipid containing one additional double bond ($^{12}C_2H_2$ vs. $^{13}C_2$: $\Delta m/z = 0.009$) [10]. An M+2 peak—two mass units heavier than the monoisotopic peak—results from the natural abundance of heavier isotopes within molecules. Resolving such overlaps typically requires a mass resolution of about 180,000 [10], with the exact value depending on the *m/z* and intensity ratio of the overlapping species. In addition, (partial) chromatographic separation with reversed-phase liquid chromatography (RPLC) can aid in resolving these overlaps, resulting in three-dimensional peak patterns as shown in Fig. 1a, which are frequently observed in lipidomics datasets.

Conventional peak integration algorithms that rely solely on two dimensions—retention time (RT) and intensity—are often inadequate for accurately resolving overlapping signals. Moreover, 2D methods can introduce significant quantification errors, even for well-separated peaks (Fig.1b), because they neglect variation in the *m/z* dimension. As a result, peaks may be truncated or signal intensities overestimated. LDA's 3D peak integration algorithm, which incorporates *m/z* alongside RT and intensity, enables more accurate quantification by adapting integration boundaries to the actual peak shape [4].

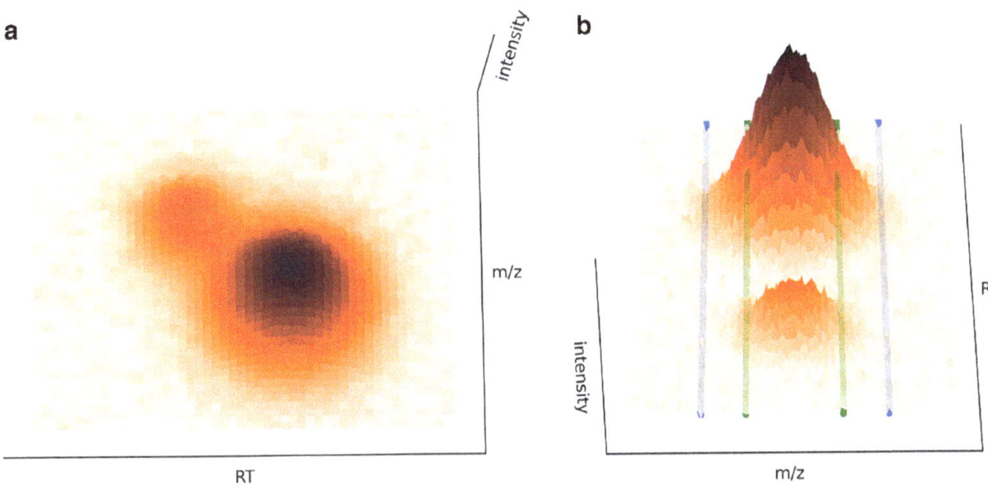

Fig. 1 Representative examples of peak patterns observed in 3D lipidomics data. (**a**) Example highlighting a case where conventional 2D peak integration using only RT and intensity would fail to achieve an accurate result. (**b**) Peak width varies with factors such as m/z and signal intensity, which requires accurate integration algorithms to adapt dynamically. Green and blue lines indicate optimal m/z integration boundaries for the lower- and higher-intensity peaks, respectively

In cases where signals cannot be separated either by *m/z* or RT, LDA's isotopic correction algorithm can still reliably estimate monoisotopic peak intensities by accounting for overlapping isotope distributions [11].

Another frequent source of isobaric overlaps stems from protonated versus sodiated adducts of the same lipid class. For instance, PC 34:1 +$[Na]^+$ and PC 36:4 +$[H]^+$ (Fig. 2), which differ by only $\Delta m/z = 0.002$ (C_2 vs. HNa), form such an isobaric pair. Although separable in theory (required mass resolution ≥600,000 [10]), this problem is often mitigated experimentally by suppressing sodium adduct formation using ammonium-based additives [12]. In contrast, isomeric overlaps, such as those between phosphatidylcholines (PCs) and phosphatidylethanolamines (PEs) with three additional carbons (e.g., PC 40:6 vs. PE 43:6, or PC 35:4 vs. PE 38:4) require orthogonal strategies for unambiguous discrimination. One such method is LDA's MS^1 peak splitting algorithm, which separates MS^1 peaks based on the distribution of MS^n spectra across the peak profile. The practical application of this strategy is illustrated using the example of an MS^1 peak (Fig. 2a) shared by PC 18:0/22:6 +$[H^+$ (Fig. 2b) and PC 18:0/20:3 +$[Na]^+$ (Fig. 2c).

A valuable complementary concept for distinguishing isomeric lipid species is the equivalent carbon number (ECN) model [13–15]. This model provides an intuitive framework for identifying false-positive identifications based on the elution order of identified species. In RPLC data, lipids within the same class typically elute earlier

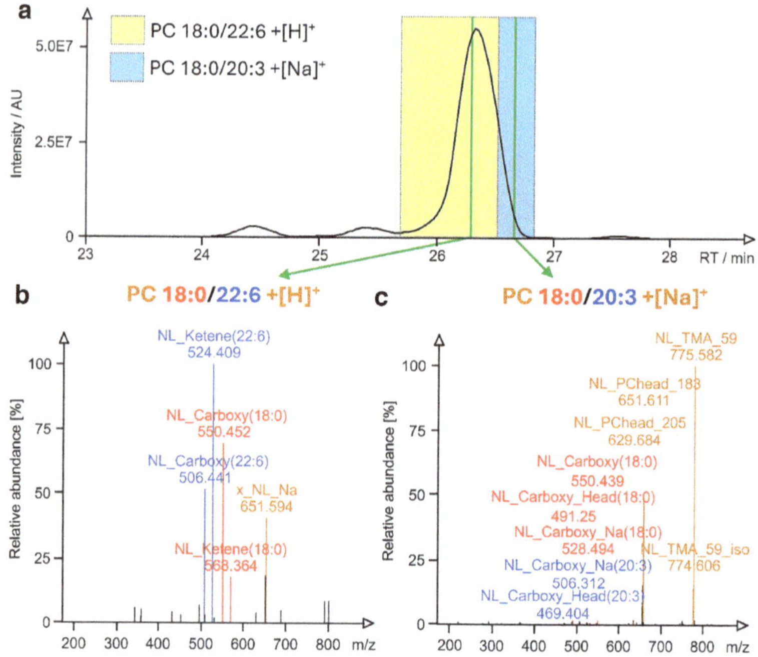

Fig. 2 Deconvolution of overlapping isobaric signals with LDA's MS[1] peak splitting algorithm. (**a**) MS[1] peak containing two isobaric lipid species: PC 18:0/22:6 +$[H]^+$ and PC 18:0/20:3 +$[Na]^+$. The isobars are neither separated by m/z nor RT. LDA separates and quantifies them by analyzing the distribution of MS^n spectra across the shared MS[1] peak. (**b**) Example MS^n spectrum showing fragment ions characteristic of PC 18:0/22:6 +$[H]^+$. Ions indicative of the head group as well as the chains 18:0 and 22:6 are highlighted in gold, red, and blue, respectively. (**c**) Example MS^n spectrum showing fragment ions indicative of PC 18:0/20:3 +$[Na]^+$. Ions characteristic of the head group as well as the chains 18:0 and 20:3 are highlighted in gold, red, and blue, respectively

with increasing number of double bonds and later with increasing carbon chain length. However, this rule is not universally applicable. Many structural features such as chain branching [16], as well as the positions and stereochemistry of double bonds, can significantly influence RT, resulting in overlapping elution windows for distinct lipid species [17]. To accommodate this complexity, LDA applies RT models with relatively broad tolerance ranges when filtering FPs that lack MS^n verification, thereby maintaining high sensitivity for correct identifications. Conversely, when RT is used to assign structural features such as double bond positions (Section 3.2) for MS^n-confirmed lipid molecular species, the software employs a dedicated algorithm capable of predicting retention times with single-digit-second accuracy [17].

The most common type of isomeric overlap in lipidomics involves lipid species that differ in the chain composition (Fig. 3).

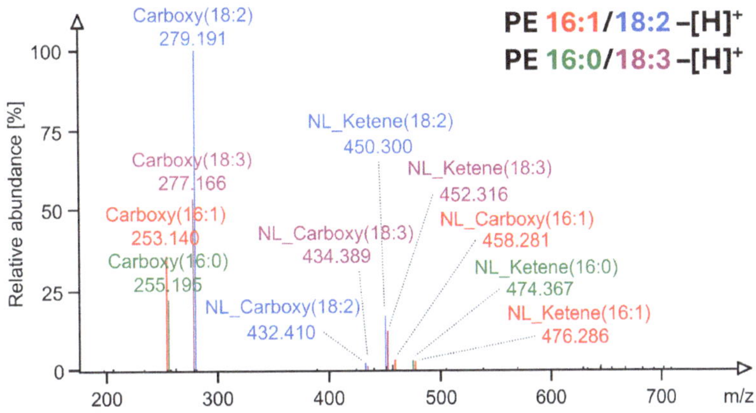

Fig. 3 Identification of coeluting lipid molecular species by LDA. Identification of two isomers of the lipid species PE 34:3 –$[H]^+$ in a multiplexed MS^n spectrum: PE 16:1/18:2 and PE 16:0/18:3. Fragment ions characteristic of the fatty acyl chains 16:1, 18:2, 16:0 and 18:3 are highlighted in red, blue, green and purple, respectively. Neutral loss fragment annotations contain the prefix "NL_"

In such cases—particularly when MS^n spectra are noisy or multiplexed—, LDA's *decision-rule*-based approach provides a distinct advantage over traditional spectral library matching. Instead of comparing entire spectra to reference entries, LDA selectively evaluates only those fragment ions that are diagnostic for each candidate lipid. This targeted strategy enhances specificity and robustness in complex datasets. Based on the intensities of fragment ions in multiplexed spectra, LDA calculates the relative abundances of the coeluting species (Fig. 3).

2 Material

2.1 Supported Operating Systems

- Windows, Linux/Unix, and macOS are supported.
- Windows (64-bit) is recommended, as it provides the most complete, out-of-the-box functionality.

2.2 Minimum System Requirements

These specifications are sufficient to install and run the software for basic workflows and smaller datasets.

- **CPU:** Dual-core processor (2 cores or more).
- **GPU:** Not required.
- **Memory (RAM):** 4 GB minimum (*see* **Note 1**).
- **Storage:** Sufficient free disk space for installation and data files.

2.3 Recommended System Requirements

For optimal performance and efficient processing of large datasets:

- **CPU:** 8-core processor or higher.
- **GPU:** Dedicated GPU with CUDA support. A GPU can significantly accelerate data analysis. Up-to-date GPU drivers are required to enable CUDA acceleration.
- **Memory (RAM):** 32 GB or more.
- **Storage:** SSD with ample free space for large datasets.

2.4 Availability

- **Download:** Installers for the latest and all previously published versions are freely available from the LDA home page: http://genome.tugraz.at/lda2
- **Available Installers:** Installers are provided for Windows, Linux/Unix, and macOS.

2.5 OS-Specific Installation Notes

The MS raw file formats mzML and mzXML are supported out-of-the-box on all OS.

2.5.1 Windows (Recommended)

- The Windows installer includes:
  - A Java Runtime Environment (JRE).
  - msconvert, enabling direct use of raw data from AB Sciex, Agilent Technologies, Bruker Daltonics, and Thermo Fisher Scientific
- Waters instruments: for direct raw data conversion, installation of Mass++ is required.
- Except for direct conversion of Waters raw data, no additional software is required. Installation is completed by simply running the installer.

2.5.2 Linux/Unix and macOS

- On Linux/Unix, Java JRE 1.8 must be installed (if not already present). For macOS, it is recommended to install Java JRE 1.8 manually as well as to ensure compatibility, as the default JRE provided by macOS has not been fully tested. It can be downloaded from the Oracle website (do not use OpenJDK).
- msconvert with vendor libraries for raw data conversion is not included.

3 Methods

3.1 General LDA Workflow

Analysis of lipidomics MS data with LDA comprises the following steps (Fig. 4a): After launching the application (*see* **Note 2**), several input files have to be provided for compound identification and quantification: (1) **MS input files**, (2) a **mass list**, (3) **settings**; and for the analysis of MS^n data, (4) **decision rules**, and (5) **chain lists** are additionally required. The MS files and mass list are selected via the input fields in the *Quantitation* tab of LDA's graphical user interface (GUI, Fig. 4b), whereas the remaining input files are specified in the *Settings* tab (Fig. 4c). A brief description of each required input file type is provided below:

1) For **MS input files**, LDA supports mzML, mzXML and chrom files (*see* **Note 3**) on all major operating systems (Windows, Linux/Unix and macOS). On Windows platforms, vendor-specific MS data from AB Sciex, Agilent Technologies, Bruker Daltonics and Thermo Fisher Scientific are also supported directly, as the Windows installers include msconvert [18].

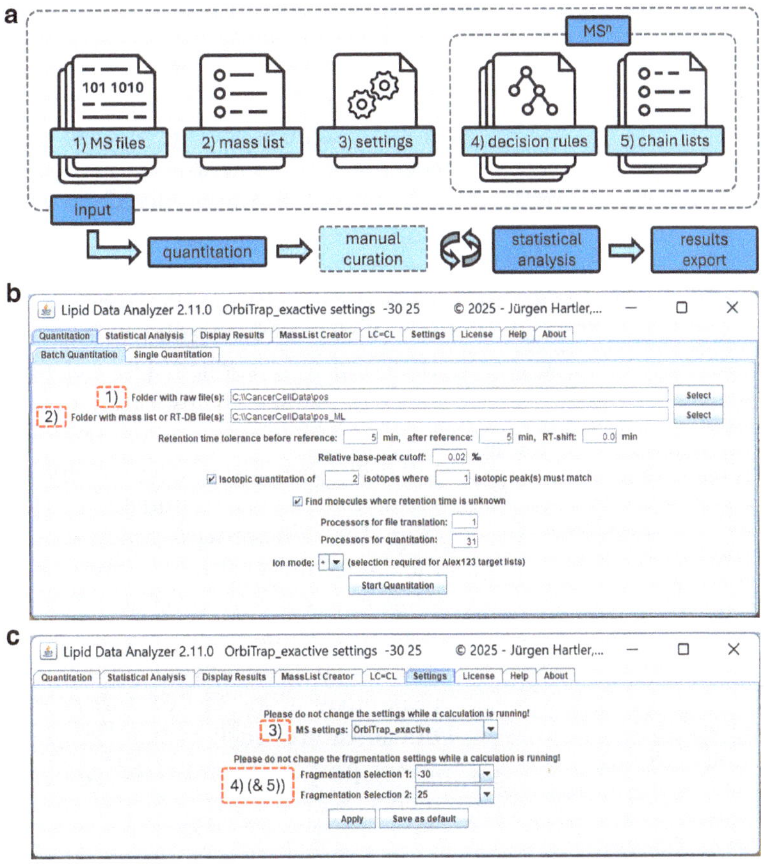

Fig. 4 Essential steps in the analysis of lipidomics experiments with LDA. (**a**) Lipid analysis with LDA requires several input files: (1) MS files, which can be provided as mzML, mzXML, LDA's chrom format, or vendor-specific file formats from AB Sciex, Agilent Technologies, Bruker Daltonics, or Thermo Fisher Scientific. In addition, LDA requires (2) a mass list, (3) settings, and for the analysis of MS^n data additionally (4) decision rules and (5) chain lists. (**b**) Graphical user interface (GUI) for LDA's lipid identification and quantitation routine with input fields for MS files and mass lists. (**c**) GUI for the selection of settings and decision rules – which implicitly specify the chain lists to be applied

2) The **mass list** defines the search space for MS^1 peak detection. LDA provides several preconfigured mass lists that can be used directly; these are located in the *exampleMassList* folder within the LDA installation directory. Sect. 3.1.1 describes in detail how to modify an existing mass list or create a new one tailored to the specific requirements of a given study.
3) LDA's **settings** can be customized for any MS instrument and experimental setup. A range of preconfigured settings files for various MS platforms are included with LDA and can be found in the *properties* folder within the LDA installation directory. Sect. 3.1.2 provides tips and best practices for configuring these settings to achieve optimal performance.
4) **Decision rules and chain lists** are required only for the analysis of MS^n data. LDA provides preconfigured decision rule sets for multiple MS setups, which are located in the *fragRules* folder within the LDA installation directory. Sect. 3.1.3 describes how these decision rule sets can be adapted and optimized to meet specific analytical requirements. The GUI allows for the selection of up to two decision rule packages for lipid annotation (Fig. 4c), typically corresponding to positive and negative ion modes. Chain lists (available in the *fattyAcids* folder within the LDA installation directory) are selected implicitly through the chosen decision rules, as each rule specifies the associated chain lists to be applied. These chain lists can be easily extended (e.g., to include chains longer than 30 carbon atoms), as they are maintained in a conventional Excel file. Once all required input files have been specified, lipid identification and quantification in LDA can be initiated by clicking "Start Quantitation" (Fig. 4b). This process generates one result file per MS input file (*see* **Note 4**). Results can be inspected individually in the *Display Results* tab or collectively visualized in the *Statistical Analysis* module (Sect. 3.1.4). Both interfaces offer tools for manual curation, including the removal of FPs and manual peak integration. The *Statistical Analysis* module offers additional features such as standardization, absolute quantitation, and export of summarized results across multiple measurements (Sect. 3.1.5).

3.1.1 Mass Lists and the MassList Creator

LDA's MS^1 quantitation module operates based on a predefined reference list of target ions, which may include lipid molecules as well as other metabolites. This "search space" is specified in LDA's Excel-based *mass list* format (Fig. 5). Several default mass lists are distributed with LDA and are located in the *exampleMassList* folder within the LDA installation directory. An LDA mass list file can comprise one or multiple sheets, each representing a distinct analyte class (*see* **Note 5**). The format supports the specification of MS^1 *m/z*

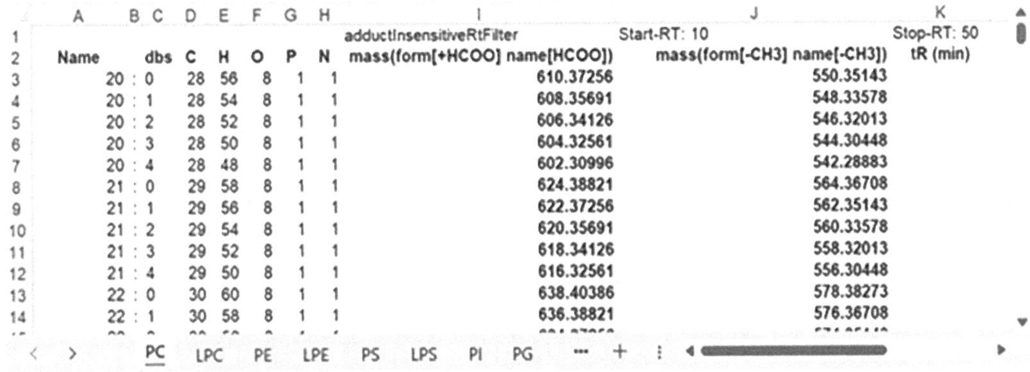

	A	B	C	D	E	F	G	H	I	J	K
1									adductInsensitiveRtFilter	Start-RT: 10	Stop-RT: 50
2	Name		dbs	C	H	O	P	N	mass(form[+HCOO] name[HCOO])	mass(form[-CH3] name[-CH3])	tR (min)
3	20	:	0	28	56	8	1	1	610.37256	550.35143	
4	20	:	1	28	54	8	1	1	608.35691	548.33578	
5	20	:	2	28	52	8	1	1	606.34126	546.32013	
6	20	:	3	28	50	8	1	1	604.32561	544.30448	
7	20	:	4	28	48	8	1	1	602.30996	542.28883	
8	21	:	0	29	58	8	1	1	624.38821	564.36708	
9	21	:	1	29	56	8	1	1	622.37256	562.35143	
10	21	:	2	29	54	8	1	1	620.35691	560.33578	
11	21	:	3	29	52	8	1	1	618.34126	558.32013	
12	21	:	4	29	50	8	1	1	616.32561	556.30448	
13	22	:	0	30	60	8	1	1	638.40386	578.38273	
14	22	:	1	30	58	8	1	1	636.38821	576.36708	

PC LPC PE LPE PS LPS PI PG

Fig. 5 Example illustrating LDA's MassList format. The first row defines RT filtering and other options to define the search space. The second row contains headers for defining the columns. This example contains the following columns from left to right: analyte name ("Name"), number of double bonds ("dbs"), element columns ("C", "H", "O", "P", "N"), mass columns ("mass(...)"; *see* **Note 5**) and the column retention time ("tR (min)"). LDA's MassList format requires the columns "Name", "tR (min)", at least one element column, and at least one "mass(...)" column. Mass columns feature three adduct parameters: 'form' defines the chemical formula of the adduct, 'name' defines the adduct name primarily used to identify the corresponding decision rules, and the optional 'charge=' parameter must be specified for multiply charged ions

values for any metabolite with any adduct ion, various oxidation types and states, and stable isotope labeling. In addition, retention time constraints can be specified to refine peak detection.

While the mass list is primarily used for MS^1 analysis, the adducts and analyte classes it defines also play a critical role in guiding LDA's MS^n analysis routine (*see* **Note 5**).

To facilitate the creation of custom mass lists, LDA includes the *MassList Creator* tool, which provides an intuitive interface for generating mass lists comprising any lipid class. For non-lipid metabolites, it is recommended to use one of the templates provided in the *exampleMassList* folder within the LDA installation directory as a starting point. The *MassList Creator* ensures compliance with LDA's formatting requirements and additionally supports alternative output formats (*see* **Note 6**).

To compute all possible *m/z* values for each lipid subclass, the MassList Creator utilizes *chain lists*, which are otherwise only needed for MS^n analysis (Sect. 3.1.3). The tool supports lipids containing any number of hydrophobic chains, including sphingoid bases and fatty acyl, alkyl, or alkenyl chains. Chain lists can define chains with arbitrary carbon numbers, degrees of unsaturation, and additional modifications such as oxidation or stable isotope labeling.

3.1.2 LDA Settings—Tips, Tricks and Troubleshooting

LDA is a highly configurable tool that can be optimized across all stages of data analysis. Appropriate parameter settings can significantly improve performance, compatibility with MS platforms, and accuracy of lipid identification. While the LDA user manual provides comprehensive coverage of all available settings, this section focuses on those most relevant to frequently asked questions.

LDA distinguishes between three types of settings: *global* settings, which are defined during installation; *general* settings stored in the *.settings* file (*see* **Note 7**); and MS platform-specific settings stored in "*.properties*" files (*see* **Note 8**).

Improperly configured global settings are the most common cause of failed analyses (*see* **Note 9**). Instructions for modifying these values after installation are provided in **Note 10**. Among the general settings, *useCuda* is particularly relevant for most users. When this option is set to "true" and a compatible GPU is available, LDA can offload parts of the lipid annotation process to a GPU, resulting in a significant performance increase. It is essential to ensure that GPU drivers are up to date, as outdated drivers may lead to errors or even cause the software to crash.

Substantial performance gains can also be achieved by optimizing MS platform-specific settings. In particular, adjusting MS file parsing for the available hardware (*see* **Note 11**), tuning the noise filter (*see* **Note 12**), and configuring peak integration parameters (*see* **Note 13**) can improve analysis speed. Of note, the MS platform-specific settings also define parameters for 3D signal integration. Whether further optimization is required can be assessed by inspecting LDA results in the 3D viewer (Fig. 6). If the current settings appear suboptimal for a given MS setup or hardware configuration, users are encouraged to contact the development team for support. For general advice on troubleshooting LDA, *see* **Note 2**.

3.1.3 MS^n Analysis—Decision Rules and Chain Lists

For the analysis of MS^n data, LDA requires *decision rules* and in addition, for analytes containing more than one hydrophobic chain, *chain lists*. Decision rules formalize expert knowledge of lipid fragmentation behavior for each analyte class and adduct combination (*see* **Note 5**) in a structured format that LDA can evaluate systematically. This enables reliable and highly adaptable automated interpretation of MS^n spectra across lipid subclasses and instrument setups. An LDA mass list defines molecule masses to be searched in the dataset. In contrast, chain lists define which hydrophobic chains—if any—are allowed within those compounds. In decision rules, these hydrophobic chains are used to calculate fragment ions that may be detected in MS^n spectra. Detection of these fragments enables identification of the chain composition and, where supported by experimental evidence, determination of *sn*-positions and additional structural features such as C=C positions (Sect. 3.2) and oxidations (Sect. 3.3). The algorithm uses two types of rules:

- **Fragment rules**: Fragments are defined by name, formula, charge, MS level, oxidation level, and a *mandatory* status.
- **Intensity rules**: For defining relationships between fragment intensities with a *mandatory* parameter.

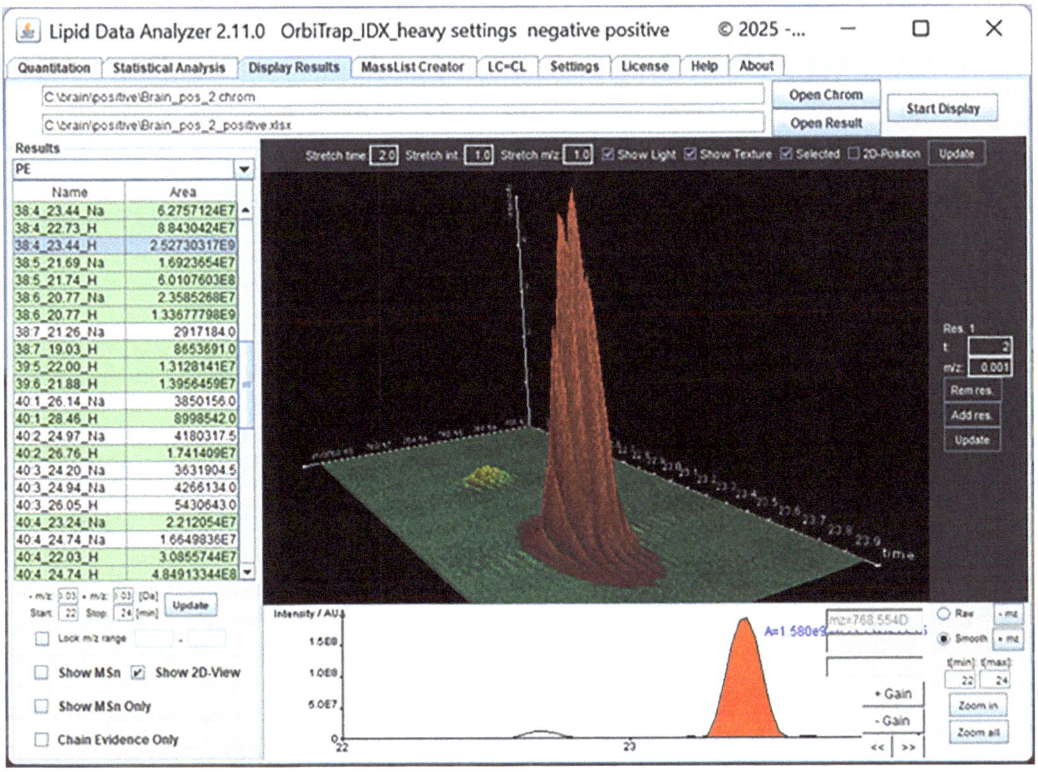

Fig. 6 Visualization of results. The *Display Results* tab provides features for data visualization and curation. In the provided example, the views relevant for the inspection of signal integration in 3D (center right) and 2D (bottom right) are shown for a selected lipid (PE 38:4 $[+H]^+$)

Each rule set has up to four sections: 1. *[GENERAL]*, which defines the analyte class and general algorithm settings; 2. *[HEAD]* (optional), defining rules related to the head group; 3. *[CHAINS]* (optional), for rules related to identifying the hydrophobic chains; 4. *[POSITION]* (optional), defining rules for positional assignment. This modular structure facilitates lipid identification at the respective level of structural detail supported by the MS data (Table 1).

The *mandatory* parameter determines how rule compliance affects annotation:

- **true**: The fragment or intensity condition must be observed.
- **false**: Optional; not always observed.
- **other**: Characteristic of a different analyte class and/or adduct.
- **class** (only in [CHAINS]): A chain fragment required to confirm the lipid species level (e.g., PI 38:4).

Rules marked **true** or **class** are required for identification at the relevant level of structural resolution, while those marked as **false** or **other** are optional and may provide supporting evidence. Fragments

Table 1

Structural elucidation with LDA at the example of PI 18:0/20:4(n-6). As more experimental evidence becomes available, LDA can report greater structural detail using the corresponding shorthand notation [5, 8]. LDA reports only the highest structural level supported by the available evidence

Data type	Stage of structural identification with LDA	Attainable level of structural resolution	Resulting annotation
MS^1	lipid species candidate; detection based on 3D algorithm and isotope distribution check	lipid species	PI 38:4
MS^n	verification of lipid species with [HEAD] rules	lipid species	PI 38:4
MS^n	assignment of chains with [CHAINS] rules	lipid molecular species	PI 18:0_20:4
MS^n	assignment of *sn*-positions with [POSITION] rules	*sn*-position resolved	PI 18:0/20:4
MS^n & RT	assignment of ω-position with LC=CL (Section 3.2)	(*sn*- and) ω-position resolved	PI 18:0/20:4(n-6)

labeled **other** specify fragment ions characteristic for another analyte class that may overlap; they are excluded from spectrum coverage calculation for the relevant analyte but may be used in intensity rules to distinguish similar classes of analytes with high specificity.

To optimize decision rules for a specific MS platform, it is recommended to start with the preconfigured *noIntensity* rule sets and adjust the rules as needed (*see* **Note 14**). LDA's GUI supports review and refinement of decision rules by providing direct visual feedback in the form annotations on user-selected spectra (Fig. 7) along with detailed textual explanations of annotation outcomes. This interactive feedback streamlines optimization of the text-based decision rule sets (*see* **Note 15**). For comprehensive guidance on rule configuration and available GUI functionalities, users are referred to the LDA user manual.

3.1.4 Data Visualization and Statistical Analysis

LDA provides a comprehensive set of tools for data visualization and basic statistical analyses. For visualization and manual correction of peak integration, chrom files must be stored in the same directory as their corresponding quantification result files (*see* **Notes 3 and 4**).

Datasets can be explored and curated through several interactive visualization panels including a 3D viewer, a 2D chromatogram viewer and an MS_n spectrum viewer. The 3D viewer enables detailed data inspection across the three dimensions: intensity, *m/z*, and RT. The 2D chromatogram viewer displays chromatographic peaks over RT and supports manual peak integration through multiple algorithms, including ASAPRatio [19] as well as the custom-developed

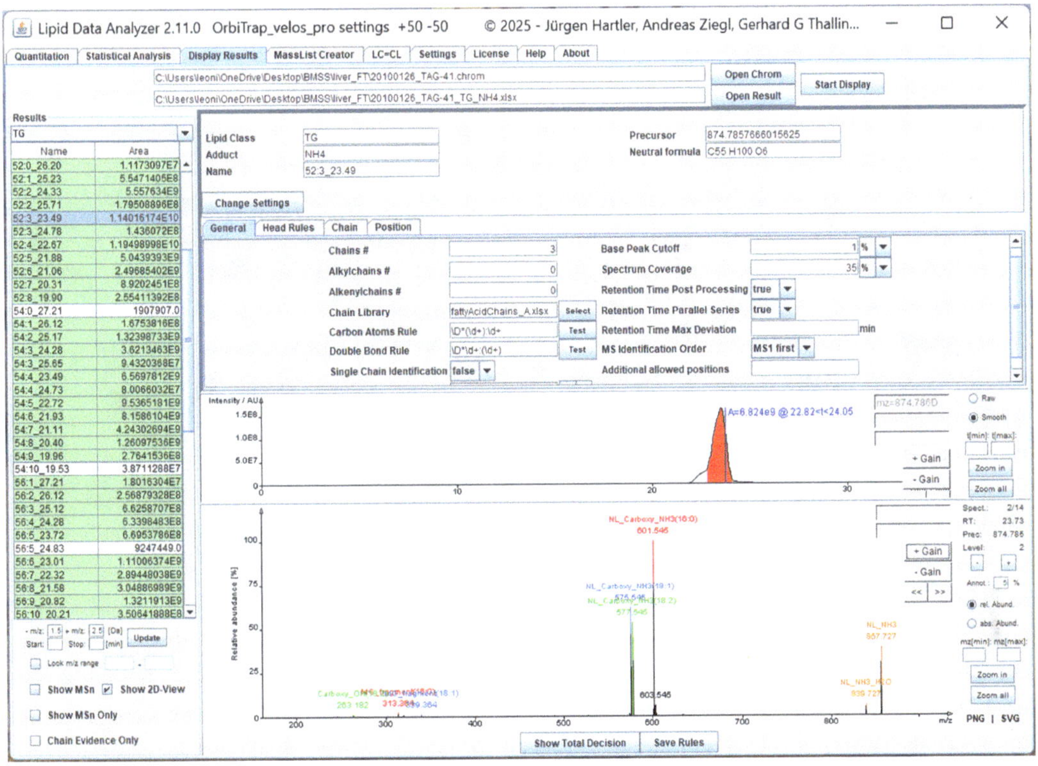

Fig. 7 LDA's GUI for review and refinement of decision rule sets at the example of TG 52:3. The interface available in the *Display Results* tab allows for adjusting any fragment or intensity rules. Feedback is provided in the form of direct visual annotation of user-defined spectra, and detailed reasons for any negative decision in text form

Greedy and 3D algorithms [4]. The MS^n spectrum viewer facilitates visualization of annotated MS^n spectra for both validation of identifications and decision rule refinement.

Statistical analysis is performed within a dedicated module that enables sample grouping, multiple normalization strategies, and absolute quantification when standard concentrations and dilution factors are provided. Quantitative results can be visualized as heatmaps (Fig. 8), illustrating relative abundance changes across samples or groups. Bar charts further aid in comparing lipid abundances across either individual analytes or sample sets. Importantly, the statistical analysis interface also enables manual curation at the group level, enabling users to adjust peak integration and remove FPs across multiple measurements simultaneously.

3.1.5 Result Summaries

LDA enables the export of result summaries across samples or groups directly from the statistical analysis interface (Fig. 8). Multiple export options are available, including customizable filters, selectable annotation levels and several output formats. These include the standardized mzTab-M format [20], which is well suited for submission to

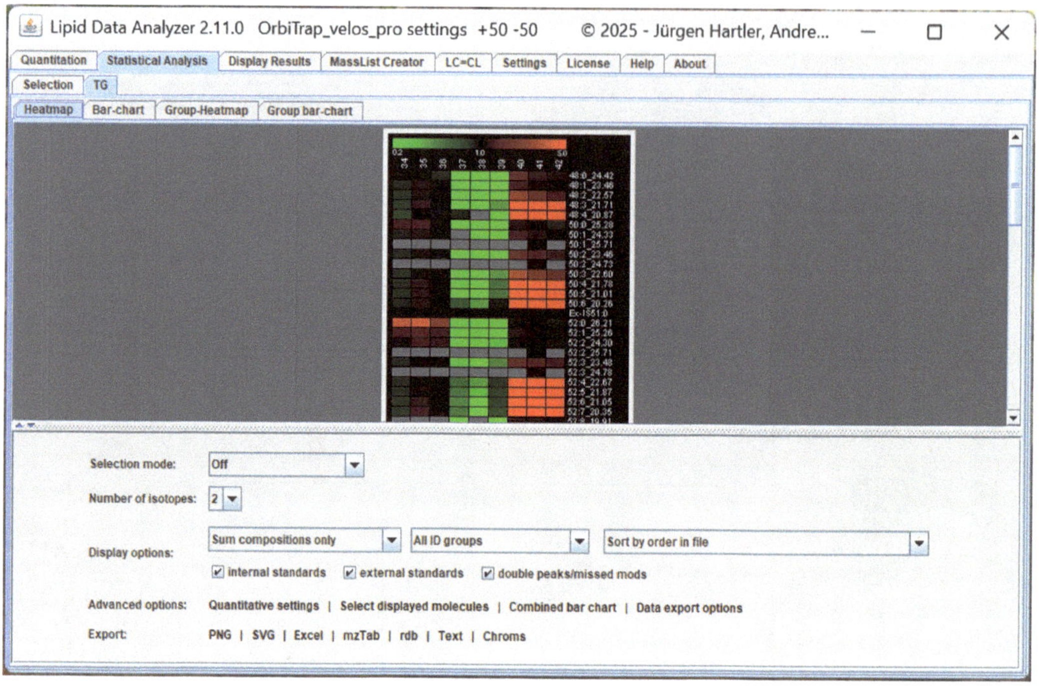

Fig. 8 Heatmap view of LDA's statistical analysis module. In the heatmap, samples/groups are per default organized horizontally and analytes vertically. The heatmap values are ratios to the median for each analyte over all samples/groups. Green, black, and red indicate values lower, equal, and higher than the median, respectively. Analytes undetected in a sample are displayed in gray. In the lower panel of this view, several options for further graphs, data curation, and export are available

public repositories; the rdb format, which is particularly suitable for downstream, script-based analysis, as it contains all analyte classes present in the original provided result files including extensive additional information that may be used for quality control; and Excel as well as plain text formats, which provide a compact overview for the identifications associated with the currently selected analyte class.

3.2 LDA's C=C Localizer (LC=CL)

Elucidating C=C positions in lipid molecules is a major focus of research, as this structural feature can dramatically influence a lipid's biophysical properties, its interactions with proteins and other biomolecules, and its role in metabolism and cellular signaling [21, 22]. LC=CL is the first software tool to extract and accurately translate FA-specific C=C-position data embedded in routine RPLC-MS/MS data into lipid structures. A key advantage of this approach is its suitability for all MS instruments capable of MS2, irrespective of the MS analyzer or ion activation method used. LC=CL's strategy also retains the full sensitivity and quantifiability of the corresponding MS method. This development was made possible by demonstrating that differences in terminal C=C positions (ω-positions) for a given lipid molecular species result in discernible RT shifts in routine RPLC measurements.

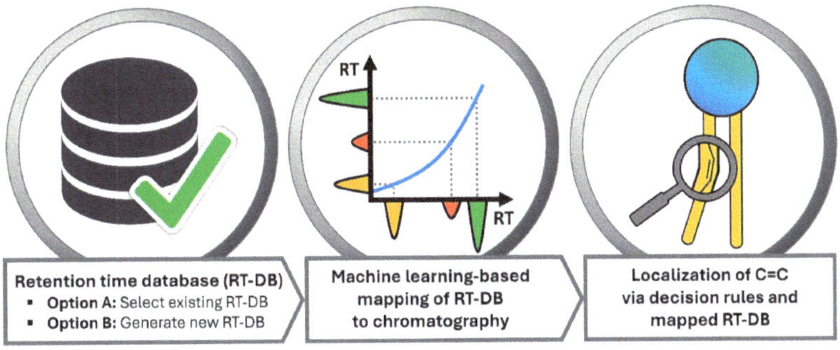

Fig. 9 LC=CL analysis pipeline, adapted from [17]. The LC=CL workflow starts with either selecting an existing RT-DB or generating a new one with LC=CL's dedicated module. The elution profile of the selected RT-DB is mapped to the relevant chromatographic conditions. Finally, LC=CL automatically identifies lipid C=C positions using LDA's decision rules and the mapped RT-DB

LC=CL's workflow consists of three main steps (Fig. 9). First, a retention time database (RT-DB) is selected to serve as a reference for the ω-position-specific elution profiles of complex lipids. While a comprehensive, experimentally validated RT-DB is provided, LDA's LC=CL module also supports the creation of custom RT-DBs. Second, a machine learning algorithm ensures accurate RT prediction for ω-resolved lipids across diverse chromatographic conditions. Third, automated ω-localization is fully integrated into LDA's lipid identification and quantification pipeline, applying RT-based ω-position annotations to lipid molecular species confirmed by decision rule sets.

3.3 Oxidized Lipids

Roles of oxidized lipids in various physiological and pathological processes, including inflammation and a number of age-related and chronic diseases, have been reported [23]. To address the analytical challenges associated with their identification and annotation, LDA has been extended accordingly [24]. The tool supports oxo modifications, but also any other modifications such as Cl, NO_2, etc., and combinations thereof.

To enable analysis of modified compounds in LDA, an adapted mass list (Sect. 3.1.1) and chain lists (Sect. 3.1.3) that detail the modifications of interest have to be provided (*see* **Note 16**). For oxidized lipids, in addition, LDA requires the file *modconfig.xml*, which defines the elemental configuration of relevant modifications. This file can be adapted if additional modifications need to be introduced.

3.4 Metabolite Identification

LDA has been successfully applied to the analysis of metabolomics data [25–28]. The general analysis workflow described in Sect. 3.1 and its subsections is equally applicable to both lipidomics and metabolomics, with one notable exception: the MassList Creator is not suitable for metabolites that lack the regular building blocks characteristic of lipids. In such cases, an LDA mass list—containing the relevant *m/z* values and, optionally, RT windows—must be manually adapted. Lipid-specific information such as the number of double bonds in the hydrophobic chains is optional in LDA's mass list and can be omitted. Similarly, the definition of sections in decision rules that only apply to lipid hydrophobic chains (*[CHAINS]* and *[POSITION]*) can be omitted; only [GENERAL] and [HEAD] sections are relevant for other metabolite groups too. For grouping metabolites into individual classes in the mass list and for decision rules, the same rules as for lipids apply (*see* **Note 5**).

3.5 LDA's Command Line Interface

Lipid identification and quantification by LDA can also be started via a command-line interface (CLI), enabling straightforward integration into automated pipelines and scripting options. Before running LDA's CLI, the relevant settings must be configured in LDA's GUI once. The CLI can then be invoked by executing either *ldaCmd.bat* (Windows) or *ldaCmd.sh* (Unix-based systems) located in LDA's installation directory.

4 Notes

1. If available RAM is insufficient for loading all results simultaneously for statistical analysis, cutoff settings can be used to analyze lipid classes individually by setting the cutoff for all other classes to 99%.
2. LDA can be started in two ways (the required files for both options are located in the root of the installation folder of LDA). The first one is a conventional executable ("Lipid Data Analyzer.exe" on Windows), while the second launches LDA via a console window ("Lipid Data Analyzer.bat" on Windows, "LipidDataAnalyzer.sh" on Linux/Unix and macOS). The console-based startup is recommended, as it provides additional progress and troubleshooting information. Additionally, the absence of errors in the console output suggests that LDA requires more time to complete the ongoing task.
3. Quantitation and data visualization in LDA are based on its internal chrom file format. After a quantitation has been started, data provided in any other supported MS file format (mzML, mzXML, or vendor-specific formats) is first converted into the internal chrom file format.
4. LDA stores the quantitation results as individual Excel files located in the same folder as the corresponding MS files. Each result file is named after the MS file (without file extension), followed by an underscore and the name of the mass list used.

5. Analyte classes in an LDA mass list should be grouped into individual Excel sheets based on similar MS^n fragmentation behavior, i.e., PC, PE, PI, etc. should each be defined in a separate sheet. This is required because each decision rule set for MS^n spectra interpretation applies to one specific adduct associated with one sheet in the selected mass list. Decision rules in LDA are stored as text files with the extension ".frag. txt" in the "fragRules" directory of the LDA installation. Each rule file corresponds to a unique combination of analyte class and adduct ion. The rule file name must exactly match the combination of the name of the corresponding sheet in the Excel mass list and the adduct name specified in the column header "mass(...)" of that sheet (Fig. 5). For example, the file "PC_Na.frag.txt" contains rules for phosphatidylcholine (PC) with the $[Na]^+$ adduct. The corresponding mass list sheet must be named "PC" and include the column header "mass(form[Na] name[Na])." Similarly, the file "HexCer_-2H.frag.txt" contains rules for hexosylceramide (HexCer) that is twice deprotonated: $[2H]^{2-}$. The associated sheet must be named "HexCer" and include the column header "mass(form[-H2] name[-2H] charge=2)." If a matching ".frag.txt" file is not found for a given sheet and adduct, LDA will skip MS^n interpretation for that analyte and perform MS^1-level identification only.
6. For straightforward integration with other software applications, the MassList Creator supports multiple output formats beyond the standard LDA MassList format. Users requiring additional formats are encouraged to contact the development team.
7. General settings are stored in the "*.settings*" file in LDA's installation directory. This file specifies LDA settings that are MS platform independent and can be edited with any text editor.
8. MS platform-specific settings are stored in *.properties* files in the properties folder located in LDA's installation directory (defined during the installation process). An individual "*.properties*" file is required for each MS platform. Users of an MS platform that is not yet available are encouraged to contact the development team.
9. Default memory settings for execution are defined during installation of LDA with the parameters -*Xms* and -*Xmx*. The -*Xms* value specifies the initial amount of memory allocated at startup, while -*Xmx* sets the maximum memory the application is allowed to use. If LDA exceeds the allocated memory limit (-*Xmx*), it will terminate. To avoid this, it is essential to assign values appropriate for the number and size of the input

files. As a general guideline, *-Xmx* should be set to approximately 2 GB less than the total available system memory, and *-Xms* should be set to about half of *-Xmx*. For example, on a system with 16 GB of RAM, recommended values (in MB) would be *-Xmx* = 14336m and *-Xms* = 7168m (default values after installation are *-Xmx* = 8192m and *-Xms* = 2048m).

10. To adapt the memory settings after installation, simply open the "LipidDataAnalyzer.bat" (Windows) or "LipidDataAnalyzer.sh" file (Linux/Unix or macOS) in the LDA installation directory with any text editor and manually adjust the values for *-Xms* and *-Xmx*. When running the LDA executable file without the console window, adjust the values in the "*Lipid Data Analyzer.vmoptions*" file in the installation directory.

11. By default, the MS file parsing settings are configured conservatively to ensure LDA can run also on resource-constrained systems. However, users working with more powerful hardware can accelerate processing by adjusting the parameter *maxFileSizeForChromTranslationAtOnce*. For large MS files—often several GB in size—LDA processes the data in multiple iterations to remain within available memory limits. The more iterations are required, the longer the translation process will take. To reduce the number of iterations and speed up processing, this parameter can be increased, provided sufficient memory is available. A recommended guideline is to set it to approximately 10% of the value assigned to *-Xmx*. For example, with *-Xmx* = 14g, a suitable value for this parameter would be 1400 (this value has to be provided in MB).

12. LDA features a noise filter for the annotation of MS^n spectra. However, the specificity of lipid annotations by LDA's decision rule-based algorithm is largely robust in the presence of noise. Therefore, especially for very noisy data, a simple way to effectively reduce processing time is to disable LDA's noise filter. This filter is activated when an MS^n spectrum contains more signals than the threshold specified by the parameter *ms2MinIntsForNoiseRemoval*. This approach was chosen as many MS instruments feature a noise filter algorithm already. If an additional noise filter is applied on such data, relevant information may be removed. Thus, LDA's noise filter should be activated only if the instrument's filter failed (which can happen for single spectra). Noise unexpectedly present in a spectrum can be detected by a disproportional increase in the number of signals. Hence, this threshold value specifies the number of signals that must be present to activate the noise filter. To completely deactivate the noise filter, set this parameter to a very high number, e.g., *ms2MinIntsForNoiseRemoval*= 1000000, in the respective "*.properties*" file.

13. The setting *use3D* enables LDA's 3D chromatographic integration algorithm. This results in more accurate peak integration, especially for complex samples with many overlapping isobars. However, for samples where such complexity is not expected, or where processing speed is of higher priority, it is advisable to disable *use3D*. This reverts peak integration to a faster 2D mode, using only intensity and RT, and can substantially reduce overall computation time.
14. LDA's GUI (Fig. 4c) includes a *noIntensity* option available for all MS platforms. This mode uses a set of unspecific decision rules designed for initial MS^n spectrum analysis. Notably, none of the rules in the *noIntensity* set are marked as mandatory = true, and intensity information is ignored. This ensures that no MS^n identifications are prematurely excluded, providing a broad starting point for defining more stringent rules to eliminate FPs that are encountered during analysis.
15. LDA's decision rules are stored as *.frag.txt* files in the *fragRules* folder in LDA's installation directory and can be edited with any text editor. The currently active decision rules are located directly in the *fragRules* folder and will be overwritten when "Save as default" in the *Settings* tab is clicked (Fig. 4c). The permanent decision rules are stored in the instrument-specific subfolders.
16. If the detection of oxidized species is required, one additional column needs to be present in LDA's mass list and chain lists with the header *PSM*. This column defines for each analyte or chain which relevant modifications to look for. For example, an entry *";O1;O2;O3;Cl1;Cl1,O1"* in the PSM column would tell the software that unmodified compounds, up to three oxidations, a chlorination, and the combination of one chlorination and one oxidation are of interest. Preconfigured and ready-to-use mass lists and chain lists for the analysis of oxidized lipids are provided with LDA.

Acknowledgments

The authors gratefully acknowledge the financial support from the University of Graz and Graz University of Technology.

References

1. Yetukuri L, Ekroos K, Vidal-Puig A, Orešič M (2008) Informatics and computational strategies for the study of lipids. Mol BioSyst 4(2):121–127. https://doi.org/10.1039/B715468B
2. Han X, Gross RW (2022) The foundations and development of lipidomics. J Lipid Res 63(2):100164. https://doi.org/10.1016/j.jlr.2021.100164
3. Bielow C, Mastrobuoni G, Orioli M, Kempa S (2017) On mass ambiguities in high-resolution shotgun lipidomics. Anal Chem 89(5):2986–2994. https://doi.org/10.1021/acs.analchem.6b04456

4. Hartler J, Trötzmüller M, Chitraju C, Spener F, Köfeler HC, Thallinger GG (2011) Lipid data analyzer: unattended identification and quantitation of lipids in LC-MS data. Bioinformatics 27(4):572–577. https://doi.org/10.1093/bioinformatics/btq699
5. Hartler J, Triebl A, Ziegl A, Trötzmüller M, Rechberger GN, Zeleznik OA, Zierler KA, Torta F, Cazenave-Gassiot A, Wenk MR, Fauland A, Wheelock CE, Armando AM, Quehenberger O, Zhang Q, Wakelam MJO, Haemmerle G, Spener F, Köfeler HC, Thallinger GG (2017) Deciphering lipid structures based on platform-independent decision rules. Nat Methods 14(12):1171–1174. https://doi.org/10.1038/nmeth.4470
6. Skotland T, Ekroos K, McDonald J, Ahrends R, Liebisch G, Sandvig K (2024) Pitfalls in lipid mass spectrometry of mammalian samples – a brief guide for biologists. Nat Rev Mol Cell Biol 25(10):759–760. https://doi.org/10.1038/s41580-024-00758-4
7. Lamp LM, Hartler J (2022) MS-RIDD paves the way toward routine double bond localization in mass spectrometry-based lipidomics. Commun Chem 5:181. https://doi.org/10.1038/s42004-022-00805-1
8. Liebisch G, Fahy E, Aoki J, Dennis EA, Durand T, Ejsing CS, Fedorova M, Feussner I, Griffiths WJ, Köfeler H, Merrill AH, Murphy RC, O'Donnell VB, Oskolkova O, Subramaniam S, Wakelam MJO, Spener F (2020) Update on LIPID MAPS classification, nomenclature, and shorthand notation for MS-derived lipid structures. J Lipid Res 61(12):1539–1555. https://doi.org/10.1194/jlr.S120001025
9. Hartler J, Armando AM, Trötzmüller M, Dennis EA, Köfeler HC, Quehenberger O (2020) Automated annotation of sphingolipids including accurate identification of hydroxylation sites using MSn data. Anal Chem 92(20):14054–14062. https://doi.org/10.1021/acs.analchem.0c03016
10. Züllig T, Köfeler HC (2021) High resolution mass spectrometry in lipidomics. Mass Spectrom Rev 40(3):162–176. https://doi.org/10.1002/mas.21627
11. Hohenwallner K, Lamp LM, Peng L, Nuske M, Hartler J, Reid GE, Rampler E (2024) FAIMS shotgun lipidomics for enhanced class- and charge-state separation complemented by automated ganglioside annotation. Anal Chem 96(30):12296–12307. https://doi.org/10.1021/acs.analchem.4c01313
12. Brugger B, Erben G, Sandhoff R, Wieland FT, Lehmann WD (1997) Quantitative analysis of biological membrane lipids at the low picomole level by nano-electrospray ionization tandem mass spectrometry. Proc Natl Acad Sci U S A 94(6):2339–2344. https://doi.org/10.1073/pnas.94.6.2339
13. White JB, Trim PJ, Salagaras T, Long A, Psaltis PJ, Verjans JW, Snel MF (2022) Equivalent carbon number and interclass retention time conversion enhance lipid identification in untargeted clinical lipidomics. Anal Chem 94(8):3476–3484. https://doi.org/10.1021/acs.analchem.1c03770
14. Ovčačíková M, Lísa M, Cífková E, Holčapek M (2016) Retention behavior of lipids in reversed-phase ultrahigh-performance liquid chromatography–electrospray ionization mass spectrometry. J Chromatogr A 1450:76–85. https://doi.org/10.1016/j.chroma.2016.04.082
15. Fauland A, Köfeler H, Trötzmüller M, Knopf A, Hartler J, Eberl A, Chitraju C, Lankmayr E, Spener F (2011) A comprehensive method for lipid profiling by liquid chromatography-ion cyclotron resonance mass spectrometry. J Lipid Res 52(12):2314–2322. https://doi.org/10.1194/jlr.D016550
16. Tanno H, Sassa T, Sawai M, Kihara A (2021) Production of branched-chain very-long-chain fatty acids by fatty acid elongases and their tissue distribution in mammals. Biochim Biophys Acta Mol Cell Biol Lipids 1866(1):158842. https://doi.org/10.1016/j.bbalip.2020.158842
17. Lamp LM, Murawska GM, Argus JP, Armando AM, Talmazan RA, Pühringer M, Rampler E, Quehenberger O, Dennis EA, Hartler J (2025) Computationally unmasking each fatty acyl C=C position in complex lipids by routine LC-MS/MS lipidomics. Nat Commun 16:7277. https://doi.org/10.1038/s41467-025-61911-x
18. Chambers MC, Maclean B, Burke R, Amodei D, Ruderman DL, Neumann S, Gatto L, Fischer B, Pratt B, Egertson J, Hoff K, Kessner D, Tasman N, Shulman N, Frewen B, Baker TA, Brusniak MY, Paulse C, Creasy D, Flashner L, Kani K, Moulding C, Seymour SL, Nuwaysir LM, Lefebvre B, Kuhlmann F, Roark J, Paape R, Suckau D, Hemenway T, Huhmer A, Langridge J, Connolly B, Chadick T, Holly K, Eckels J, Deutsch EW, Moritz RL, Katz JE, Agus DB, MacCoss M, Tabb DL, Mallick P (2012) A cross-platform toolkit for mass spectrometry and proteomics. Nat Biotechnol 30(10):918–920. https://doi.org/10.1038/nbt.2377
19. Li XJ, Zhang H, Ranish JA, Aebersold R (2003) Automated statistical analysis of protein abundance ratios from data generated by stable-isotope dilution and tandem mass spectrometry. Anal Chem 75(23):6648–6657. https://doi.org/10.1021/ac034633i
20. Hoffmann N, Rein J, Sachsenberg T, Hartler J, Haug K, Mayer G, Alka O, Dayalan S, Pearce JTM, Rocca-Serra P, Qi D, Eisenacher

M, Perez-Riverol Y, Vizcaíno JA, Salek RM, Neumann S, Jones AR (2019) mzTab-M: a data standard for sharing quantitative results in mass spectrometry metabolomics. Anal Chem 91(5):3302–3310. https://doi.org/10.1021/acs.analchem.8b04310
21. Hayashi D, Mouchlis VD, Dennis EA (2021) Omega-3 versus Omega-6 fatty acid availability is controlled by hydrophobic site geometries of phospholipase A_2s. J Lipid Res 62:100113. https://doi.org/10.1016/j.jlr.2021.100113
22. Dennis EA (1987) Regulation of eicosanoid production: role of phospholipases and inhibitors. Nat Biotechnol 5(12):1294–1300. https://doi.org/10.1038/nbt1287-1294
23. Vieira SA, Zhang G, Decker EA (2017) Biological implications of lipid oxidation products. J Am Oil Chem Soc 94(3):339–351. https://doi.org/10.1007/s11746-017-2958-2
24. Krettler CA, Hartler J, Thallinger GG (2020) Identification and quantification of oxidized lipids in LC-MS lipidomics data. Stud Health Technol Inform 271:39–48. https://doi.org/10.3233/SHTI200072
25. Züllig T, Zandl-Lang M, Trötzmüller M, Hartler J, Plecko B, Köfeler HC (2020) A metabolomics workflow for analyzing complex biological samples using a combined method of untargeted and target-list based approaches. Metabolites 10(9):342. https://doi.org/10.3390/metabo10090342
26. Zandl-Lang M, Züllig T, Holzer M, Eichmann TO, Darnhofer B, Schwerin-Nagel A, Zobel J, Haidl H, Biebl A, Köfeler H, Plecko B (2025) Multi-omics profiling in spinal muscular atrophy (SMA): investigating lipid and metabolic alterations through longitudinal CSF analysis of Nusinersen-treated patients. J Neurol 272:183. https://doi.org/10.1007/s00415-025-12909-4
27. Zandl-Lang M, Züllig T, Trötzmüller M, Naegelin Y, Abela L, Wilken B, Scholl-Buergi S, Karall D, Kappos L, Köfeler H, Plecko B (2022) Changes in the cerebrospinal fluid and plasma lipidome in patients with Rett syndrome. Metabolites 12(4):291. https://doi.org/10.3390/metabo12040291
28. Pferschy-Wenzig E-M, Ortmann S, Atanasov AG, Hellauer K, Hartler J, Kunert O, Gold-Binder M, Ladurner A, Heiß EH, Latkolik S, Zhao Y-M, Raab P, Monschein M, Trummer N, Samuel B, Crockett S, Miao J-H, Thallinger GG, Bochkov V, Dirsch VM, Bauer R (2022) Characterization of constituents with potential anti-inflammatory activity in Chinese Lonicera species by UHPLC-HRMS based metabolite profiling. Metabolites 12(4):288. https://doi.org/10.3390/metabo12040288

Chapter 13

Prediction, Targeting, and Annotation of Oxidized Lipids Using LPPtiger2 Software

Zhixu Ni and Maria Fedorova

Abstract

The epilipidome, a subset of the natural lipidome arising from enzymatic and non-enzymatic lipid modifications, remains largely unexplored. Within this emerging class, oxidized complex lipids have raised considerable interest due to their diverse biological functions, including the modulation of inflammation, cell fate decisions, and the execution of programmed cell death. However, the discovery and annotation of these typically low-abundant yet structurally diverse lipid species present significant analytical challenges, often necessitating advanced bioinformatics tools. Here, we present a computational pipeline powered by LPPtiger2 software, designed for the comprehensive discovery, detection, and annotation of complex oxidized lipids within the context of a defined lipidome. The LPPtiger2 hybrid workflow offers a robust solution for high-quality epilipid profiling by integrating a predictive algorithm with a semi-targeted experimental protocol. Using a knowledge-based in silico epilipidome prediction algorithm, it generates a highly customized, sample-specific search space prior to data acquisition. This approach transforms the conventional untargeted lipidomics pipeline into a semi-targeted workflow that selectively focuses on predicted epilipid precursors. Such specificity enhances LPPtiger2-supported annotation of modified epilipids through improved sensitivity, superior MS/MS spectral quality, and a tailored lipid search space.

Keywords Oxidized lipids, Epilipidomics, Lipid annotations, Software, Lipidomics

1 Introduction

Lipids are hydrophobic or amphiphilic molecules that serve diverse biological roles, including energy storage, organization of membrane architecture, and signal transduction. This functional versatility arises from the extensive structural diversity of the lipidome, which currently encompasses over 50,000 experimentally characterized species, with thousands more predicted. Beyond their canonical forms, lipids are subject to a wide range of enzymatic and non-enzymatic modifications, such as phosphorylation, hydroxylation, halogenation, and glycosylation, that introduce small chemical moieties capable of altering lipid function. These modifications give rise to the epilipidome, a vast and chemically heterogeneous subset of the lipidome,

Michael Witting (ed.), *Computational Methods and Data Analysis for Metabolomics*, Methods in Molecular Biology, vol. 3063,
https://doi.org/10.1007/978-1-0716-5452-1_13,

implicated in the regulation of biological functions [1, 2]. Despite its potential biological significance, the systematic discovery and annotation of the epilipidome remain in their infancy. Oxygenated lipids represent some of the most extensively studied epilipid species to date, with oxylipins, oxygenated derivatives of free fatty acids, routinely analyzed through targeted mass spectrometry (MS) [3]. Targeted liquid chromatography (LC)–MS analysis facilitates sensitive and specific detection of oxylipins; however, it requires definition of targeted analytes. The identification of complex oxidized lipids such as oxygenated glycerophospholipids (oxGPLs), cholesteryl esters (oxCEs), and triacylglycerides (oxTGs), is more challenging due to their high structural diversity and context-dependent formation, which together result in a vast chemical space that exceeds the capacity of targeted analytical approaches.

Over the years, various approaches have been developed for the detection and annotation of complex oxidized lipids [4–8]. Generally, these strategies fall into two categories: targeted methods optimized for a limited set of selected epilipids, and untargeted workflows designed to enhance epilipidome coverage, often coupled with dedicated computational algorithms. LPPtiger2 software integrates strengths from both approaches, offering a semi-targeted hybrid workflow that leverages an innovative in silico prediction algorithm prior to data acquisition [4]. LPPtiger2 computational pipeline is based on four tasks—(1) in silico epilipidome prediction based on the composition of studied lipidome, which is used for (2) generation of inclusion lists for semi-targeted analysis, followed by (3) the annotation of detected oxidized lipids using pre-defined fragmentation rules, and (4) review of epilipid annotations (Fig. 1). Computational part is supported by three LC–MS experiments, each which specific aim: (A) untargeted LC–MS/MS analysis of unmodified lipidome to be used for in silico prediction,

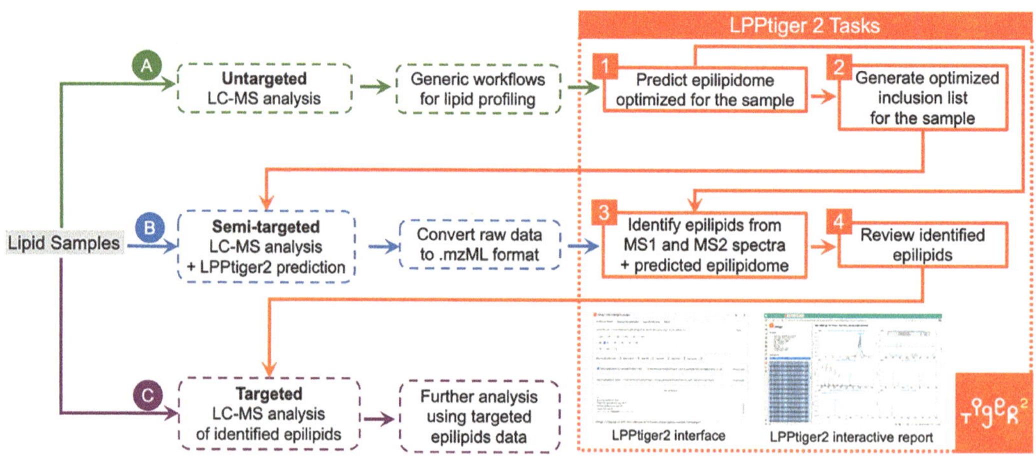

Fig. 1 An overview of hybrid LPPtiger2 workflow for epilipidomics analysis covering LC–MS/MS experiments (**A–C**) and the corresponding computational tasks (1–4) performed by LPPtiger2 software

(B) semi-targeted LC–MS/MS analysis using LPPtiger2 generated inclusion list of predicted epilipids, and (C) targeted LC–MS/MS for relative quantification of annotated oxidized species. Here, we present the design and application of an LPPtiger2-supported workflow for the prediction, software-guided targeting, and annotation of complex oxidized lipids, including oxGPL, oxCE, and oxTG species.

2 Materials

2.1 Wet Lab Requirements

LC–MS/MS data must be acquired through wet-lab experiments complimenting multiple steps of the hybrid workflow (Fig. 1A–C). A representative study demonstrating the LPPtiger2-assisted analytical and computational pipeline for annotation and relative quantification of oxidized complex lipids in human blood plasma samples, including full details of the corresponding wet-lab methodologies, has been published by Criscuolo et al. [4].

2.2 Hardware Requirements

LPPtiger2 is a cross-platform, open-source software tool designed for broad accessibility and usability across computing environments. It operates efficiently on standard laptop configurations equipped with at least four CPU cores and 16 GB of RAM. For more demanding lipid annotation workflows, systems with eight or more CPU cores and a minimum of 32 GB RAM are recommended to ensure optimal performance. Given the substantial size of raw and processed LC–MS datasets, as well as the volume of output files generated, a minimum of 10 GB of available disk space is advised to accommodate data storage requirements.

2.3 Dry-Lab Requirements

2.3.1 LPPtiger2 Installation

The latest release of the LPPtiger2 software, along with comprehensive tutorial materials, is available via the GitHub repository of the Lipid Metabolism: Analysis and Integration (LMAI) laboratory (https://github.com/LMAI-TUD/lpptiger2/releases/). For Windows users, a ready-to-use zip package (LPPtiger2_RC_win10_x64_exe.zip) is provided, containing a click-to-run executable file. No additional installation is required. To launch LPPtiger2, users simply extract the contents of the Zip archive and double-click the included LPPtiger2.exe file, which opens the main graphical user interface. For other operating systems, users must install Python with the required dependencies and run the source-code version of LPPtiger2. A step-by-step installation guide, along with example input and output files for a demonstrative session, is included in the tutorial package (LPPtiger2_tutorial_pack.zip), available on the release page. For clarity, the following instructions refer specifically to the Windows platform.

2.3.2 Proteowizard Installation and Raw Files Conversion into mzML Format

To convert raw MS files to the mzML format, users on Windows platforms are advised to download the latest version of ProteoWizard from the official website (https://proteowizard.sourceforge.io/) and follow the installation instructions provided. For users operating on other systems, a Docker-based version of ProteoWizard is available.

2.3.3 Selection of Representative Lipid Lists for Epilipidome Prediction

To initiate the LPPtiger2 workflow, users must first prepare a curated list of lipids selected for in silico oxidation. This list should ideally include representative lipid species spanning a range of acyl chain lengths and degrees of unsaturation. It is recommended that these lipids be selected based on results from untargeted lipidomics analyses performed on the same biological sample (Fig. 1A). While users are free to employ their preferred approaches to generate this list, several general guidelines are suggested to ensure comprehensive and biologically relevant coverage (*see* **Note 1**).

3 Methods

3.1 Epilipidome In Silico Prediction

3.1.1 Parameter Selection

In the example workflow, users are encouraged to perform predictions for each lipid class separately. A representative screenshot of the LPPtiger2 interface, illustrating example parameters for the generation of oxidized phosphatidylcholine (PC) species, is provided in Fig. 2.

The key parameters include the following:

List of FA (.csv)

This file configures the list of fatty acyl chains to be used for in silico oxidation (exemplified in Table 1; *see* **Note 2** for more details). Isomeric lipid species can be distinguished within LPPtiger2 using standardized SMILES strings and descriptive remarks. For example, both the ω-3 and ω-6 isomers of FA 18:3 are supported. Users may also extend the lipid list by incorporating custom structures, such as odd-chain fatty acids or species with modified degrees of unsaturation. However, to ensure compatibility with the software prediction engine, all custom SMILES entries must conform to the formatting style defined in the provided template

Select lipid subclass for in silico oxidation

Next, the user can select a lipid subclass to be used for in silico oxidation using fatty acyl combinations defined in the List of FA file.

Max modification site

This parameter defines the maximum number of modification sites used for in silico oxidation on a given fatty acyl chain.

Max total O, max OH, max keto, max OOH, max epoxy

This parameter controls the type of modifications introduced into the fatty acyl chain during in silico oxidation. By adjusting the

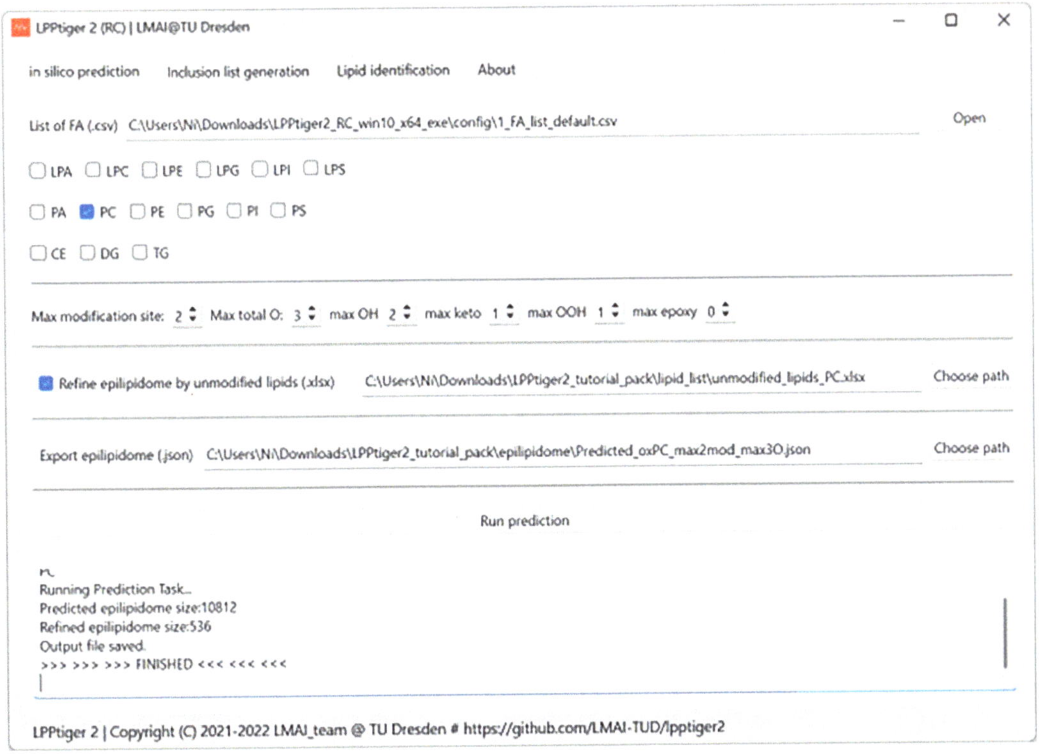

Fig. 2 The LPPtiger2 interface of the in silico prediction module is shown with parameters configured to exemplify the prediction of oxidized phosphatidylcholine (PC) species. In this example, a maximum of two modification sites and the addition of up to three oxygen atoms are allowed, constrained by combinations of hydroxyl (up to 2), keto (one), or hydroperoxyl (one) modifications

Table 1
Example of the FA list file including abbreviations, SMILES, and remarks

Residues	SMILES	Remark
FA16:0	OC(CCCCCCCCCCCCCCC)=O	FA16:0
FA16:1	OC(CCCCCCC\C=C/CCCCCC)=O	FA16:1(9Z)
FA18:0	OC(CCCCCCCCCCCCCCCCC)=O	FA18:0
FA18:1	OC(CCCCCCC\C=C/CCCCCCCC)=O	FA18:1(9Z)
FA18:2	OC(CCCCCCC\C=C/C\C=C/CCCCC)=O	FA18:2(9Z,12Z)
FA18:3	OC(CCCCCCC\C=C/C\C=C/C\C=C/CC)=O	FA18:3(9Z,12Z,15Z)
FA18:3	OC(CCCC\C=C/C\C=C/C\C=C/CCCCC)=O	FA18:3(6Z,9Z,12Z)
FA20:3	OC(CCCCCC\C=C/C\C=C/C\C=C/ CCCCC)=O	FA20:3(8Z,11Z,14Z)
FA20:4	OC(CCC\C=C/C\C=C/C\C=C/C\C=C/ CCCCC)=O	FA20:4(5Z,8Z,11Z,14Z)

allowed number and type of oxidative modifications, users can generate customized prediction sets tailored to specific sample types or experimental conditions (*see* **Note 3**).

Refine lipidome by modified lipids (.xlsx)

The list of lipid molecular species used for in silico oxidation is provided as input and must contain fatty acyl chains defined in the accompanying List of FA (.csv) file. This lipid list is typically derived from prior LC–MS/MS-based lipidomics experiments aimed at annotating unmodified lipid species present in the sample of interest (Fig. 1A). From such lipidomics profiling, one can select the most abundant, most regulated, or most PUFA-rich lipid molecular species to be used for in silico oxidation. Furthermore, one list can include multiple lipid subclasses (e.g., PC and PE lipids). In this case, by selecting the corresponding lipid class options in the LPPtiger2 interface, users can generate predicted epilipidomes for multiple lipid classes one by one (*see* **Note 4**).

Export epilipidome (.json)

This parameter specifies the directory path where the predicted epilipidome output, generated based on the selected settings, will be saved.

3.1.2 The Epilipidome JSON File

The predicted epilipidome is exported as a customized library in JSON format, containing detailed information for each lipid species, including shorthand notation, possible structures, and predicted fragmentation patterns generated through in silico fragmentation algorithms (*see* **Note 5**). A representative example of the data structure for an oxidized lipid PC(16:0_18:1<OH>) is shown in Fig. 3. Each entry in the library follows a consistent nested structure comprising general lipid information and MS/MS-specific annotations, including class-specific fragments, neutral losses, and predicted fragmentation patterns. Most predicted fragment ions are annotated with their elemental compositions. Diagnostic signals used as a "predicted fingerprint" provided as *m/z* values which may not always have a defined formula. This rich annotation makes the output suitable for advanced bioinformatics analyses and downstream tool development. To support interoperability with other formats and cheminformatics workflows (e.g., SDF, SMILES, InChI), LPPtiger2 provides a representative SMILES string for each predicted lipid structure, based on one of the possible regioisomeric forms and its constituent residues. Please note that these SMILES representations are intended for demonstrative purposes only and do not define exact structures, due to inherent uncertainty in the precise position of modifications. An example JSON output file is included in the LPPtiger2 tutorial package, allowing users to explore the full data structure and evaluate potential applications of the predicted epilipidome.

At present, some epilipids identified by LPPtiger2 may retain structural ambiguities, particularly regarding the precise positions of modifications or the stereospecific (*sn*-1/*sn*-2) localization of fatty acyl residues. During the development of LPPtiger2, none of the existing lipid nomenclature systems fully met the requirements for representing

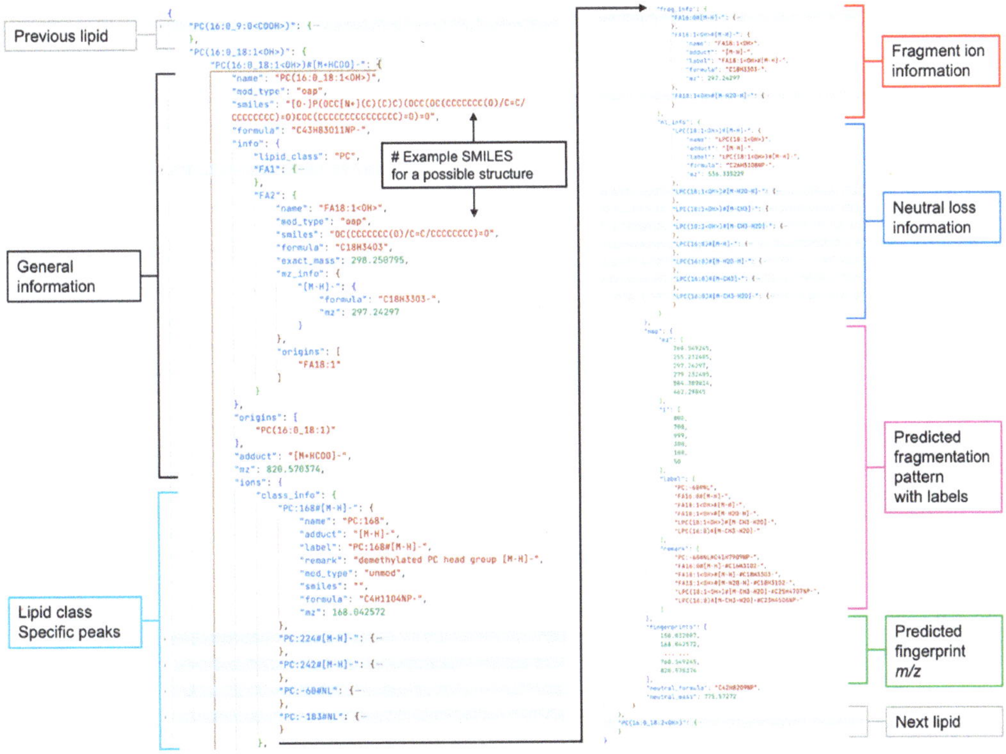

Fig. 3 Basic data structure of the predicted epilipidome in JSON format generated by LPPtiger2 exemplified for PC(16:0_18:1<OH>). Each lipid entry is organized in a nested JSON structure comprising general lipid metadata and predicted MS/MS fragmentation information, including class-specific fragments, neutral losses, diagnostic ions, and fingerprint peaks. Elemental compositions are provided for predicted fragments and neutral losses. To avoid the redundancies and reduce the complexity, the fingerprint peaks were provided in *m/z* only

such partially defined structures. To address this, a dedicated LPPtiger-specific nomenclature was adapted from the established shorthand lipid notation, but modified to accommodate oxidative modifications while maintaining a balance between readability, annotation accuracy, and computational compatibility. The core principle of this system is the use of angle brackets ("< >") to encapsulate all modifications, allowing multiple types of oxidative modifications to be annotated even when their exact positions are unknown.

For example:

PC(16:0_9:0<COOH>): A glycerophosphatidylcholine bearing esterified fatty acyl chains of 16:0 and 9:0 with unspecified *sn*-1/*sn*-2 positions, where the 9:0 chain contains a terminal carboxylic acid (–COOH) functional group.

PC(16:0_18:1<OH>): A glycerophosphatidylcholine bearing esterified fatty acyl chains of 16:0 and 18:1 with unspecified *sn*-1/*sn*-2 positions, where the 18:1 chain contains a hydroxyl (OH) functional group.

These flexible abbreviations can prevent over-reporting the positional isomers without solid MS2 spectra confirmation. Thus, in LPPtiger2, positional isomers of hydroxyeicosotetraenoid acid (HETE) are abbreviated as FA20:4<OH> to ensure the match between spectra-based evidence and the structure. When high-quality spectra with signals annotating specific positional isomers are available, one can further update the corresponding abbreviation. Some examples of position-specific fragments can be found in the Supplementary Data 2 file of a previous publication by Criscuolo et al. [4]. For example, MS2 fragment ion at *m/z* 171.1027 supports the annotation of the hydroxyl group position to carbon 9 of FA 18:1 acyl chain, and thus such a lipid can be annotated as PC(16:0_18:1< OH{9}>) in LPPtgier2 style or PC 16:0_18:1;9OH according to the shorthand notation [9].

3.2 Generation of Inclusion Lists for Semi-Targeted LC–MS Analysis

3.2.1 Inclusion List Generation

To prepare the inclusion list for semi-targeted LC–MS/MS detection of predicted oxidized lipids LPPtiger2 requires the following input files (Fig. 4):

Fig. 4 The LPPtiger2 interface of the Inclusion list generation module

Table 2
An example of the representative lipidome file to be used for in silico oxidation.

LIPID	CLASS
PC(16:0_18:1)	PC
PC(16:0_18:2)	PC
PC(16:0_20:3)	PC
PC(16:0_20:4)	PC
PC(16:0_22:4)	PC
PC(16:0_22:5)	PC
PC(16:0_22:6)	PC
PC(18:0_18:2)	PC
PC(18:0_20:3)	PC
PC(18:0_20:4)	PC
PC(18:0_22:4)	PC
PC(18:0_22:5)	PC
PC(18:1_18:2)	PC
PC(18:1_20:4)	PC
PC(O-16:0_20:4)	PC
PC(P-16:0_20:4)	PC
PC(P-18:0_20:4)	PC
PE(16:0_18:1)	PE
PE(16:0_18:2)	PE

List of unmodified lipids (.xlsx)

This is the list of unmodified lipids used for in silico oxidation, equal to the list "Refine lipidome by modified lipids (.xlsx)" used in the previous section (Table 2).

Load epilipidome (.json)

This JSON file represents the output of the epilipidome prediction and serves as a customized spectral library for the preparation of the inclusion list.

Export inclusion list (.csv)

This parameter specifies the directory path where the generated inclusion list will be saved.

3.2.2 Refinement of LPPtiger2 Generated Inclusion List for PRM and MRM Experiments

The inclusion list file generated by LPPtiger2 contains key information for MS-based targeted analysis, including the neutral molecular formula, corresponding charged formula, exact mass, and calculated *m/z* values for each predicted epilipid (Table 3). This file is intended as a basis for constructing instrument-specific inclusion lists and must be reformatted

Table 3
Example of LPPtiger2-generated inclusion list of oxPC lipids (shown top 11 rows)

Lipid_name	Neutral_formula	Neutral_ mass	Charge	Charged_ formula	mz	oxFA_name	oxFA_ neutral_ formula	oxFA_ neutral_ mass
PC(16:0_9:0 <COOH>)	$C_{33}H_{64}O_{10}NP$	665.426784	[M−H]−	$C_{33}H_{63}O_{10}NP^-$	664.418959	FA9:0 <COOH>	$C_9H_{16}O_4$	188.104859
PC(16:0_9:0 <oxo>)	$C_{33}H_{64}O_9NP$	649.431869	[M+ HCOO]−	$C_{34}H_{65}O_{11}NP^-$	694.429523	FA9:0<oxo>	$C_9H_{16}O_3$	172.109944
PC(16:0_18:1 <oxo>)	$C_{42}H_{80}O_9NP$	773.55707	[M+ HCOO]−	$C_{43}H_{81}O_{11}NP^-$	818.554724	FA18:1<oxo>	$C_{18}H_{32}O_3$	296.235145
PC(16:0_18:1 <OH>)	$C_{42}H_{82}O_9NP$	775.57272	[M+ HCOO]−	$C_{43}H_{83}O_{11}NP^-$	820.570374	FA18:1<OH>	$C_{18}H_{34}O_3$	298.250795
PC(16:0_18:1 <OOH>)	$C_{42}H_{82}O_{10}NP$	791.567634	[M+ HCOO]−	$C_{43}H_{83}O_{12}NP^-$	836.565289	FA18:1 <OOH>	$C_{18}H_{34}O_4$	314.24571
PC(16:0_18:2 <OH,oxo>)	$C_{42}H_{78}$O10NP	787.536334	[M+ HCOO]−	$C_{43}H_{79}O_{12}NP^-$	832.533989	FA18:2 <OH,oxo>	$C_{18}H_{30}O_4$	310.214409
PC(16:0_18:2 <oxo,OOH>)	$C_{42}H_{78}O_{11}NP$	803.531249	[M+ HCOO]−	$C_{43}H_{79}O_{13}NP^-$	848.528903	FA18:2 <oxo,OOH>	$C_{18}H_{30}O_5$	326.209324
PC(16:0_18:2 <oxo>)	$C_{42}H_{78}O_9NP$	771.54142	[M+ HCOO]−	$C_{43}H_{79}O_{11}NP^-$	816.539074	FA18:2<oxo>	$C_{18}H_{30}O_3$	294.219495
PC(16:0_18:2 <2OH>)	$C_{42}H_{80}O_{10}NP$	789.551984	[M+ HCOO]−	$C_{43}H_{81}O_{12}NP^-$	834.549639	FA18:2<2OH>	$C_{18}H_{32}O_4$	312.23006
PC(16:0_18:2 <OH,OOH>)	$C_{42}H_{80}O_{11}NP$	805.546899	[M+ HCOO]−	$C_{43}H_{81}O_{13}NP^-$	850.544553	FA18:2 <OH,OOH>	$C_{18}H_{32}O_5$	328.224974
PC(16:0_18:2 <OH>)	$C_{42}H_{80}O_9NP$	773.55707	[M+ HCOO]−	$C_{43}H_{81}O_{11}NP^-$	818.554724	FA18:2<OH>	$C_{18}H_{32}O_3$	296.235145

according to the requirements of the specific MS platform in use (*see* **Note 6**). The exported file provides all essential fields to support this process, allowing researchers to select only the relevant parameters, such as the neutral formula with assigned charge, or simply the *m/z* value, for instrument-specific inclusion list generation. Adduct assignment is automated based on lipid class and chemical modifications.

For example, oxPC is assigned the $[M+HCOO]^-$ adduct by default, while PC species containing a terminal carboxylic acid (–COOH) are automatically assigned $[M-H]^-$. In addition to the information on oxidized lipid precursor ion, the file includes a information (elemental composition, neutral formula, and exact mass) for the main product ion corresponding to the oxidized acyl chain, facilitating the preparation of transition list for targeted analysis (e.g., MRM) or enabling rapid inspection of PRM-derived MS/MS spectra upon data acquisition. All data provided in the inclusion list CSV file are derived directly from the predicted epilipidome JSON output described above, offering developers the flexibility to generate customized tables or integrate this information into other workflows.

3.2.3 Semi-Targeted Detection of Oxidized Lipids Using LPPtiger2 Generated Inclusion list

Experimental details of the LC–MS/MS analysis employing LPPtiger2-predicted oxidized lipids and the corresponding inclusion list-based data-dependent acquisition (semi-targeted) workflow are available in the published protocol [4].

3.3 Identify Epilipids Using LPPtiger2

3.3.1 Conversion of Raw MS Files into mzML Format

LPPtiger2 supports two primary input formats for downstream analysis: the generic mzML format, converted from individual raw LC–MS files, and pre-processed CSV files exported from Lipostar2 [8, 10] (see Fig. 5). mzML files can be generated using the MSConvert module from ProteoWizard with default settings. Selecting the 32-bit option is recommended to reduce mzML file size without compromising data quality. Alternatively, Lipostar2 users can bypass mzML conversion by directly exporting a CSV file containing preliminarily identified oxidized lipids, which is compatible with LPPtiger2. Detailed instructions for exporting Lipostar2-compatible files for use with LPPtiger2 are available in the Lipostar2 user guide, which can be found in the Lipostar2 installation folder.

3.3.2 LPPtiger2-Based Annotation of Oxidized Lipids

For lipid annotation, the converted spectra and the corresponding predicted epilipidome JSON files are required. The following set of parameters must be configured (*see* **Note 7**):

Spectra

If mzML tab is selected, a mzML file is required. If the Lipostar2 (.csv) tab is selected, then the exported CSV Lipostar2 file needs to be provided.

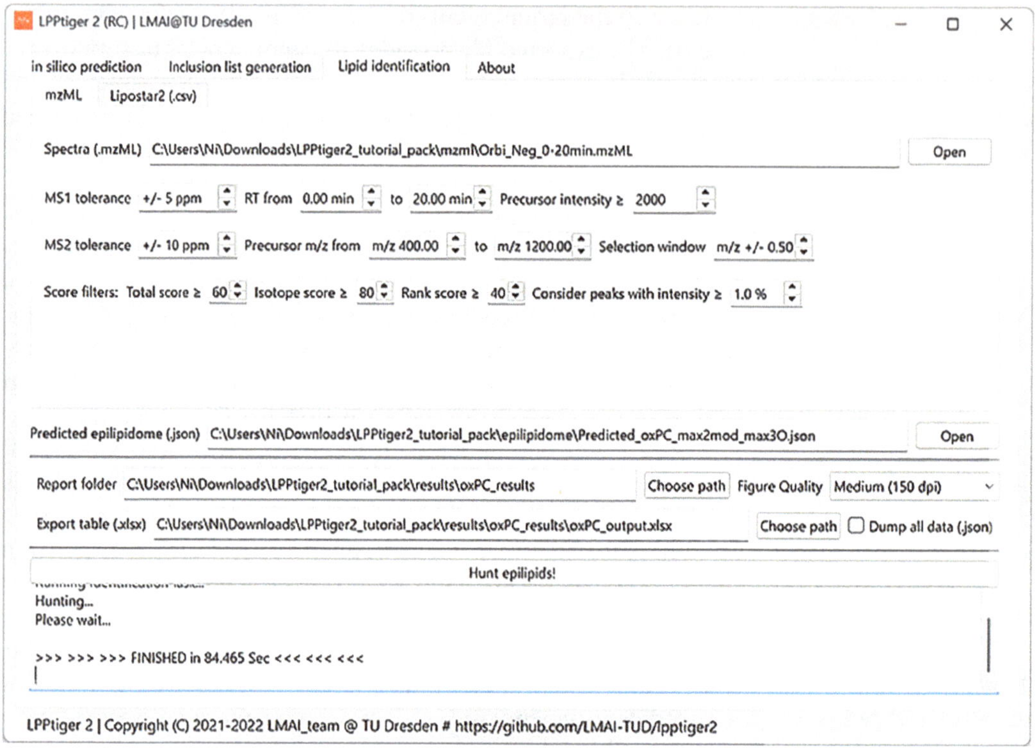

Fig. 5 The LPPtiger2 interface for annotation of oxidized complex lipids

MS1 tolerance, MS2 tolerance

The allowed mass tolerance, defined in parts per million (ppm), specifies the acceptable deviation between experimental and predicted *m/z* values for both precursor and product ions during spectral matching

RT range, Precursor m/z range

The allowed range of retention times and precursor *m/z* to be considered for the annotation.

Precursor intensity, Selection window

These parameters control the precursor ion intensity threshold at the MS1 level and enable LPPtiger2 to associate the MS2 spectra with corresponding precursor *m/z* in the MS1 level.

Total score, Isotope score, Rank score

These scores are applied to filter annotation results based on LPPtiger2's five-criteria scoring system, which integrates multiple spectral and structural features to assess annotation confidence. For more information on the scoring algorithms, please refer to the original publication by Ni et al. [11].

Consider the peak with intensity

Only MS2 peaks with relative signal intensities exceeding a defined threshold are considered for lipid identification and contribute to the scoring process.

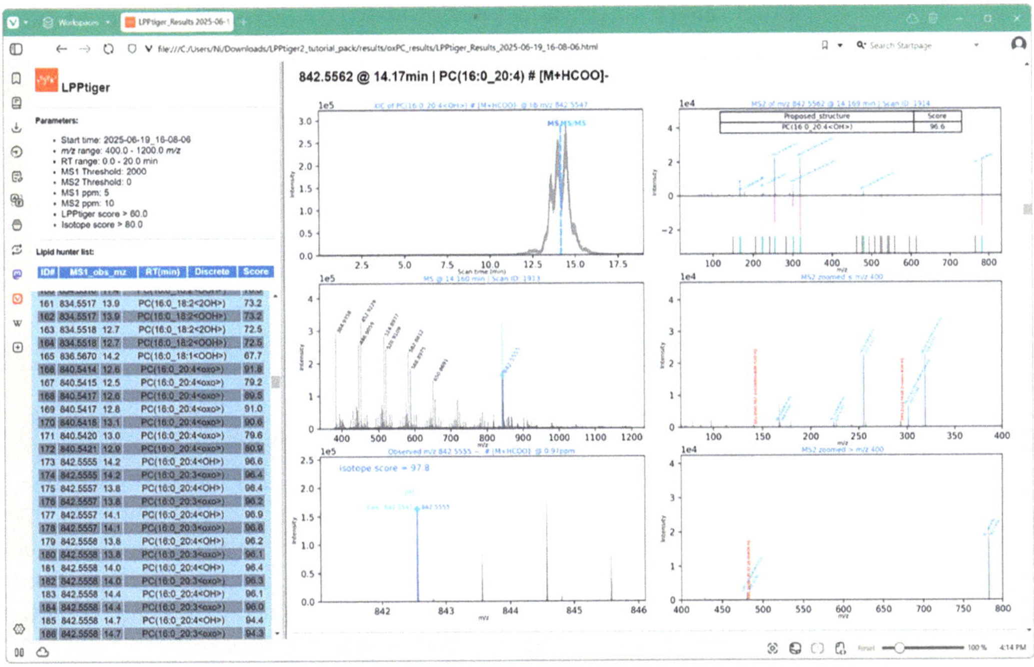

Fig. 6 Example of the LPPtiger2 interactive HTML output. The main parameters and list of identified lipids are displayed on the left, while six-panel images of annotated XIC, MS, and MS/MS spectra appear on the right. Users can click on lipid names to view the corresponding spectra

Figure quality

Users can select from three image quality settings—Low (96 DPI), Medium (150 DPI), and High (300 DPI), to suit the requirements of different projects and publication standards.

Dump all data (.json)

When enabled, this option allows LPPtiger2 to export a comprehensive JSON file containing all detailed identification scores and matched MS/MS peaks for each annotated lipid. Depending on spectral quality and the number of identifications, the resulting file may range from 100 MB to several gigabytes in size. This feature is primarily intended for advanced users developing custom data analysis pipelines or re-plotting spectra. The exported JSON provides full access to spectral data, identified fragment peaks, scoring metrics for each identification, and output image data with original *m/z* values and intensities. This level of detail ensures complete transparency in the identification process (Fig. 6).

After setting the path for input and output files, users can click on the "**Hunt epilipids!**" button to start the identification.

3.4 Review of Epilipid Annotations

LPPtiger2 produces an interactive HTML output alongside a corresponding results table in.xlsx format (Fig. 6). Identification results can be examined using any web browser, enabling users to explore annotated information of extracted ion chromatogram (XIC), MS, and MS/MS spectra across six-panel images within the HTML report (*see* **Note 8**).

In addition to the HTML files, a corresponding output table is provided in.xlsx format. This table has rich information distributed in following columns: lipid notation, proposed structures, lipid class, neutral formula, formula of detected ion, adduct, library *m/z*, observed MS1 *m/z*, mass accuracy (ppm), observed MS1 intensity, MS/MS precursor *m/z*, MS/MS scan time, scan ID, total score, isotope score, rank score, spectral match score, fingerprint score, signal-to-noise ratio, figure reference (file name of the corresponding six-panel-image), and identified peaks.

4 Notes

1. Criteria for selecting representative lipids for predictive analysis are (1) limit selection to a maximum of 50 lipids per lipid subclass; (2) prioritize lipids based on their relative abundance and biological significance; and (3) lipid nomenclature should adhere to established shorthand notation guidelines, with a modification notation enclosed within brackets.
2. The fatty acid (FA) list file comprises three columns: the name of fatty acid in the lipid shorthand notation format, the SMILES representations indicating possible double-bond positions, and annotations such as ω-3 and ω-6 isomers. The SMILES representations are essential for accurate prediction of epilipids, particularly truncated oxidized lipids. Users must ensure that all FA residues in the selected representative lipid list are included in the FA list file. Sample FA list files are provided in the "config" subfolder within the LPPtiger2 directory.
3. The maximum total oxygen count must be equal to or greater than the maximum number of oxidation sites. LPPtiger2 automatically restricts modification sites to not exceed the number of double bonds within the lipids.
4. The comprehensive predicted lipid search space derived from selected fatty acids in the configuration can become exceedingly large, compromising identification specificity [4]. Therefore, refining the predicted lipid space by focusing on selected lipid precursors markedly reduces the search scope and enhances the accuracy of epilipid identification.
5. The exported JSON file encapsulates comprehensive information on predicted epilipids and can be utilized for downstream analyses, including molecular networking. However, the SMILES strings within the JSON serve primarily as illustrative representations for calculating molecular formulas and m/z values, and should not be interpreted as definitive structures of the identified epilipids.

6. To enhance compatibility across different instruments, information on both neutral and charged precursors is provided; users must further reformat the table to meet specific instrument requirements. For certain platforms, it is advisable to remove isomeric lipids based on their molecular formula to generate a list of unique formulas or m/z values.
7. Identification parameters must be carefully aligned with the corresponding data acquisition settings, particularly MS1 tolerance, MS2 tolerance, and selection window parameters, as these directly influence the performance of identification algorithms.
8. Manual revision can enhance identification accuracy by evaluating peak shapes and intensities to refine threshold settings, followed by re-running the identification to achieve improved results. Isomeric lipid species bearing distinct modifications may be confidently distinguished through this process. Users are encouraged to prepare the identification table using the summary provided alongside the interactive HTML report.

Acknowledgments

Work in the Fedorova lab is supported by “Sonderzuweisung zur Unterstützung profilbestimmender Struktureinheiten” by the SMWK to TUD, TG70 by Sächsische Aufbaubank and SMWK, the measure is co-financed with tax funds on the basis of the budget passed by the Saxon state parliament, Deutsche Forschungsgemeinschaft (FE 1236/5-1, FE 1236/8-1, TRR 412/1 – 535081457), and Bundesministerium für Bildung und Forschung (031L0315A, DEEP_HCC and 01EJ2205A, FERROPath).

References

1. Penkov S, Fedorova M (2024) Membrane epilipidome—Lipid modifications, their dynamics, and functional significance. Cold Spring Harb Perspect Biol 16 7):a041417
2. Ni Z, Goracci L, Cruciani G, Fedorova M (2019) Computational solutions in redox lipidomics–Current strategies and future perspectives. Free Rad Biol Med 144:110–123
3. Schebb NH, Kampschulte N, Hagn G, Plitzko K, Meckelmann SW, Ghosh S, Joshi R et al. (2025) Technical recommendations for analyzing oxylipins by liquid chromatography–mass spectrometry. Sci Signal 18:887). eadw1245
4. Criscuolo A, Nepachalovich P, Garcia-Del Rio DF, Lange M, Ni Z, Baroni M, Cruciani G, Goracci L, Blüher M, Fedorova M (2022) Analytical and computational workflow for in-depth analysis of oxidized complex lipids in blood plasma. Nat Commun 13 1):6547
5. Allen-Redpath K, Aldrovandi M, Lauder SN, Gketsopoulou A, Tyrrell VJ, Slatter DA, Andrews R et al. (2019) “Phospholipid membranes drive abdominal aortic aneurysm development through stimulating coagulation factor activity. Proc Natl Acad Sci 116 16):8038–8047
6. Ni Z, Sousa BC, Colombo S, Afonso CB, Melo T, Pitt AR, Spickett CM et al. (2019) Evaluation of air oxidized PAPC: A multi laboratory study by LC-MS/MS. Free Rad Biol Med 144:156–166
7. Lu J, Guo S, Xue X, Chen Q, Ge J, Zhuo Y, Zhong H et al. (2017) Identification of a novel series of

anti-inflammatory and anti-oxidative phospholipid oxidation products containing the cyclopentenone moiety in vitro and in vivo: implication in atherosclerosis. J Biol Chem 292 13):5378–5391
8. Damiani T, Bonciarelli S, Thallinger GG, Koehler N, Krettler CA, Salihoglu AK, Korf A et al. (2023) Software and computational tools for LC-MS-based epilipidomics: challenges and solutions. Anal Chem 95 1):287–303
9. Liebisch G, Fahy E, Aoki J, Dennis EA, Durand T, Ejsing CS, Fedorova M et al. (2020) Update on LIPID MAPS classification, nomenclature, and shorthand notation for MS-derived lipid structures. J Lipid Res 61 12):1539–1555
10. Goracci L, Tortorella S, Tiberi P, Pellegrino RM, Di Veroli A, Valeri A, Cruciani G (2017) A comprehensive platform-neutral cheminformatics tool for lipidomics. Anal Chem 89 11):6257–6264
11. Ni Z, Angelidou G, Hoffmann R, Fedorova M (2017) LPPtiger software for lipidome-specific prediction and identification of oxidized phospholipids from LC-MS datasets. Sci Rep 7 1):15138

Chapter 14

Comprehensive LC–MS Metabolomics Data Processing with *notame* R/Bioconductor Package

Ville Koistinen, Retu Haikonen, Atte Lihtamo, Vilhelm Suksi, Olli Kärkkäinen, Leo Lahti and Kati Hanhineva

Abstract

Liquid chromatography–mass spectrometry (LC–MS) is widely used in metabolomics. Raw LC–MS data is relatively complex, consisting of molecular features originating not only from unique metabolites but also from redundant adducts, in-source fragments, artifacts, and impurities. It is also prone to signal intensity drift during long sequences, missing values, and false positives in statistical tests. To tackle these instrument-related and data-dependent challenges with robust pre-processing and quality evaluation tools, we have developed the notame R package bundle, which recently became available as a R/Bioconductor release and now supports SummarizedExperiment data format. It pre-processes LC–MS metabolomics data by correcting signal intensity drift, flagging potential low-quality and contaminant features, imputing missing values, and clustering features likely originating from the same metabolite. Most of the functions include default recommended values that can be modified by the user. Univariate and multivariate statistics with parametric and non-parametric alternatives and false discovery rate can be performed with notameStats package. Results and data visualizations, such as quality control figures, PCA, heatmaps, volcano plots, and feature-wise graphs, are available in notameViz package. Together, these packages contribute to a complete metabolomics data analysis workflow, connecting the phases between signal detection/alignment and metabolite identification while producing publication-ready illustrations.

Keywords Preprocessing, Statistics, Quality assurance, Data visualization, Programming

1 Introduction

Metabolomics data produced by liquid chromatography–mass spectrometry (LC–MS) instruments is rich in signals originating from various ions generated from metabolites, solvent contaminants, and artifacts from the instrument [1]. Long sequences of samples are also prone to drift in signal intensity between and within analytical batches, and necessitate careful monitoring and correction prior to entering any statistical evaluation [2]. The statistical models may be negatively impacted by missing values, either from the true missingness of the metabolite in certain samples or from missed signals during the signal

Michael Witting (ed.), *Computational Methods and Data Analysis for Metabolomics*, Methods in Molecular Biology, vol. 3063,
https://doi.org/10.1007/978-1-0716-5452-1_14,

detection process, for instance [3]. The large number of detected and aligned molecular features also affects the results from the statistical models because the risk of false positives needs to be controlled by multiple-testing correction, such as false discovery rate (FDR) [4].

Taken together, the characteristic complexities in LC–MS metabolomics data necessitate data analysis procedures beyond signal detection and alignment to allow an unbiased biological interpretation of the data. Several measures can be taken to ensure the best possible data quality. Preparing extraction blanks and quality control (QC) samples pooled from the experimental samples are key processes during sample preparation because they will enable both the quality checkup and correction of signal drift during the data analysis. Optimizing the sample preparation, configuring the chromatography, and setting the parameters for data acquisition and signal detection naturally play a crucial role in creating an outcome that has the optimal potential to be polished into a reliable and reproducible dataset [5–7].

Here, we describe *notame*, a metabolomicsworkflow [8, 9] and data analysis tool written into a package bundle [10] that operates in the R programming environment, pre-processes aligned LC–MS data, performs quality assessment, calculates statistics, and generates visualizations from the quality assessment and statistical results (Fig. 1). We will go through the wet-lab and data-related prerequisites for using *notame*, pre-processing steps including data quality control,

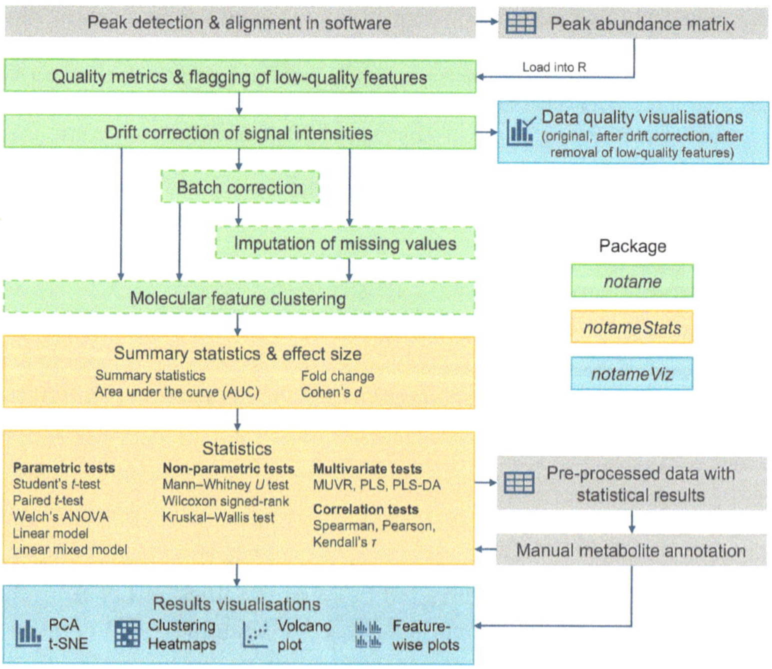

Fig. 1 General workflow of the metabolomicsdata processing in *notame* in context with the other necessary tools, such as the software for peak detection and alignment spreadsheet processing, and the manual and computational annotation of metabolites

signal drift correction and imputation of missing values, the array of statistical models (with *notameStats*), and visualization possibilities (with *notameViz*). For detailed instructions on the practical usage of these R packages, we refer to the documentation available online [10].

2 Considerations and Prerequisites for Pre-Processing

Data quality assurance needs to be considered already during the sample preparation protocol by preparing the extraction blank and QC samples [11]. The extraction blank is essentially the same extraction solvent used to extract the metabolites, and it goes through the same sample preparation steps as the experimental samples using the same containers, pipettes, filters, etc. The purpose of the extraction blank is to flag solvent contaminants and other non-biological artifacts during data processing. Solvent contaminants are typically present in the extraction solvent after contact with plastics.

QC samples are prepared by pooling aliquots of all the experimental samples (or sometimes by acquiring a commercial QC sample) to obtain a single representative sample of the study [11, 12]. The QC sample is used by *notame* to assess data quality, flag molecular features with detection issues (such as carryover ions), and correct the signal drift in relatively long sequences (more than 50 to 100 injections). At least four QC injections are needed for *notame* to perform these tasks, and they need to be injected before and after every sequence of samples (Fig. 2). We recommend injecting a QC sample every 10–12 experimental samples [8], although published methods have reported injecting a QC sample every 4–15 samples [13].

In case the LC–MS run consists of several hundreds or thousands of samples, it is necessary to run the samples in several analytical batches, each batch usually consisting of 100–300 samples [8, 14]. In long sequences, more than one vial/well may be needed for the QC sample injections. If between-batch variation is expected due to a variable sample material, a batch-specific QC sample may be

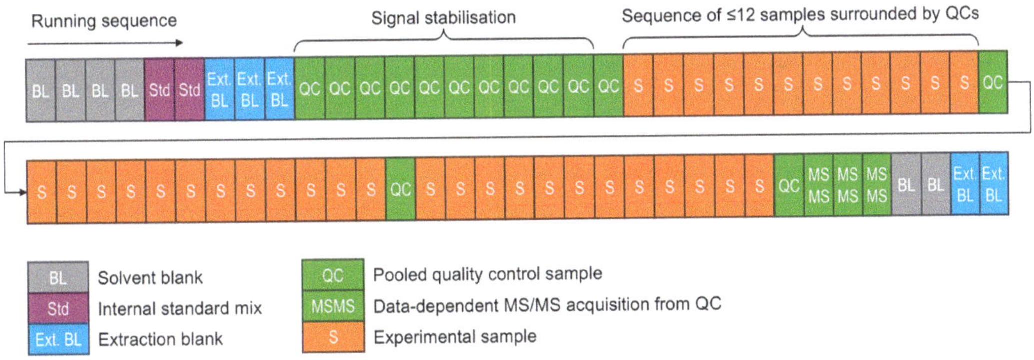

Fig. 2 An example LC–MS sequence of 36 randomized samples, including suggested solvent blank, extraction blank, internal standard, and QC injections

prepared by pooling the experimental samples from a single batch to be used for the signal drift correction [9].

It is noteworthy that in studies containing highly variable samples (e.g., from different species / biological material) or several study or treatment groups, there is a chance that a metabolite only present in a fraction of samples gets diluted in the QC sample to the extent that the peak is no longer reproducibly detected from the QC sample, resulting in being perceived as a low-quality signal during the data quality assessment (see Chap. 3). Such metabolites may nevertheless be among the most relevant for the study question. Therefore, in these types of study setups, it is not advised to discard any molecular features based on the flags given during data preprocessing but rather to consider the flags as warnings of potential issues in data quality.

The suitable input data for *notame* is a molecular feature abundance data matrix containing features as rows and samples as columns, the first columns being reserved for molecular feature metadata and the first rows for sample metadata (Fig. 3). With some modifications, this kind of output should be obtainable from most vendor and open-source metabolomics analysis software Currently, *notame* directly supports the alignment output from MS-DIAL [15] in a tab-separated.txt,.csv, or.xlsx file. Some of the metadata is mandatory for pre-processing: the molecular feature metadata must include the mass-to-charge and retention time values (for generating unique feature IDs) and the analytical mode (if more than one in the same data matrix). The sample metadata must contain the injection sequence order and a row that specifies the QCs (named "QC" by default). Because *notame* can also detect potential solvent contaminants, it is advisable to keep the data from the extraction blank injections in the

			Class	Blank EXT	Blank EXT	Blank EXT	QC	A	B	B	A
			File type	Blank	Blank	Blank	QC	Sample	Sample	Sample	Sample
			Injection order	1	2	3	4	5	6	7	8
			Batch ID	1	1	1	1	1	1	1	1
Mode	Alignment ID	Average Rt(min)	Average Mz	Filename_1	Filename_2	Filename_3	Filename_4	Filename_5	Filename_6	Filename_7	Filename_8
RP_pos	1	7.703	100.0213	381	652	513478	513478	633377	606602	701414	80532
RP_pos	2	4.212	101.0793	75952	64925	72967	830284	370514	1625446	705233	469641
RP_pos	3	12.200	103.1045	0	0	0	409235	530718	231911	573868	700366
RP_pos	4	8.052	105.0034	489165	493120	426981	276555	155464	407091	59225	256970
RP_pos	5	12.514	120.0473	2946	7953	1098	53177	769442	841850	0	642994
HILIC_pos	1	0.949	101.0211	209684	197359	267349	742067	912350	99276	201481	602960
HILIC_pos	2	3.184	110.0618	4629	6730	5207	359368	907818	134559	126817	880186
HILIC_pos	3	2.472	118.0626	4073	3772	547	232021	232021	682210	690570	53470
HILIC_pos	4	2.014	126.1025	364	2496	2553	27256	42548	157130	705456	427825
HILIC_pos	5	6.492	130.0273	968	1457	3090	394894	473404	618389	179585	503437

Fig. 3 Example of a simple data matrix structure suitable for pre-processing with *notame*, with sample metadata (blue), molecular feature metadata (green), and molecular feature abundances as raw peak areas (yellow) aligned across the extraction blank, QC, and experimental samples. The package will be looking for the corner where the molecular feature abundance data begins, which is why this data matrix structure should be followed. The metadata presented here is the minimum requirement; the data may contain as many additional sample metadata rows and feature metadata columns as necessary

data matrix. Signal abundance data from MS/MS samples should be removed at this point because of the low number of MS^1 data points used for the peak area calculation.

The following steps contain basic requirements for preparing metabolomics data for preprocessing. The up-to-date script is available at https://github.com/hanhineva-lab/notameTutorials.

1. Generate a peak abundance matrix from aligned metabolomics data (acquired from the software handling the peak detection and alignment and save it in an Excel format or as a tab-separated text or CSV file with a structure according to Fig. 3. Combine the abundance tables from the different modes and add a column specifying the analytical mode, such as `Mode` (containing the modes, e.g., RP_pos, HILIC_neg). Remove columns containing abundances from solvent blanks (keep extraction blanks), internal standards, and MS/MS injections, and any average or SD values.
2. Install the latest Bioconductor release of *notame*, *notameStats*, and *notameViz*. The versions of *notame*, *notameStats*, and *notameViz* used to run the commands described in this chapter are 1.0.1, 1.0.0, and 1.0.0, respectively, under R version 4.5.2 and Bioconductor version 3.22. To test the script, it is possible to use the `toy_notame_set` data object available within the package.

```
if (!requireNamespace("BiocManager", quietly =
TRUE)) install.packages("BiocManager")
BiocManager::install("notame")
BiocManager::install("notameViz")
BiocManager::install("notameStats")
```

3. Load the R packages required to run the scripts in this workflow and set the path for the working directory. Create a `figures` subfolder for the visualizations.

```
library(notame)
library(notameViz)
library(notameStats)
library(dplyr)
library(tidyr)
ppath <- "C:/File_path/"
setwd(ppath)
dir.create("figures")
```

4. (a)
Load your dataset containing all the modes (replace `data.xlsx` with the correct file name) into R while ensuring that the `split_by` argument corresponds with the name of the column containing the analytical mode information. If the data consists of only one analytical mode, this needs to be

specified with the name argument instead of using split_by, e.g., name = HILIC_pos. This command also generates the SummarizedExperiment objects from the data.

```
se <- import_from_excel(file = file.path(ppath,
"data.xlsx"), sheet = 1, split_by = "Mode")
```

(b) If the modes (in this example, four modes named HILIC_neg.xlsx, etc.) are in separate spreadsheets, load them separately instead while ensuring they have the same run sequence (the same number and order of samples in the columns). It is recommended to merge them as well at this point into one SummarizedExperiment object.

```
modesList <- c("HILIC_neg", "HILIC_pos", "RP_
neg", "RP_pos")
modes <- list() for (mode in modesList)
{   # Read single mode, set it in the list
modes[[mode]] <-
import_from_excel(file = file.path(ppath, mode,
".xlsx"), name = mode) }
se <- merge_notame_sets(object = modes)
```

(c) Alternatively, to test the script with the in-built dataset, load toy_notame_set into the environment.

```
data(toy_notame_set)
se <- toy_notame_set
```

5. Classify the data as (metabolite) abundances. Create any necessary missing columns for the pheno and feature data, clean the object, and split it into a list of SummarizedExperiment objects by mode using fix_object. The split into different analytical modes (if present) is recommended for the data preprocessing steps.

```
names(assays(se)) <- "abundances"
modes <- fix_object(object = se, split_data = TRUE,
assay.type = "abundances")
```

6. Ensure that the objects contain all the data in the correct form. The output from names should be a list of the modes, and the output from sapply should be the same list with "SummarizedExperiment" as the class label.

```
names(modes) sapply(modes, class)
```

3 Data Quality Assessment and Flagging of Low-Quality and Contaminant Features

3.1 Calculation of Quality Metrics

The quality of metabolomics data should be routinely assessed because of the "rubbish in, rubbish out" principle. The quality checkup may reveal issues in the data that may require consideration during the results interpretation, corrective measures, or even the samples to be re-run in the instrument. The data quality of each molecular feature is assessed by *notame* with three main quality metric calculations originally defined by Broadhurst et al. [12]: the detection rate, relative standard deviation (RSD), and *D*-ratio. The detection rate is calculated before drift correction, and the quality metrics after the drift correction (Chap. 4) to assess whether the drift correction has improved the quality of some of the features to an acceptable level. The detection rate is determined in two ways: within the QC samples (default recommended requirement: detected in ≥ 70% of the QC samples) and within the experimental samples, divided into user-defined study groups (default recommended requirement: detected in ≥ 50% of samples within at least one study group). The detection rate function in *notame* is called with the command `flag_detection(object)`, where `object` is replaced with the name of the data object that contains the metabolomics data.

The RSD and *D*-ratio are calculated as their parametric and non-parametric (robust) versions (RSD* and *D*-ratio*) because the relative abundances in metabolomics data do not always have a linear relationship with the actual concentrations. RSD, in this case, is the ratio of the standard deviation and mean of the QC sample peak areas. The RSD parameter in *notame* is given as a direct ratio, whereas otherwise, RSD is often reported as a percentage. The formula for calculating the non-parametric RSD* is shown in below.

$$RSD^{*}_{QC} = \frac{1.4826 \times MAD_{QC}}{median_{QC}}.$$

Here, the standard deviation is replaced by a scaled median absolute deviation (MAD), which is defined as $MAD = median(|m_{QC} - median(m_{QC})|)$, multiplied BY a scaling factor, and the mean is replaced by the median [12].

The *D*-ratio (from ***dispersion ratio***) describes the statistical dispersion (spread) of the QC samples in relation to that of the experimental samples. The parametric *D*-ratio is the standard deviation of the peak areas in the QC samples divided by the standard deviation of the peak areas in the experimental samples. In the non-parametric *D*-ratio*, the standard deviation is replaced by MAD for both QC and experimental samples (below). Again, the *D*-ratio and its non-parametric form are given as ratios but are often otherwise used as percentages.

$$D\!-\!ratio^{*} = \frac{MAD_{QC}}{MAD_{sample}}.$$

3.2 Flagging Low-Quality and Contaminant Features

A molecular feature that passes the detection rate criteria is considered of good quality if it fulfills at least either of the following (default) criteria: (1) RSD* < 0.2 and *D*-ratio* < 0.4 and (2) RSD < 0.1, RSD* < 0.1, and *D*-ratio < 0.1. The quality metrics are calculated for every molecular feature and reported in the data matrix output. Those molecular features that do not pass the detection rate criteria in the QC samples are flagged in a separate column (*Flag*) as *Low QC detection*, and those not passing the detection rate criteria in the study groups are flagged as *Low group detection*. Molecular features not fulfilling the other quality metrics are flagged as *Low quality*. The low-quality molecular features are retained in the final pre-processed data matrix but excluded from the first round of the missing value imputation to prevent the low-quality signals from introducing noise to the missing value imputation. The function to calculate the quality metrics is called with the command `flag_quality(object)`, and the default criteria described above can be modified with the argument `condition`.

The potential contaminant features in the metabolomics data originate from various sources, such as additives during sample preparation, ghost peaks from derivatization, impurities in solvents and other reagents, and contaminants (such as plasticizers) from the laboratory equipment [16]. These features appear as signals with a relatively high abundance in the extraction blanks compared to the instrument noise level and/or the experimental samples. The metabolomics community has yet to reach a consensus on the exact criteria for flagging or removing molecular features that are likely contaminants [12]. Based on a survey conducted by the Metabolomics Quality Assurance and Quality Control Consortium (mQACC) [17], González-Domínguez et al. recently suggested that a molecular feature is flagged as a potential contaminant if its mean abundance in the experimental samples is not more than three times the mean abundance in the blanks [16]. MS-DIAL has an optional in-built blank filter, and by default, a feature gets flagged if its maximum abundance across the experimental samples is less than five times the average abundance in the blank samples [15]. As Fig. 4 shows, the selection of the descriptor in calculating the sample-to-blank ratio can have a considerable impact on whether a molecular feature gets flagged. This may become crucial when the feature is detected at high levels only in a small percentage of the experimental samples, which results in mean and median values that may become too low to fulfill criteria based on these descriptives. However, molecular features that exhibit high levels in a single study group may be among the most relevant metabolites for the study question and therefore should not be dismissed.

In *notame*, the blank filter is called with the command `flag_contaminants(object)`, and it allows the user to define the descriptives (`mean`, `median`, or `max` for `blank_type` and

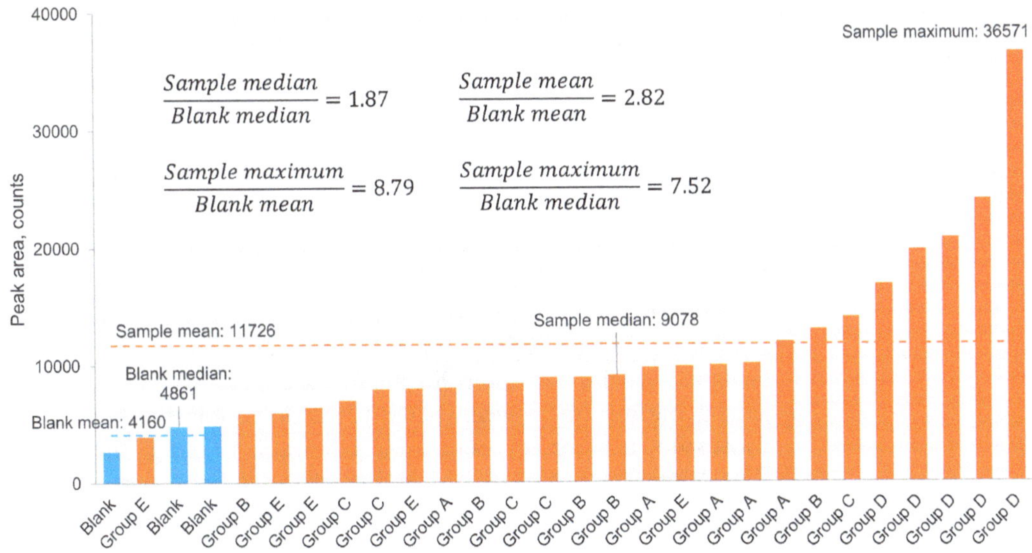

Fig. 4 Mock data on the abundance of one molecular feature in blanks and 25 experimental samples, divided into five study groups, and the various descriptors calculated from the samples. A relatively high abundance of the molecular feature in a single study group, even though indicating that it at least partially originates from the biological samples, means that the feature gets flagged as a contaminant if the sample median/blank median > 3 or sample mean/blank mean > 3 is used as the criterian

`sample_type`) used for calculating the sample-to-blank ratio, as well as the threshold for the minimum ratio for the feature *not* to be flagged as a contaminant (with the argument `flag_thresh`). The potential contaminants are, by default, flagged as *Contaminant* in a separate *Flag* column in the output data matrix and not removed from the data.

4 Drift and Batch Correction

LC–MS suffers from a characteristic drift in measured responses, such as signal intensity, retention time, and mass accuracy, because the samples are in direct contact with the chromatographic column and other components of the measurement platform [14]. The issue may become evident already in sequences of tens of injections, requiring the use of QC samples. Whereas the retention time and mass-to-charge drift often remain within acceptable limits to be handled by the signal detection and alignment functions, the potential drift in signal intensity needs a dedicated normalization method to be accounted for. The simplest methods to normalize metabolomics data are based on the total ion chromatogram (TIC) or individual metabolites considered stable enough throughout the samples; however, they are not recommended because molecular features tend to exhibit individual drift patterns, and these methods do not consider potential differences in the overall metabolite composition and intensity in different study groups [2]. Modern drift correction techniques are based on curve fitting, such as LOESS [14] or cubic spline regression [2] applied on the QC samples, which work as "beacons" based on which the intensities of the experimental samples are then normalized.

If the sample sequence contains more than 100–300 samples, it usually becomes necessary to analyze them in several batches (uninterrupted sequences), creating an additional challenge for managing data integrity [13]. While certain approaches treat the analytical batches separately for the whole data analysis process [18], thus avoiding the between-batch variation problem. However, in most study designs, it is nevertheless desirable to merge the batches for optimal statistical power [19] and the metabolite identification process. Several methods for batch correction exist for metabolomics data, and many of them rely on QC samples for correcting the signal abundances, including R packages *BatchCorrMetabolomics* [13] and *batchCorr* [2]. As suggested in Chap. 2, preparing separate within-batch and between-batch QCs will ensure that any potential differences between samples in each batch do not interfere with the batch correction procedure.

In *notame*, a feature-wise cubic spline regression is fitted to the peak areas of the QC samples [8] to correct for the signal intensity drift within one analytical batch. A smoothing parameter is included to avoid overfitting the model in case of individual outlier QC samples. The smoothed regression line is then used to correct the signals of the experimental samples. The drift correction is performed in *notame* by running the function `correct_drift(object)`. By default, the algorithm applies the correction on $\log_{10}$-transformed abundances, which better follow the assumptions of the cubic spline regression model and do not result in negative corrected values, which would reject the correction result. The smoothing parameter, by default, is chosen by leave-one-out cross-validation for each molecular feature from a range of 0.5–1.5. An optional quality control is performed by setting `check_quality = TRUE` and `condition = "RSD_r < 0 & D_ratio_r < 0"` (for example), which means that the drift correction result is only kept if (in this example) both RSD* and *D*-ratio* decrease during the drift correction. In case the sequence has several batches and a batch-specific QC was prepared, the batch-specific QC injections can be used for the drift correction instead of the QC pooled from the whole study samples.

An array of visualizations describing the data quality and metabolite profile overview is generated in *notame* with the `visualizations()` function, which can be applied to any object during the pre-processing to assess and compare the impact of each processing stage (i.e., drift and batch correction merging the analytical modes) on the metabolomics data. The exact types of visualizations created by this function depend on the content and type of the object. The following visualizations are generated regardless of the object type:

- histograms of the distribution of the quality metrics and flags,
- Boxplots of the samples (containing all peak areas) in the injection order,
- PCA scores plot of the samples colored by injection order,
- t-SNE plot of the samples colored by injection order,
- Dendrogram of the samples arranged by hierarchical clustering, and
- Heatmap of the Euclidean distances between samples arranged by hierarchical clustering.

In addition, the following visualizations are generated under certain conditions:

- Hexagonal binned (hexbin) PCA and t-SNE plots colored by the injection order if the object contains more than 60 samples,
- Density curves of the Euclidean distances between the experimental and QC samples, respectively, if the object contains QC samples, and
- Histograms of the *p*-values from a linear regression of features against injection order in the experimental and QC samples, respectively, if the object contains QC samples.

Furthermore, if the object has a `group`, `time`, and/or `subject` column, the PCA, t-SNE, and sample boxplots will be modified to accommodate these metadata. These visualizations are practical for visually inspecting whether the drift correction and flagging of low-quality features have indeed improved the data quality. In addition, they can reveal quality issues that are beyond the algorithms to correct, such as the evaporation of samples located on a certain plate, which causes an overall increase in signal intensities in a group of samples without affecting the adjacent QC samples. Figures 5 and 6 show examples of the data quality visualizations generated before and after the drift correction and flagging of features based on their quality.

7. Take several processor cores into use while leaving two cores for other tasks. This will reduce the computing time for the quality check, drift correction, imputation, and statistics. Once the computationally intensive tasks are done, release the cores by running `register(SerialParam())`.

```
library(BiocParallel) register(SnowParam(workers=
snowWorkers() - 2))
```

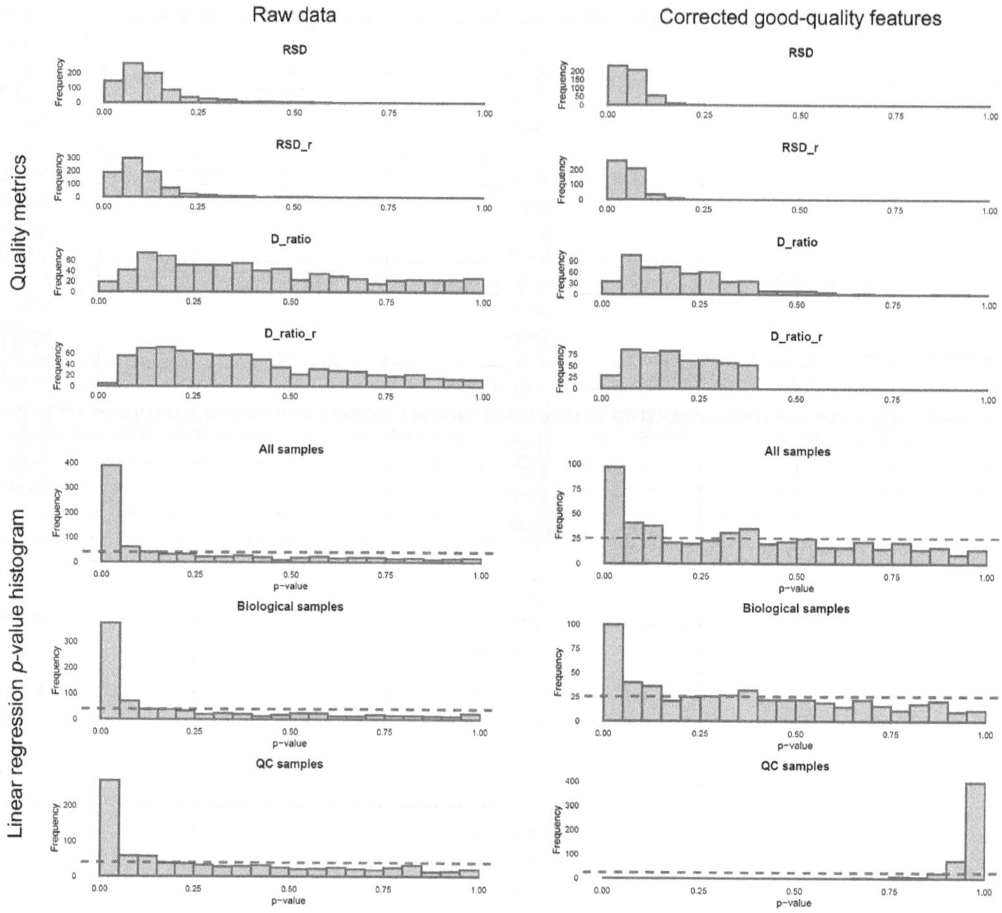

Fig. 5 Histograms of the quality metrics (from `plot_quality()`) and the *p*-values from the linear regression model results (molecular features against injection order; from `plot_p_histogram()`) before and after the quality assessment and drift correction of a test dataset in the HILIC positive mode. Generally, the quality assessment and drift correction should move the quality metrics distribution toward lower values, and the *p*-value distribution of the biological samples closer to the uniform distribution, and the *p*-value distribution of the QC samples toward 1

8. Perform the contaminant flagging, quality check, and drift correction on each mode separately within a loop and create visualizations of the various phases of the process in the `figures` folder of the working directory. The detection limit is by default set to 0.7 (≥70%) of the QC samples and 0.5 (≥50%) of the experimental samples in at least one group (defined by `Group`). If the study does not contain repeated measurements, the `time` and `id` parameters can be set to `NULL`.

```
processed <- list()
for (i in seq_along(modes)) {
name <- names(modes)[i]
mode <- modes[[i]]
```

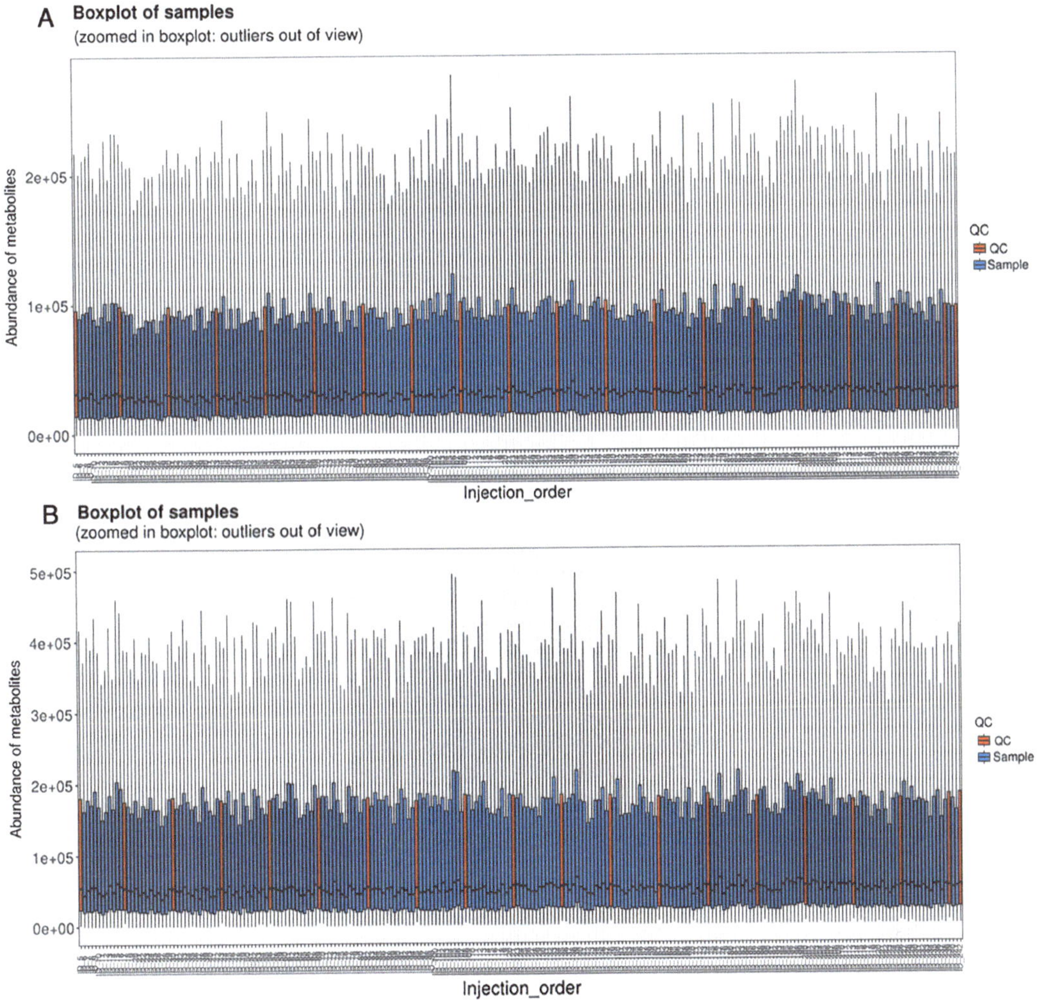

Fig. 6 Boxplots of the molecular feature abundances in all samples analyzed in one of the analytical modes (HILIC positive), arranged based on the randomized injection order. (**A**) Raw abundances of all detected features; (**B**) drift-corrected abundances of the good-quality features

```
mode <- mark_nas(mode, value = 0) # Set all zero
abundances to NA
mode <- flag_detection(mode, group = "Group") # Flag
features with low detection rate
mode <- flag_contaminants(mode, blank_col = "QC",
blank_label = "Blank", blank_type = "mean", sample_
type = "max", flag_thresh = 5) # Flag contaminant
features
mode <- mode[, mode$QC!= "Blank"] # Remove blanks
save_QC_plots(mode, prefix = paste0(path, "figures/",
name, "_ORIG"), perplexity = 5, group = "Group",
time = "Time", id = "Subject_ID", color = "Group")
# Visualize data before drift correction
corrected <- correct_drift(mode) # Correct drift
```

```
save_QC_plots(corrected, prefix = paste0(path,
"figures/", name, "_DRIFT"), perplexity = 5, group
= "Group", time = "Time", id = "Subject_ID", color
= "Group") # Visualize data after drift correction
corrected <- corrected %>%
assess_quality() %>%
flag_quality() # Flag low-quality features
save_QC_plots(corrected, prefix = paste0(path,
"figures/", name, "_CLEAN"), perplexity = 5, group
= "Group", time = "Time", id = "Subject_ID", color
= "Group") # Visualize data after removal of low-
quality features
processed[[i]] <- corrected # Save iterated results
}
```

9. Merge the processed data (modes) and create visualizations of the complete data.

```
merged <- merge_notame_sets(object = processed)
save_QC_plots(merged, prefix = paste0(path,
"figures/_FULL"), group = "Class", time = NULL, id
= NULL, color = "Class")
```

10. Remove the QC sample data and save visualizations once more without the QC samples.

```
merged_no_qc <- drop_qcs(merged)
save_QC_plots(merged_no_qc, prefix = paste0(path,
"figures/FULL_NO_QC"), group = "Group", time =
"Time", id = "Subject_ID", color = "Group")
```

5 Imputation of Missing Values

Missing values are a typical irregularity in nontargeted LC–MS data and can occur for multiple reasons. The value may be missing completely at random (MCAR), which essentially means that there is zero correlation between the missing and observed values [20]. In metabolomics, this can result from random issues or errors during ionization or data acquisition [21]. Missing at random (MAR) means that the missingness depends on the observed molecular features and may be due to inaccuracies in peak detection and deconvolution. These are often handled by several metabolomics software, such as MS-DIAL, through a separate gap-filling process. Missing not at random (MNAR), on the other hand, indicates a potential true biological missingness, which can be explained simply by the absence of the metabolite in a sample or the concentration not reaching the limit of detection, in which case the data often appear as left-truncated [22]. There can be several plausible biological reasons for a large proportion of MNAR values, such as drug or dietary metabolites that exist only in samples from exposed individuals/animals/cells or species-specific metabolites in a dataset containing samples from several species. These biological reasons behind missingness are worth considering when interpreting data with imputed values. In some applications where the aim is to find samples with a less common metabolite pattern, such as in exposomics or pharmaceutical studies, the imputation method (or the use of imputation altogether) needs to be carefully considered and tested to avoid introducing false detections of metabolites, which we have illustrated in Fig. 7.

By default, *notame* performs imputation of missing values by the random forest method [22]. It is accessed by `impute_rf()`, which is a wrapper function around the *missForest* R package. Because the random forest method introduces a random element into the data, the reproducibility of the imputation step can be ensured by setting a seed number by `set.seed()`. A simple imputation of a value other than zero can be performed with `impute_simple()`. In our typical workflow, where metabolites of interest are expected to be present in the majority of samples, the imputation process is performed three times: first, a random forest imputation for the good-quality features only, then a simple imputation for those remaining features with more than 50% missing values, and lastly, a random forest imputation for the rest of the features. However, this process is evaluated separately for every dataset.

11. Set a seed number for reproducibility, such as 10.

```
set.seed(10)
```

12. Perform the first round of imputation with the random forest method on the good-quality features.

```
imputed <- impute_rf(object = merged_no_qc, all_
features = FALSE)
```

13. Perform the second round of imputation by imputing 1 for the remaining features with more than 70% missing values. This value may need adjustment based on the type of study. This is to avoid issues with further data analysis steps, such as transformation, normalization, and certain statistical models.

```
imputed <- impute_simple(object = imputed, 1, na_
limit = 0.5)
```

14. Perform the third round of imputation with the random forest method for the remaining features.

```
imputed <- impute_rf(object = imputed, all_features
= TRUE)
```

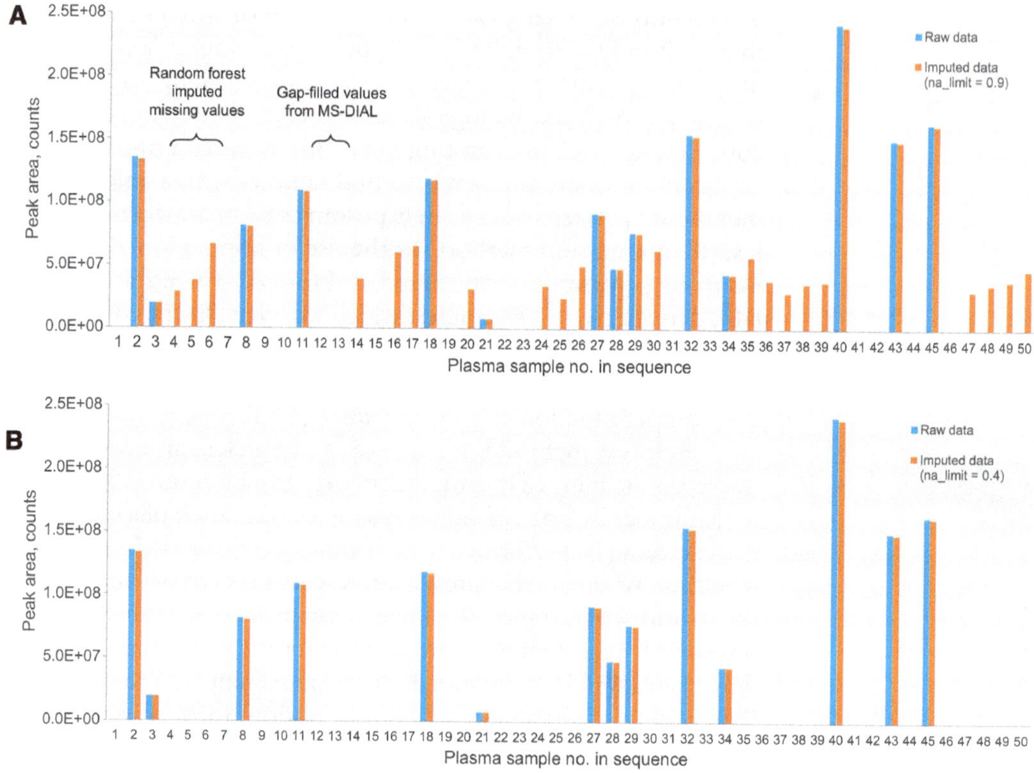

Fig. 7 **(A)** An example of random forest imputation creating unwanted artifact peaks in the data, showing the abundances of a drug metabolite in a subset of plasma samples from a clinical intervention study. Blue bars represent the raw signal abundances (as peak areas) derived from MS-DIAL after gap filling, and the orange bars represent the data after preprocessing with default settings in *notame*, including the drift correction and random forest imputation of missing values. The imputation process has generated relatively high abundances to replace the missing values, whereas the gap-filled abundances from the original raw data remain close to the instrument noise level. **(B)** The issue can be resolved by lowering the na_limit to 0.4 or less (meaning that, since in this case missing values were 40% of the samples, the feature was imputed with 1 and not chosen for random forest imputation)

6 Batch Correction

Batch correction (between-batch correction), if needed, is performed after the (within-batch) drift correction by using the function `normalizeBatches` from the *batchCorr* R package [2], also available via Bioconductor with SummarizedExperiment support. The batch correction (i.e., between-batch normalization is performed on common molecular features between the batches by normalizing their signal areas based on the QC samples or the sample median. The normalization based on QC samples is performed if the QC samples are considered sufficiently representative of the samples using the heuristics described in more detail by Brunius *et al.* [2]; otherwise, the sample median is used.

15. Perform batch correction if the samples were analyzed in more than one batch. By default, the QC samples are named as "QC".

```
batch_corrected <-
batchCorr::normalizeBatches(peakTableCorr =
imputed, batches = "Batch", sampleGroup = "QC",
refGroup = "QC", population = "all", assay.type =
"abundances", name = "normalized")
```

7 Molecular Feature Clustering

Any metabolite detected by LC–MS typically produces several molecular features in the dataset. This redundancy originates from detecting the same metabolite in multiple analytical modes, the metabolite generating various typical adducts during the ionization, such as $[M + H]^+$, $[M + Na]^+$, and $[M - H_2O + H]^+$, and the presence of isotopic peaks, in-source fragments, multiply charged ions, oligomers, and unusual adducts, such as those with other co-eluting metabolites [1]. Peak detectionsoftware usually bundles the isotopic peaks to produce only one molecular feature with the isotopic distribution given in the metadata; however, other redundant signals often remain separate in the data matrix and may be connected via information in a separate metadata column. There is an additional feature in *notame* for clustering the remaining redundant molecular features, which is based on two assumptions: firstly, the Spearman correlation between the abundances of the molecular features across all the samples is high, and secondly, they co-elute in the liquid chromatography. The algorithm will then recursively identify clusters assumed to represent the same metabolite by clustering features that (1) fulfill these two assumptions and (2) are highly correlated to a sufficient number (e.g., 80%) of other features within the cluster, also called the relative degree. Thus, features having high correlation to only a small proportion of the features in the cluster will be dropped (see supplementary material in Klåvus et al. 2020 for a detailed description of the algorithm [8]).

Function `cluster_features` is called in *notame* to perform the feature clustering. The result is an additional column in the data matrix, which contains the cluster assigned for each molecular feature and is named after the molecular feature that has the highest median abundance within the cluster, assumed to represent the base peak. Thus, no molecular features will be removed from the data by the function. By default, the retention time window for co-elution is one second (or 1/60 min), the threshold for Pearson correlation is 0.9, and the threshold for the relative degree required by each node during the feature clustering is 0.8 (i.e., the two assumptions are fulfilled with at least 80% of the other features in the cluster).

16. Optional step: Cluster the molecular features. Use the same time unit as the retention time information in the data.

```
clustered <- cluster_features(object = batch_
corrected, rt_window = 1/60, corr_thresh = 0.9,
d_thresh = 0.8)
```

8 Statistics

Several summary statistics, effect size metrics, and statistical models can be calculated with *notame*. In addition to the raw *p*-value, the statistical models return a false discovery rate (FDR)-corrected *p*-value, known as *q*-value, which by default uses the Benjamini–Hochberg FDR on the good-quality molecular features. Adjusting the *p*-values for multiple testing is important in metabolomics, where often thousands of consecutive tests are performed separately on each molecular feature, making the data prone to false positive results.

8.1 Summary Statistics and Effect Sizes

The summary statistics are calculated with function `summary_statistics` and they include the mean, standard deviation (SD), median, MAD (see Chap. 3), and quartiles calculated for each molecular feature. The area under the curve (AUC) can be calculated for each study subject and feature in repeated-measures studies using the function `perform_auc`. For the effect size, fold changes are calculated with function `fold_change`, and Cohen's *d* values with the function `cohens_d`. Whereas fold change is the direct ratio between the group averages and an intuitive way to assess relative differences, Cohen's *d* will also consider whether the effect size is significant in terms of the variance within the groups. The benefit of Cohen's *d* in metabolomics analysis is that when the variation between the samples is taken into account, both endogenous metabolites and metabolites from external exposures are treated more equally. With fold change, metabolites from external exposures easily get very high FC, even though their biological significance could be low, and the variance between samples is very high. In contrast, endogenous metabolites under tight homeostatic control (i.e., having low variance) can receive low FC even though this small change can have a large biological importance.

The execution of the following steps depends on the type of data and study question in hand.

17. Calculate the summary statistics for each molecular feature. Optionally, one or several columns can be given as a character vector for groupwise summary statistics. In the upcoming

steps, replace the `object` with the name of the preferred pre-processed object, such as `imputed` or `clustered`.

```
summary_statistics <- summary_statistics(object =
object, grouping_cols = NULL)
```

18. Calculate the area under the curve (AUC) for each feature and study subject in studies where samples from several time points are included. The resulting object contains the AUC values for each subject, and it can be used for statistics, such as *t*-test (Step 21).

```
aucs <- perform_auc(object = object, time =
"Time", subject = "Subject_ID", group = "Group")
```

19. Calculate fold changes between the study groups. If there are more than two study groups, the function will, by default, calculate the fold changes between all possible pairs.

```
fc <- fold_change(object = object, group =
"Group")
```

20. Calculate Cohen's *d* between the study groups. This can be done for two specified groups at a time. If id (for subject ID) and time are given, Cohen's *d* will be computed for the change in time.

```
cohens_d <- cohens_d(object = object, group =
"Group", id = "Subject_ID", time = "Time")
```

8.2 Univariate Tests

The user can choose parametric and non-parametric univariate tests depending on the type of data and study question. Nontargeted metabolomics data often do not fulfill all three assumptions required for the reliable use of parametric tests: normality, homoscedasticity (the equality of variances), and independence of the sample abundances, in which case a non-parametric test would produce more robust results [23]. On the other hand, non-parametric tests are not as powerful as the parametric ones, and especially after the adjustment of *p*-values for multiple testing, may miss biologically significant differences.

When comparing two groups, *notame* uses Welch's *t*-test (to manage unequal variances) as the parametric test and Mann–Whitney *U*-test as the non-parametric alternative. These can be accessed with functions `perform_t_test` and `perform_non_parametric`, respectively. For paired samples, a paired *t*-test or Wilcoxon signed-rank test as a non-parametric alternative can be performed using the same two functions via the addition of attribute `is_paired = TRUE` (see Steps 23 and 24).

21. Perform Welch's *t*-test between two or more study groups. In case more than two study groups exist, all possible pairs will

be tested separately. This function, and several other statistical functions, use a formula interface, where `Feature` is replaced by the ID of the molecular feature in each iteration. If the group variances are known to be equal, a Student's *t*-test can be performed using the same function by setting the attribute `var.equal = TRUE`.

```
t_test_results <- perform_t_test(object = object,
formula_char = "Feature ~ Group")
```

22. Perform Mann–Whitney *U*-test between the two study groups.

```
mann_whitney_results <- perform_non_
parametric(object = object, formula_char = "Feature
~ Group")
```

23. Perform a paired parametric *t*-test between two study groups.

```
paired_t_results <- perform_t_test(object = object,
formula_char = "Feature ~ Group", is_paired = TRUE,
id = "Subject_ID")
```

24. Perform a paired non-parametric *t*-test (Wilcoxon signed-rank test) between the two study groups.

```
wilcoxon_results <- perform_non_parametric(object =
object, formula_char = "Feature ~ Group", is_paired
= TRUE, id = "Subject_ID")
```

Comparisons between more than two groups can be performed with Welch's ANOVA (parametric) using the function `perform_oneway_anova()` and Kruskal–Wallis test (non-parametric) with `perform_kruskal_wallis()`. If the comparison involves adjustment for fixed effects, such as age or sex, a linear model can be fitted separately for each feature using the function `perform_lm()`, which uses the `stats` package. If random effects, such as subject ID in repeated-measures study settings, need to be included in the model as well, a linear mixed model should be fitted instead by using `perform_lmer()`, which uses the `lme4` package.

25. Perform Welch's ANOVA to compare the averages of two or more study groups.

```
oneway_anova_results <- perform_oneway_anova(object
= object, formula_char = "Feature ~ Group")
```

26. Perform Kruskal–Wallis test to compare the averages of two or more study groups.

```
Kruskal_wallis_results <- perform_kruskal_
wallis(object = object, formula_char = "Feature ~
Group")
```

27. Perform a linear model to study whether the molecular features predict the difference in selected fixed effects, such as group and time, as well as their interaction. An equivalent formula would be `"Feature ~ Group * Time"`.

```
lm_results <- perform_lm(object = object, formula_
char = "Feature ~ Group + Time + Group:Time")
```

28. Perform a linear mixed model to study whether the molecular features predict the difference in selected fixed effects, such as group and time, and the contribution of random effects, such as subject ID. The confidence interval for the parameters is, by default, calculated with the Wald method, which is suitable for large datasets with normally distributed effects. In other cases, `profile` or `boot` (bootstrapped) confidence interval should be considered.

```
lmer_results <- perform_lmer(object = object,
formula_char = "Feature ~ Group + Time + (1 |
Subject_ID)", ci_method = "Wald")
```

8.3 Multivariate Tests

Multivariate tests available via *notame* include a MUVR (multivariate modeling with minimally biased variable selection in R), a wrapper around *MUVR2* package, and partial least squares methods, including partial least squares (PLS), sparse PLS (SPLS), and partial least squares discriminant analysis (PLS-DA), a wrapper around *mixOmics* package. MUVR is a supervised learning algorithm for selecting a set of molecular features most likely related to the research outcome using PLS and random forest modeling [24]. The multivariate approaches also include dimensionality reduction methods, such as PCA and t-SNE, and hierarchical clustering, which will be introduced in Chap. 9.

29. Perform MUVR to select relevant molecular features for the target variable `y` to predict. The number of iterations (`nRep`) is recommended to be at least 30.

```
rf_model <- muvr_analysis(object = object, y =
"Group", nRep = 30, method = "RF")
```

30. Perform PLS to identify a stable set of metabolites that best explain the outcome `y`, while controlling overfitting.

```
pls_opt <- mixomics_pls_optimize(object = object,
y = "Group", ncomp = 3, nrepeat = 5)
```

Manual annotation of metabolites may need to be performed before performing some of the statistics (such as correlation between metabolites) and generating certain visualizations described in the following steps (such as heatmaps). In addition, to ensure that there is no redundancy in the final list of reported metabolites, representative ions may need to be chosen at this point if the same metabolite has been identified multiple times as several adducts and/or in several analytical modes. Although algorithms, such as the molecular feature clustering in *notame* and MS-CleanR [25], aim at producing a clean list of molecular features representing unique metabolites, there may still be a need to manually verify this for the identified metabolites. If metabolite annotation is performed after the data preprocessing and statistics, it may be necessary to load and merge additional columns from an edited Excel file into *notame* for adding the metabolite names for visualization purposes.

31. Combine the statistics results (needed for manual annotation of metabolites) into the main preprocessed object. In this example, results from the Cohen's *d*, Mann–Whitney *U*-test, and linear model were chosen. Export the combined data object into an Excel table.

```
with_results <- join_rowData(imputed, cohens_d)
with_results <- join_rowData(with_results,
mann_whitney_results)
with_results <- join_rowData(with_results,
lm_results)
write_to_excel(with_results, file = paste0(path,
"imputed_statistics.xlsx"))
```

32. Perform manual annotation of metabolites and add manual metabolite ID (column `Curated_ID`), MSI ID level (column `ID_level`), and manually chosen representative ion columns (column `Representative_ion`, where the representative ions have been marked with `x`). Remove the extra rows, so that Feature_ID etc. become the column names (top rows). Save the Excel file under a new name, such as `curated_data.xlsx`.

33. Join the manual annotation results from the Excel file (via `openxlsx` package) as extra data columns to the `with_results` object. Ensure that there is at least one redundant column common for both datasets, such as the unique Feature ID generated by *notame*.

```
extra_data <- openxlsx::read.xlsx(paste0(path,
"curated_data.xlsx")) colnames(extra_data) extra_
data <- extra_data[, c("Feature_ID", "Curated_ID",
"ID_level", "Representative_ion")] with_results <-
join_rowData(with_results, extra_data)
```

34. Generate an object containing only the representative ions of each annotated metabolite.

```
annotated <- with_results[!is.na(rowData(data_
added)$Representative_ion) & rowData(with_
results)$Representative_ion == "x",]
```

8.4 Correlation Tests

Correlation tests in *notame* can be performed between metabolites themselves or between metabolites and other given variables, such as clinical measurements, dietary data, or even abundances from other omics datasets. The statistic can be chosen between Pearson's r (the result equaling that from standardized linear regression), Spearman's ρ, and Kendall's τ (also known as Kendall rank correlation) by using the `method` argument passed to `corr.test` and choosing between `pearson`, `spearman`, or `kendall`. The choice of the statistical method depends on the data and research question: Pearson correlation measures the linear relationship between variables that are assumed to be normally distributed (which is often not the case in typical metabolomics data), whereas Spearman's ρ and Kendall's τ are non-parametric rank correlation tests that do not require assumptions of normality. Spearman's ρ often gives conservative results, whereas Kendall's τ may be especially suitable for small sample sizes with unknown distribution of metabolite abundances [26]. In the current versions of the *notame* packages, the correlation results objects can be used for heatmap visualizations.

35. Calculate Spearman correlation coefficients and p-values between the annotated metabolites. For Pearson correlation and Kendall's τ, use `pearson` and `kendall`, respectively, as `method`.

```
correlations <- perform_correlation_tests(object
= annotated, x = rownames(annotated), method =
"spearman")
```

36. Calculate Kendall's τ correlation coefficients and p-values between the annotated metabolites and two sample information variables: time and injection order. `Time` needs to be converted to `numeric`.

```
annotated$Time <- as.numeric(annotated$Time)
correlations_variables <- perform_correlation_
tests(object = annotated, x = rownames(annotated),
y = c("Time", "Injection_order"), method =
"kendall")
```

9 Results Visualizations

Informative and impactful visualizations are key in most types of research. They are particularly important in areas such as metabolomics and multi-omics, where large amounts of complex data need to be presented concisely and understandably for different purposes, such as quality inspection (see Chap. 4), evaluation of different data analysis steps, and communication of the results [27]. Metabolomics data is often visualized on different levels of detail, from overviews like PCA to metabolite-specific plots. This chapter describes the main results visualization options available in *notame*.

9.1 PCA and t-SNE

Principal component analysis (PCA) and *t*-distributed stochastic neighbor embedding (t-SNE) are unsupervised dimensionality reduction methods aimed at simplifying complex data into orthogonal components that best explain the variability within the dataset. For example, they can reveal class differences or outlier samples in the data unbiasedly [28]. In practice, the scores of each sample and the two chosen components (by default, the first and second components) are mapped into a two-dimensional plot for a visual assessment of the differences between the metabolite profiles of each sample. The main difference between PCA and t-SNE is that PCA is a linear method that aims to preserve as much of the global variance in the data as possible, whereas t-SNE is a nonlinear method that better preserves local structures and clusters similar data points together [29]. Both methods can thus complement each other in finding patterns from metabolomics data.

Both PCA and t-SNE are generated automatically with the `visualizations()` function; however, they can be accessed directly by `plot_pca()` and `plot_tsne()` for a custom plot, where the components, distance scale, color and shape of the data points, and the use of density curves can be chosen (Fig. 8). They are wrappers for *pcaMethods* and *Rtsne* packages, respectively. The PCA loadings plot can be generated with `plot_pca_loadings()`. In addition, both PCA and t-SNE can be plotted with arrows for data with several observations in time for each study subject and as hexbin plots, where the components are plotted as hexagonal bins, and the color of each bin is summarized from the samples. The dedicated functions for these are `plot_pca_arrows()`, `plot_tsne_arrows()`, `plot_pca_hexbin()`, and `plot_tsne_hexbin()`.

37. Plot a PCA including the first two principal components, unit variance scaling, density curves, `Group` defining the color, and `Time` defining the shape.

```
plot_pca(object = object, pcs = c(1, 2), scale
= "uv", density = TRUE, title = NULL, color =
"Group", shape = "Time")
```

38. Plot an arrow t-SNE including the first two principal components, unit variance scaling, density curves, the color defining grouping variable 1, the time point for constructing the arrows, and subject ID. If the sample size is large and the image becomes crowded, it is possible to show the groups in separate faceted plots by adding + `facet_wrap(~Group)`. Note that the perplexity (default: 30) may need to be adjusted in studies

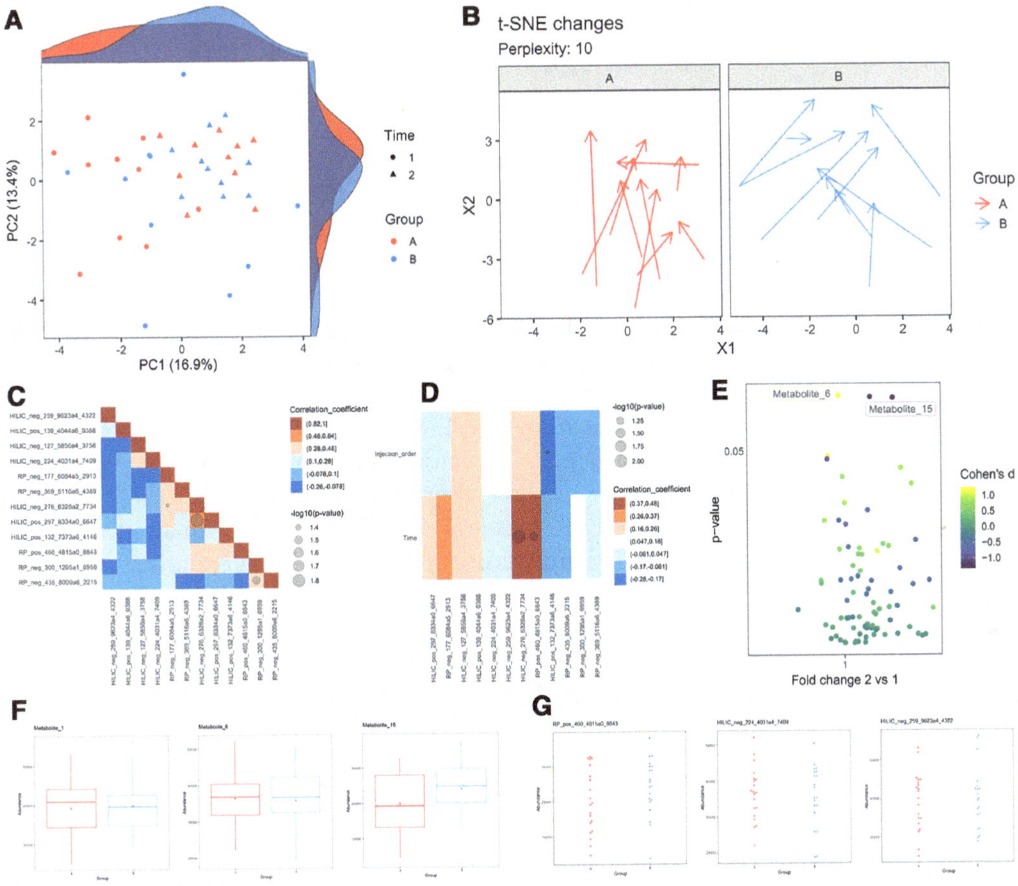

Fig. 8 Data visualizations from the `toy_notame_set` object. (**A**) PCA generated with the script example given in Step 37, and (**B**) faceted arrow t-SNE generated with the script in Step 38. (**C**) Heatmap of the metabolite–metabolite correlation results (Step 39) and (**D**) correlations between a set of clinical markers and identified metabolites (Step 40). (**E**) A volcano plot showing the differential metabolites (based on fold change and *p*-value) from a pairwise comparison between the study groups (Step 41). Annotated metabolites with $p < 0.05$ have been labeled and the dots colored based on the Cohen's *d* value of the change in time between the same study groups. Note that "2 vs 1" refers to group 1 being the "control" (i.e., the denominator when calculating the fold change), and thus fold change values higher than 1 and positive Cohen's *d* values indicate a higher abundance in group 2. (**F**) Boxplots of the abundances of selected identified molecular features labeled according to the user-given metabolite name (Step 42). (**G**) Beeswarm plots of the same molecular features with the unique Feature ID automatically generated by *notame* during the preprocessing (Step 43). The images were exported to SVG format, and additional editing was carried out in Inkscape, including removing white background elements except for the plot area and adding a gray background

where the number of samples is less than 100 [the perplexity should be no higher than (number of samples − 1)/3].

```
plot_tsne_arrows(object = imputed, scale = "uv",
perplexity = 10, color = "Group", time = "Time",
```

```
subject = "Subject_ID", arrow_style = arrow(angle
= 30, length = unit(0.15, "inches"), ends =
"last", type = "open"), line_width = 0.75) +
facet_wrap(~Group)
```

9.2 Clustering and Heatmaps

Heatmap combined with clustering is a powerful means to visualize metabolomics data because it can intuitively reveal metabolite-level patterns in the results. Each cell within the heatmap will be colored according to a variable, such as the correlation coefficient, effect size, or normalized peak area (e.g., as *z*-scores). The correlation coefficient can be calculated between each metabolite, in which case only the lower triangular part of the heatmap needs to be plotted (Fig. 8C), or between metabolites and other variables, such as clinical measurements (Fig. 8D). If the heatmap shows test statistics, the *p*-value (or *q*-value) can be plotted as well, represented by asterisks or circles (where the circle diameter can be used to signify the value). It is important to choose both an intuitive and color-blind-friendly color palette for the heatmap. If the data is centered around a meaningful zero, such as log-transformed ratios, *z*-scores, and correlation coefficients, a diverging red–white–blue palette (such as `RdBu` from `RColorBrewer` used by ***notame***) is usually the optimal choice.

39. Generate an effect heatmap with hierarchical clustering from the correlation results (object `correlations` generated in Step 35) between the annotated metabolites.

```
plot_effect_heatmap(correlations, x = "X", y
= "Y", effect = "Correlation_coefficient", p =
"Correlation_P_FDR", point_size_range = c(2, 8),
discretize_effect = TRUE, breaks = 7, clustering =
TRUE, lower_tri = TRUE)
```

40. Generate an effect heatmap with hierarchical clustering from the correlation results (object `correlations` generated in Step 36) between annotated metabolites and a set of other variables.

```
plot_effect_heatmap(correlations_variables, x =
"X", y = "Y", effect = "Correlation_coefficient", p
= "Correlation_P_FDR", point_size_range = c(2, 8),
discretize_effect = TRUE, breaks = 7, clustering =
TRUE, lower_tri = TRUE)
```

9.3 Volcano Plot

Volcano plot visualizes the differences between two study groups by plotting the molecular features and identified metabolites based on their effect size and statistical significance of a pairwise comparison performed previously (see Chap. 8.2). Typically, the x-axis is an effect size measure, such as fold change or Cohen's *d*, and the y-axis is the *p*-value, often in a negative logarithmic scale. As a result, there will be a vertical "zero-effect" center in the image, and typically, the metabolites are scattered on both sides of the center with the most significantly differential metabolites located toward both upper corners of the plot, giving an impression of a volcanic eruption (Fig. 8E). The correct interpretation of the plot requires that it is clear what the effect size is measuring; i.e., a positive effect size could mean a higher metabolite abundance in treatment vs control or species A vs B. In *notame*, the plot is generated with the function `volcano_plot()`. The columns containing the effect size and *p*-value information need to be specified. In addition, the x-axis can be $\log_2$-transformed, the zero effect point centered, and identified metabolites labeled.

41. Generate a volcano plot from a pairwise comparison by using the Cohen's *d* (Step 20) and *p*-values from the linear mixed model (Step 28), log2-transform the x-axis, center the zero-effect point, label those metabolites with manually curated identifications and FDR < 0.05, and make custom axis and legend labels.

```
volcano_plot(object = with_results, x = "B_vs_A_
FC", p = "GroupB.Time2.p.value", color = "B_vs_A_2_
minus_1_Cohen_d", log2_x = TRUE, center_x_axis =
TRUE, label = "Curated_ID", label_limit = 0.05,
title = NULL) +
labs(x = "Fold change 2 vs 1", y = "p-value", color
= "Cohen's d")
```

9.4 Feature-Wise Plots

Feature-wise visualizations show the abundance of an individual molecular feature in the samples across study groups and/or time. For instance, they can present the groupwise interquartile range (boxplot), the abundance of each sample (beeswarm plot), change in time (individual or group line plot), and the relationship between a feature and another measurement (scatterplot) (Fig. 8F and G).

42. Save boxplots of all features by group, with "Curated ID" column used to label the metabolites, into a single PDF file.

```
save_group_boxplots(object = object, file_path =
"./figures/group_boxplots.pdf", format = "pdf", x =
"Group", title = "Curated_ID", color = "Group")
```

43. Save beeswarm plots of uniquely identified features (see Steps 33 and 34) by group into separate PNG files.

```
save_beeswarm_plots(object = annotated, file_path
= "./figures/beeswarm_plots/", format = "png", x =
"Group", color = "Group")
```

10 Conclusions

Using an aligned LC–MS data matrix as such without considering signal drift, contamination, and peak quality compromises the integrity of metabolomics data. With *notame*, we aim to provide a user-friendly workflow of established data analysis tools to process aligned metabolomics data. It uses quality metrics, correction tools, statistics, and visualizations to support the user in finding relevant ways to interpret the results while providing ways to critically assess the LC–MS data and optimize its quality.

The release of *notame* within Bioconductor has connected the package to a wide data ecosystem and active community, which enables streamlined multi-omics analyses utilizing the SummarizedExperiment data container format for research as well as training purposes [30]. This will also allow us to reduce overlapping features and integrate *notame* functionalities with other omics and Bioconductor packages utilizing SummarizedExperiment, such as *R4MS* (https://www.rformassspectrometry.org/), *mia* [31], and *SCE* [32]. We are constantly developing *notame* to provide new and expanded features to further facilitate metabolomics data analysis. For example, a "traffic light system" could highlight those good-quality features that are most relevant based on a statistical analysis of the study groups and the presence of acquired MS/MS data, and thus have higher priority for manual annotation.

Acknowledgments

Anton Klåvus (University of Eastern Finland) and Jussi Paananen (University of Eastern Finland) are acknowledged as the developers of the first version of *notame*. Hany Ahmed (University of Turku) is acknowledged for providing some of the demonstration data. Artur Sannikov (University of Turku) is acknowledged for technical assistance and review of the R script. We acknowledge funding from the following agencies: European Union's Horizon 2020 Research and Innovation Programme (grant agreements No. 952914, No. 101060247, and No. 101000309), Jane and Aatos Erkko Foundation, and Lantmännen Research Foundation.

References

1. Sindelar M, Patti GJ (2020) Chemical discovery in the era of metabolomics. J Am Chem Soc 142(20):9097–9105
2. Brunius C, Shi L, Landberg R (2016) Large-scale untargeted LC-MS metabolomics data correction using between-batch feature alignment and cluster-based within-batch signal intensity drift correction. Metabolomics 12:1–13
3. Do KT, Wahl S, Raffler J, Molnos S, Laimighofer M, Adamski J, Suhre K, Strauch K, Peters A, Gieger C (2018) Characterization of missing values in untargeted MS-based metabolomics

data and evaluation of missing data handling strategies. Metabolomics 14:1–18

4. Peluso A, Glen R, Ebbels TM (2021) Multiple-testing correction in metabolome-wide association studies. BMC Bioinform 22:1–18
5. Albóniga OE, González O, Alonso RM, Xu Y, Goodacre R (2020) Optimization of XCMS parameters for LC–MS metabolomics: an assessment of automated versus manual tuning and its effect on the final results. Metabolomics 16:1–12
6. Roca M, Alcoriza MI, Garcia-Cañaveras JC, Lahoz A (2021) Reviewing the metabolome coverage provided by LC-MS: focus on sample preparation and chromatography – -a tutorial. Anal Chim Acta 1147:38–55
7. Defossez E, Bourquin J, von Reuss S, Rasmann S, Glauser G (2023) Eight key rules for successful data-dependent acquisition in mass spectrometry-based metabolomics. Mass Spectrom Rev 42(1):131–143
8. Klåvus A, Kokla M, Noerman S, Koistinen VM, Tuomainen M, Zarei I, Meuronen T, Häkkinen MR, Rummukainen S, Farizah Babu A (2020) "Notame": workflow for non-targeted LC–MS metabolic profiling. Metabolites 10(4):135
9. Zarei I, Eloranta A-M, Klåvus A, Väistö J, Lehtonen M, Mikkonen S, Koistinen VM, Sallinen T, Haapala EA, Lintu N (2024) Eight-year diet and physical activity intervention affects serum metabolites during childhood & adolescence: a nonrandomized controlled trial. iScience 27(7), 110295.
10. GitHub. (2026). https://github.com/hanhineva-lab/notame.
11. Dudzik D, Barbas-Bernardos C, García A, Barbas C (2018) Quality assurance procedures for mass spectrometry untargeted metabolomics. a review. J Pharm Biomed Anal 147:149–173
12. Broadhurst D, Goodacre R, Reinke SN, Kuligowski J, Wilson ID, Lewis MR, Dunn WB (2018) Guidelines and considerations for the use of system suitability and quality control samples in mass spectrometry assays applied in untargeted clinical metabolomic studies. Metabolomics 14:1–17
13. Wehrens R, Hageman JA, van Eeuwijk F, Kooke R, Flood PJ, Wijnker E, Keurentjes JJ, Lommen A, van Eekelen HD, Hall RD (2016) Improved batch correction in untargeted MS-based metabolomics. Metabolomics 12:1–12
14. Dunn WB, Broadhurst D, Begley P, Zelena E, Francis-McIntyre S, Anderson N, Brown M, Knowles JD, Halsall A, Haselden JN (2011) Procedures for large-scale metabolic profiling of serum and plasma using gas chromatography and liquid chromatography coupled to mass spectrometry. Nat Protoc 6(7):1060–1083
15. Tsugawa H, Ikeda K, Takahashi M, Satoh A, Mori Y, Uchino H, Okahashi N, Yamada Y, Tada I, Bonini P (2020) A lipidome atlas in MS-DIAL 4. Nat Biotechnol 38(10):1159–1163
16. González-Domínguez Á, Estanyol-Torres N, Brunius C, Landberg R, González-Domínguez R (2024) QC omics: Recommendations and Guidelines for Robust, Easily Implementable and Reportable Quality Control of Metabolomics Data. Anal Chem 96(3):1064–1072
17. Evans AM, O'Donovan C, Playdon M, Beecher C, Beger RD, Bowden JA, Broadhurst D, Clish CB, Dasari S, Dunn WB (2020) Dissemination and analysis of the quality assurance (QA) and quality control (QC) practices of LC–MS based untargeted metabolomics practitioners. Metabolomics 16(10):113
18. Goh WWB, Wang W, Wong L (2017) Why batch effects matter in omics data, and how to avoid them. Trends Biotechnol 35(6):498–507
19. Han W, Li L (2022) Evaluating and minimizing batch effects in metabolomics. Mass Spectrom Rev 41(3):421–442
20. Little RJ, Rubin DB (2019) Statistical analysis with missing data, vol 793. John Wiley & Sons
21. Wei R, Wang J, Su M, Jia E, Chen S, Chen T, Ni Y (2018) Missing value imputation approach for mass spectrometry-based metabolomics data. Sci Rep 8(1):663
22. Kokla M, Virtanen J, Kolehmainen M, Paananen J, Hanhineva K (2019) Random forest-based imputation outperforms other methods for imputing LC-MS metabolomics data: a comparative study. BMC Bioinform 20:1–11
23. Vinaixa M, Samino S, Saez I, Duran J, Guinovart JJ, Yanes O (2012) A guideline to univariate statistical analysis for LC/MS-based untargeted metabolomics-derived data. Metabolites 2(4):775–795
24. Shi L, Westerhuis JA, Rosén J, Landberg R, Brunius C (2019) Variable selection and validation in multivariate modelling. Bioinformatics 35(6):972–980
25. Fraisier-Vannier O, Chervin J, Cabanac G, Puech V, Fournier S, Durand V, Amiel A, André O, Benamar OA, Dumas B (2020) MS-CleanR: a feature-filtering workflow for untargeted LC–MS based metabolomics. Anal Chem 92(14):9971–9981
26. Di Giovanni N, Meuwis M-A, Louis E, Focant J-F (2023) Correlations for untargeted GC× GC-HRTOF-MS metabolomics of colorectal cancer. Metabolomics 19(10):85
27. Mildau K, Ehlers H, Meisenburg M, Del Pup E, Koetsier RA, Ortega LRT, de Jonge NF, Singh KS, Ferreira D, Othibeng K (2025) Effective

data visualization strategies in untargeted metabolomics. Nat Prod Rep 42(6):982–1019
28. Worley B, Powers R (2013) Multivariate analysis in metabolomics. Curr Metab 1(1):92–107
29. Belkina AC, Ciccolella CO, Anno R, Halpert R, Spidlen J, Snyder-Cappione JE (2019) Automated optimized parameters for T-distributed stochastic neighbor embedding improve visualization and analysis of large datasets. Nat Commun 10(1):5415
30. Drnevich J, Tan FJ, Almeida-Silva F, Castelo R, Culhane AC, Davis S, Doyle MA, Geistlinger L, Ghazi AR, Holmes S (2025) Learning and teaching biological data science in the Bioconductor community. PLOS Comput Biol 21(4):e1012925
31. Borman T, Benedetti G, Muluh G, Raulo A, Valderrama B, Sannikov A, Peschel S, Liu Y, Hindström R, consortium O (2025) Orchestrating microbiome analysis with Bioconductor. bioRxiv:2025.2010. 2029.685036
32. Amezquita RA, Lun AT, Becht E, Carey VJ, Carpp LN, Geistlinger L, Marini F, Rue-Albrecht K, Risso D, Soneson C (2020) Orchestrating single-cell analysis with Bioconductor. Nat Methods 17(2):137–145

Check for updates

Chapter 15

Metabolomics Data Analysis with TIGER

Rui Wang-Sattler and Siyu Han

Abstract

TIGER, a non-parametric method, was developed to address technical variations (e.g., plate and batch effects) in targeted and non-targeted metabolomics datasets. It integrates the random forest (RF) algorithm into a flexible ensemble learning framework, combining multiple base models with a meta-model. These base models are trained using diverse RF hyperparameter combinations, eliminating the need for manual hyperparameter tuning. This chapter highlights practical considerations for using TIGER effectively, including incorporating quality control (QC) samples into study design. When QCs are unavailable, randomly selected samples can be remeasured to facilitate cross-kit corrections. To optimize processing time, highly correlated metabolites from QC samples are selected to train the base models, with weights assigned via an exponential decay function. TIGER employs relative standard deviation (RSD) and mean absolute percentage error (MAPE) as key metrics to ensure models generalize well to unseen data while avoiding overfitting. Additionally, the developed dynamic website has been demonstrated with raw and TIGER-normalized data, enabling performance evaluation and visualization of longitudinal patterns of metabolites or metabolite ratios. This platform demonstrates TIGER's ability to accurately normalize data and uncover trends across three time points spanning a decade. With its proper application, TIGER stands to be a powerful tool for metabolomics studies.

Keywords Metabolomics, Targeted metabolomics, Non-targeted metabolomics, Data-preprocessing, Batch correction, Cross-kit correction, ML algorithms

1 Introduction

TIGER (**t**echnical variation el**i**mination for metabolomics data usin**g** ensemble learning archit**e**ctu**r**e) integrates the random forest (RF) algorithm into a versatile ensemble learning framework [1]. The R package, TIGER (https://github.com/HAN-Siyu/TIGER), has been downloaded over 8000 times and has demonstrated superior performance in addressing technical variations in metabolomics datasets, as validated by an independent study [2]. A dynamic website (https://han-siyu.github.io/TIGER_web/) offers an interactive interface to visualize and compare summarized results from the KORA S4, F4, and FF4 studies [3–6], supporting the exploration of TIGER's reliability and reproducibility [1].

Michael Witting (ed.), *Computational Methods and Data Analysis for Metabolomics*, Methods in Molecular Biology, vol. 3063, https://doi.org/10.1007/978-1-0716-5452-1_15,

This chapter introduces key concepts and practical considerations for using TIGER effectively, including the integration of quality control (QC) samples into study design and remeasuring samples for cross-kit corrections. While examples focus on targeted metabolomics datasets, TIGER also supports non-targeted datasets, such as the P20 Negative dataset (used as a demo dataset for the systematic error removal using the random forest [SERRF] algorithm) [7], and the Amide dataset (included in the WaveICA R package) [8]. Overall, TIGER is a flexible tool applicable to any type of quantitative dataset with QC data, making it suitable for diverse omics applications.

2 Materials

2.1 Incorporating Required QC Data into Study Design for TIGER

To effectively correct batch and plate effects (referred to as "plate" in the following text) using TIGER, it is essential to include at least three identical QC samples per plate in the study design. These QC samples serve as training data for the model and are critical for ensuring the reliability of plate effect corrections. Along with randomizing subject samples, QC samples must be carefully included in the study design to ensure robust model training. TIGER uses these QC samples to split datasets into training and validation folds (*see* **Note 1**).

Additionally, raw quantitative data at the individual level are required, including key identifiers such as sample ID, sample type, and plate ID. These data points are critical for TIGER's ability to identify and adjust for plate effects. For detailed guidance, please refer to the user manual for the TIGERr package.

For example, TIGER was applied to metabolomics datasets from the KORA FF4. The dataset was analyzed using the biocrates AbsoluteIDQ™ p180 kits (referred to as p180) across 29 plates as described in detail [1]. The raw data included:

- One each of QC1, QC2, and QC3 per plate.
- Five identical QC samples per plate [9].
- Raw metabolite values for 2218 study participants.

The QC samples were used to train the model, while QC1, QC2, QC3, and subject samples were used to test the models [1].

It is important to double-check the validity of training models on QC samples and to test them on subject samples (*see* **Note 2**), although usually this is the case. For example, in the KORA FF4 study, with an example of metabolite PC aa C36:4, the variation of subject samples alongside QC samples (Fig. 1A) supports this validity.

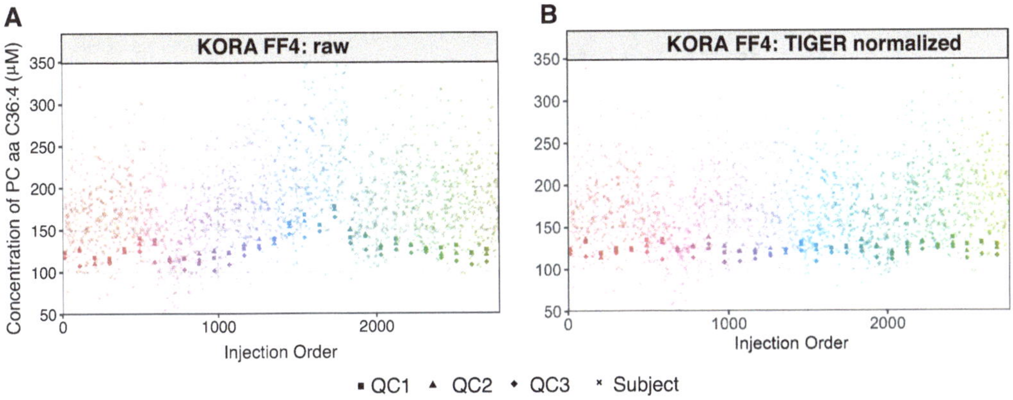

Fig. 1 Validation of training models using QC and subject samples before and after TIGER normalization. Plot A shows the variation of subject samples alongside QC samples with an example of metabolite PC aa C36:4 in the KORA FF4 study. This plot is part of Fig. 1 in the TIGER paper. Plot B presents TIGER normalized data for the same metabolite

2.2 Remeasured Data Enabling Cross-Kit Correction or Longitudinal Dataset Correction

In the absence of identical QC samples measured on each plate, randomly selected samples (ideally five per plate) can be remeasured using targeted analytical kits to serve as QC samples. This approach enables cross-kit corrections using TIGER.

For example, TIGER supports across-kit data harmonization, as demonstrated in the longitudinal datasets of the baseline KORA S4 study and its two follow-up studies, F4 and FF4. Data from S4 and FF4 were quantified using the same p180 kits, while data from F4 were quantified using the biocrates AbsoluteIDQ™ p150 kits (referred to as p150) [10–12].

To enable cross-kit correction, 288 samples were selected from the original 38 p150 kit plates and remeasured alongside FF4 samples using four p180 kits. Additionally, 22 identical QC samples were included in the remeasurement experiment [1]. The data of KORA FF4's QC samples can be found in TIGER's R package.

3 Methods

3.1 The Key Component of TIGER

TIGER combines multiple base models with a meta-model (Fig. 2A). The base models are trained with QC samples using varied combinations of hyperparameters of the RF algorithm (Fig. 2B), aiming to adjust the raw observed metabolite values to approach target values. The target value serves as a reference, which is the mean value of each metabolite across all QC samples. The meta-model assigns weights to the base models using an exponential decay function (Fig. 2C), calculated based on the relative error between prediction and target values in the validation dataset [1]. After training, TIGER can be

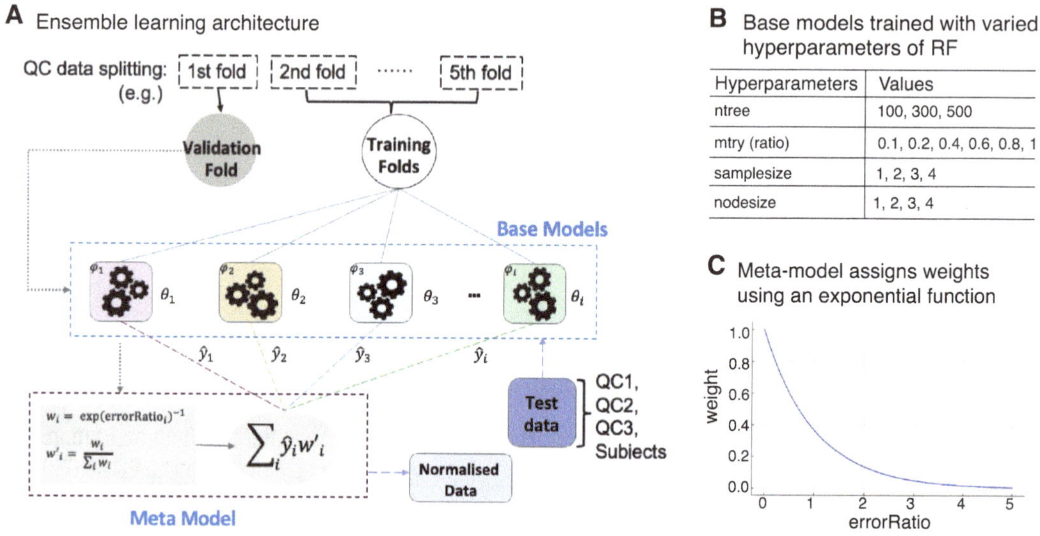

Hyperparameters	Values
ntree	100, 300, 500
mtry (ratio)	0.1, 0.2, 0.4, 0.6, 0.8, 1
samplesize	1, 2, 3, 4
nodesize	1, 2, 3, 4

Fig. 2 Key concepts of TIGER. (**A**) The ensemble learning architecture consists of multiple base models and one meta-model. QC samples are split into training and validation folds to train the base models, which are then combined by the meta-model. (**B**) Each base model is trained with a different combination of hyperparameters of the random forest (RF) algorithm. This table corresponds to Table S2 (b) in the TIGER paper [1]. (**C**) The meta-model assigns weights to base models using an exponential decay function, with models having lower error ratios receiving higher weights

applied to test samples (typically subject samples), making their values approximate the values expected under variation-free conditions.

By training base models with various hyperparameter combinations, TIGER is free from the need for hyperparameter tuning (*see* **Note 3**) and, to some extent, resistant to hyperparameter-induced overfitting. For instance, the RF algorithm includes four hyperparameters (Fig. 2B). Each base model represents one unique combination of values from these four hyperparameters. This design comes at the cost of increased computational complexity, but ultimately leads to more robust performance.

The meta-model assigns weights to each base model using an exponential decay function. Models with low error ratios are assigned higher weights, while models with larger error ratios are assigned smaller weights to prevent bias (*see* **Note 4**). By incorporating all base models into the data correction process, this ensemble learning strategy further enhances the robustness of TIGER.

3.2 Cleaning Data Before Running TIGER

To ensure data reliability, metabolites with higher intra- and inter-plate variation (based on QC samples) or a higher percentage of missing values should be discarded. Additionally, metabolites with values less than 50% of the limits of detection (LOD) for their corresponding plates should also be discarded (*see* **Note 5**).

In the KORA study, cross-kit corrections were considered across data from three time points. The following five QC steps were applied for metabolite selection:

1. **Overlap**: metabolites must overlap between the p150 and p180 kits.
2. **LOD threshold**: at least 50% of its measured sample values are above the LOD for the corresponding plates.
3. **Missing values**: metabolites with less than 10% missing values were retained.
4. **Variation threshold**: the median relative standard deviation (RSD) should be less than 25% (also known as the coefficient of variation, CV). The code below shows how to compute the RSD.
5. **Correlation**: Spearman correlation coefficients between KORA F4 (remeasured) and F4 (original) must be greater than 0.5. This can be computed using R's cor() function.

The following code can be used to compute RSD:

```
### R code: compute RSD
# compute RSD for one metabolite (one column in a
dataset)
compute_RSD <- function(one_meta) {
val_RSD <- sd(one_meta, na.rm = TRUE) / (mean(one_
meta, na.rm = TRUE) + 1e-7) # add 1e-7 to the
denominator to avoid division by zero
val_RSD
}
# Use apply function to compute RSD on the whole
dataset:
RSD_values <- sapply(your_meta_dataset[your_meta_
columns], compute_RSD)
# your_meta_columns indicates the columns of your
metabolite values.
### end of R code: compute RSD
```

Following these steps, 103 metabolites were retained for analysis out of the 194 investigated. Working with reliably measured metabolites is crucial, not only for accurate results but also to save processing time during model training.

3.3 Run TIGER for Technical Variation Removal

TIGER can estimate technical variation only based on plate labels, treating plate effects as the primary source of technical variation. Users can additionally visualize QC values against injection order or well position to assess whether their datasets are also affected by well positions or injection orders. If pronounced well position effects or temporal drifts are observed, TIGER allows users to remove these effects by specifying well position values via the col_position parameter or injection order values via the col_order (see code in example).

We recommend applying these corrections only when there is clear evidence of well position effects or temporal drifts, to prevent TIGER from becoming too aggressive and over-correcting the data.

By default, the target value is the mean value computed on QC samples across all batches, but median is also supported by setting targetVal_method as "median." Median can sometimes achieve better performance if the dataset contains very strong technical variation. Additionally, the target value can also be computed for each batch by setting targetVal_batchWise = TRUE (see code in example). In this case, each batch has its own target value for each metabolite. TIGER will first adjust intra-batch variation and then do additional normalization across all batches. Using the batch-wise target values, TIGER is able to adjust strong temporal drifts in the dataset Amide. However, this setting is recommended only for datasets in which both intra- and inter-batch effects are highly pronounced, as the results under this setting may be less robust than the default setting.

Important: We advise users to select the above-mentioned settings based on the severity of technical variation in their data, and to always validate the results, regardless of the technical variation removal methods used, to ensure that batch effects have been removed while meaningful biological variation is preserved. An intuitive evaluation can be performed by comparing the data variation or error ratio before and after batch effect removal (see Sect. 3.4).

TIGER requires additional computational time, particularly when analyzing large datasets with many variables. TIGER supports parallel computing to improve processing efficiency. This can be set via the parallel.cores parameter (see code in example). To further optimize performance, the models for each metabolite are fitted on its highly correlated metabolites (correlation coefficient > 0.5). If no highly correlated metabolites are found, the top-ranking metabolites based on correlation will be selected. By default, TIGER uses groups of five to ten highly correlated metabolites for each metabolite in the dataset to fit the base models and regress out technical variation, which clearly reduces processing time. Comparable results were observed when fitting the model using only five to ten highly correlated metabolites versus using the whole dataset on the KORA FF4 dataset. However, under the same test environment, training the model with the full dataset required approximately 5.3 times more processing than the default setting (*see* **Note 6**).

The minimum and maximum number of correlated metabolites can be defined using the selectVar_minNum and selectVar_maxNum parameters (see code in example). The correlated metabolites can be selected based on correlation or partial correlation coefficients using the parameter selectVar_corType. By default, TIGER uses the Pearson method to compute correlation coefficients, which focuses on linear relationships between variables. Users can choose the Spearman method by setting the parameter selectVar_corMethod =

"Spearman," which focuses on a monotonic relationship based on the rank order of the data. TIGER can yield reliable results under both settings. However, computing partial correlation coefficients on a large dataset can be very time-consuming.

Important: TIGER accepts datasets with NA (missing values) or zeros. However, correlation cannot be computed if data have fewer than two valid pairs, which means that for a given metabolite, if there are fewer than two valid paired observations with another metabolite in the QC samples, the correlation coefficient between these two metabolites will be NA. If selectVar_minNum = 5, each metabolite needs to have valid correlation coefficients with at least five other metabolites. If selectVar_minNum is not reached, TIGER will assume that a reliable correction for technical variation cannot be performed and will throw an error. In this case, users may either exclude metabolites lacking sufficient observations or, alternatively, select relevant variables based on domain knowledge and provide them to TIGER via the selectVar_external parameter.

The following code shows how to run TIGER for batch effect removal on one dataset:

```
### R code: run TIGER for batch effect removal
# install TIGER from CRAN
install.packages(TIGERr) # version on CRAN
# or install TIGER from GitHub (with the most
recent updates)
if (!library("devtools", logical.return = T)) {
install.packages("devtools")
}
devtools::install_github("HAN-Siyu/TIGER")
# load TIGER
library(TIGERr)
# import your data. Here we import demo dataset:
data(FF4_qc)
# train samples are the samples used to estimate
technical variation, which are generally QC samples
# test samples are those in which you want to
remove technical variation,
# typically the subject samples or samples used for
benchmarking,
# in this example, we use samples labeled as "QC"
as training samples,
# and the remaining samples as test samples:
train_samples <- FF4_qc[FF4_qc$sampleType ==
"QC",]
test_samples <- FF4_qc[FF4_qc$sampleType!= "QC",]
# run TIGER to remove batch effect:
test_norm <- run_TIGER(
test_samples = test_samples,
train_samples = train_samples,
col_sampleID = "sampleID", # column name for sample ID
# indicate the samples are QC samples or subject
samples (also useful
```

```
# when the train samples contain different types of
QC samples):
col_sampleType = "sampleType", # column name for
sample type
col_batchID = "plateID", # column name for batch/
plate ID
# optional, column name for injection order values:
col_order = "injectionOrder",
# optional, column name for well position values:
col_position = "wellPosition",
# parameters for computing the target values:
targetVal_method = "mean", # or use "median"
targetVal_batchWise = FALSE, # if use batch-wise
target value?
# parameters for selecting highly correlated
variables:
selectVar_corType = "cor", # use "pcor" for partial
correlation
selectVar_corMethod = "pearson", # "pearson" or
"spearman"
selectVar_minNum = 5, # at least 5 variables will
be selected
selectVar_maxNum = 10, # select no more than 10
variables
# indicates how many cores will be used for
parallel computing
# (if -1, all CPU cores detected on the current
host will be used):
parallel.cores = 2)
### end of R code: run TIGER for batch effect
removal
```

A major innovation and highlight of TIGER is its ability to perform data correction across multiple datasets, which is particularly valuable for longitudinal studies. This is achieved through defining target values. As described in the previous text, the target value acts as a theoretical reference point that the model aims to approximate. Each metabolite has one target value. In cross-sectional datasets, the target value is by default defined as the mean across all QC samples. To enable adjustment across multiple datasets, the same set of target values can be used to adjust all datasets, allowing the model to align data from different datasets to a comparable level. Using TIGER's compute_targetVal( ) function, the target values may be calculated as the mean of all QC samples across all datasets, or as the mean from QC samples of a single dataset, depending on which reference the users want the model to approximate. This strategy can also be applied to harmonize datasets obtained from different measurement kits.

The previous example showed how to remove technical variation for a cross-sectional dataset using the target values automatically computed from the training samples. The next example demonstrates how to correct and align datasets generated by different kits. This example also reflects how we adjust for technical variation introduced by the p150 and p180 kits in the KORA S4, F4, and FF4 datasets.

```
### R code: run TIGER for datasets alignment
data(FF4_qc)
# again, use samples labeled as "QC" as training
samples
train_samples <- FF4_qc[FF4_qc$sampleType ==
"QC",]
# Suppose we align two datasets, one containing QC1
and QC2 samples,
# and the other containing QC2 and QC3 samples
test_subset1 <- FF4_qc[FF4_qc$sampleType %in%
c("QC1", "QC2"),]
test_subset2 <- FF4_qc[FF4_qc$sampleType %in%
c("QC2", "QC3"),]
# compute target values
# here, we use values computed from train_samples
as reference,
target_val <- compute_targetVal(
QC_num = train_samples[-c(1:5)], # numeric
metabolite values
sampleType = train_samples$sampleType, # sample
types
targetVal_method = "mean", # use default setting
targetVal_batchWise = FALSE, # disable batch-wise
mode
# if targetVal_batchWise = TRUE, batch/plate ID is
required
# (this argument will be ignored if not using
batch-wise mode)
batchID = FF4_qc$plateID)
# adjust the first test set
test_align_1 <- run_TIGER(
test_samples = test_subset1,
train_samples = train_samples,
col_sampleID = "sampleID",
col_sampleType = "sampleType",
col_batchID = "plateID",
col_order = "injectionOrder",
col_position = "wellPosition",
targetVal_external = target_val # pass computed
target values)
# adjust the second test set
test_align_2 <- run_TIGER(
test_samples = test_subset2,
train_samples = train_samples,
col_sampleID = "sampleID",
col_sampleType = "sampleType",
col_batchID = "plateID",
col_order = "injectionOrder",
col_position = "wellPosition",
targetVal_external = target_val # use the same
target values)
### end of R code: run TIGER for datasets alignment
```

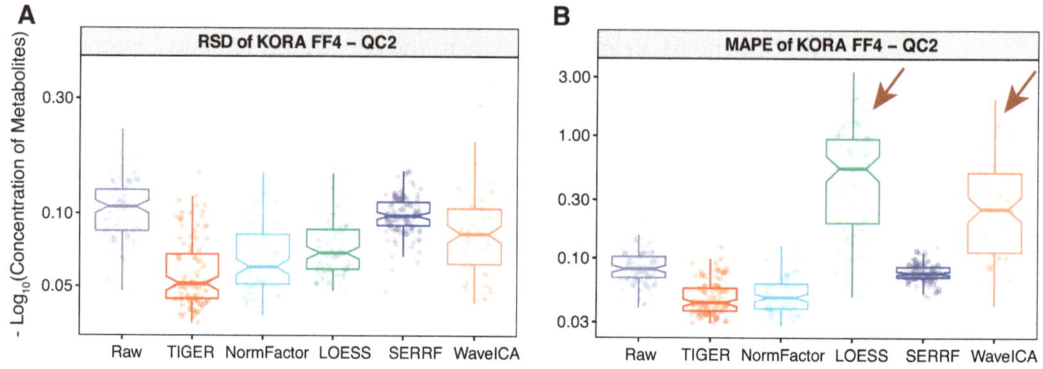

Fig. 3 RSD and MAPE of QC2 of the KORA FF4 study. (**A**) Relative standard deviation (RSD) for raw data, TIGER, NormFactor, LOESS, SERRF, and WaveICA of QC2 from the KORA FF4 study. (**B**) Mean absolute percentage error (MAPE) for the same methods, demonstrating their performance in reducing technical variations. Black arrows highlight specific differences in performance across methods. Both plots are presented in the TIGER paper as part of Fig. 3 [1]

TIGER does not require hyperparameter tuning, but it does offer some options to adjust the training behaviors of its models. These are technical settings intended for users interested in evaluating or customizing the ensemble architecture implemented in TIGER. Readers interested in further technical details may refer to the TIGER documentation.

3.4 Evaluating Models Using Both RSD and MAPE

Traditional machine learning models often lack statistical metrics, such as *p* values, which are commonly used in hypothesis testing. To address this and evaluate the performance of models in TIGER, RSD, and mean absolute percentage error (MAPE) were implemented as key metrics [1].

- RSD measures the variability of predictions relative to the mean and is expressed as a percentage. Lower RSD values indicate that the model's predictions are consistent.
- MAPE quantifies the mean absolute difference between predicted and actual values as a percentage of the actual values. A lower MAPE reflects higher prediction accuracy.

Ideally, both RSD and MAPE should be low, as this indicates a model that is both consistent and accurate, as demonstrated with TIGER, normalization factor (normFactor) [13, 14], and SERRF [7] (Fig. 3). However, when a model achieves a low RSD but has a high MAPE, as observed with LOESS (local polynomial regression) [15] and WaveICA [8] (Fig. 3A, B), it suggests overfitting or over-correction. In such cases, the model may perform well on training data but fails to generalize to unseen data, thereby limiting its practical application (*see* **Note 7**).

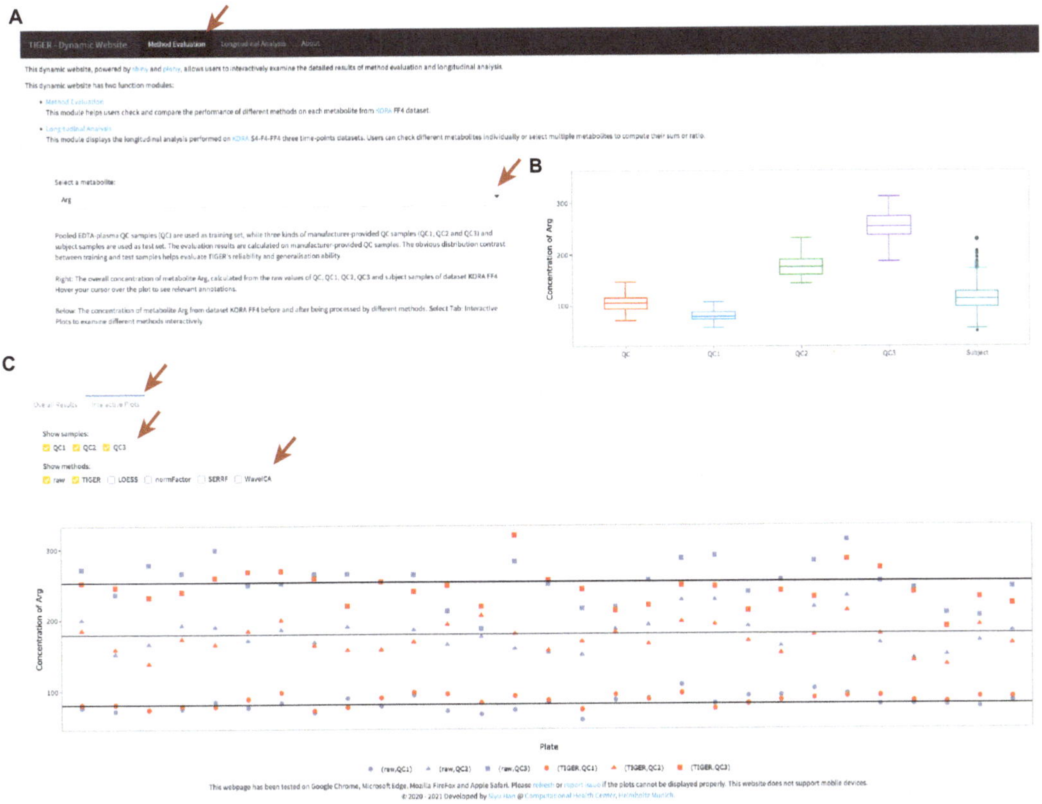

Fig. 4 Screenshot from the TIGER dynamic website illustrating method evaluation with Arginine (Arg) as an example. (**A**) The "Method Evaluation" section is highlighted with a red arrow. This section provides a brief explanation of the selected method and its evaluation. (**B**) A dropdown menu allows users to select a metabolite of interest from the 103 analyzed metabolites, indicated with a red arrow. Boxplots display the concentration of Arg across raw values of QC, QC1, QC2, QC3, and subject samples across plates from the KORA FF4 study. (**C**) A scatter plot shows "Interactive plots" across QC1, QC2, QC3, and TIGER-normalized values of Arg. Here, four additional methods (e.g., LOESS, normFactor, SERRF, and WaveICA) can also be selected to compare their performance

3.5 The TIGER Dynamic Website Illustrating Method Evaluation and Longitudinal Analysis

A dynamic website (https://han-siyu.github.io/TIGER_web/) was developed to visualize TIGER's performance and explore longitudinal analysis from the KORA S4, F4, and FF4 studies. This platform provides an interactive interface to compare results across methods, focusing on both method evaluation and longitudinal analysis.

The website enables users to assess TIGER's ability to correct plate effects and compare its performance with other commonly used methods, such as LOESS, normFactor, SERRF, and WaveICA.

As an example, in the "Method Evaluation" section (Fig. 4), the metabolite Arginine (Arg) is used to demonstrate the platform's functionality, highlighting how TIGER effectively normalizes data and facilitates comparisons across methods.

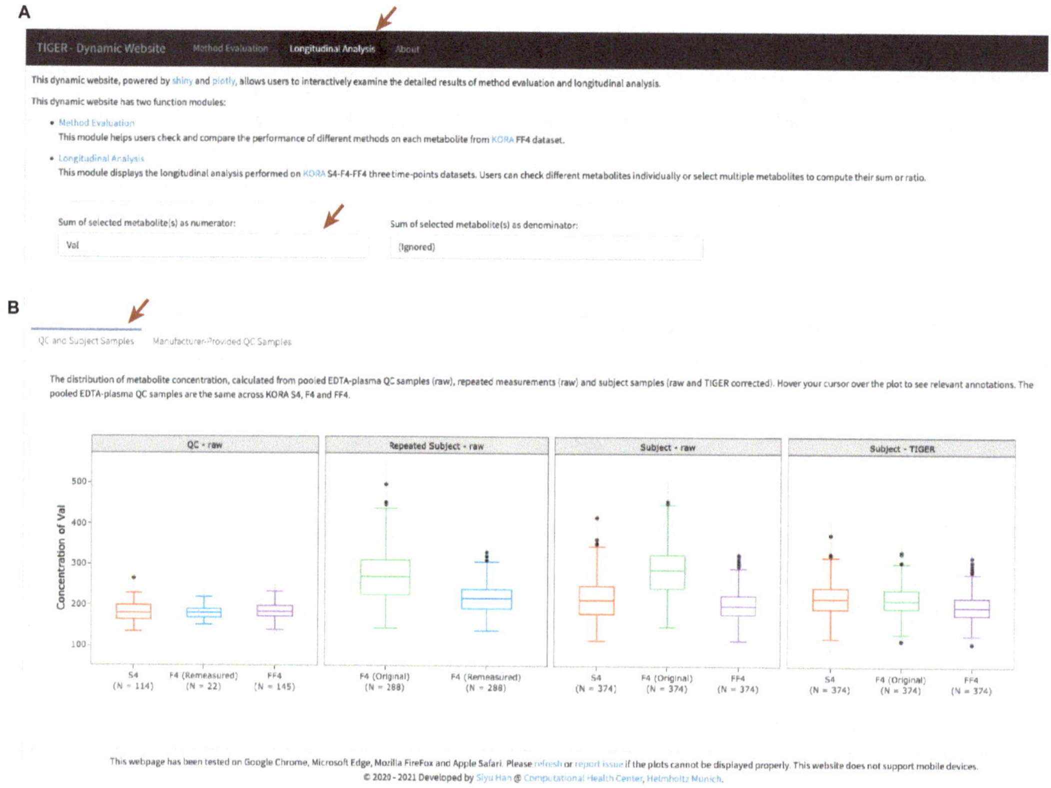

Fig. 5 Screenshot from the TIGER dynamic website showing longitudinal analysis with Valine (Val) as the selected metabolite. (**A**) This figure displays the "Longitudinal Analysis" section as indicated with a red arrow. The dropdown menu allows users to select a metabolite of interest from the 103 available metabolites using Valine as an example. (**B**) By selecting "QC and Subject samples," the plot compares raw values of QC, original and remeasured KORA F4 data, and raw and TIGER-normalized subject sample results across the KORA S4, F4, and FF4 of longitudinal studies

The dynamic website not only serves as a tool to visualize method evaluation but also offers insights into the reproducibility and reliability of results. By integrating longitudinal data, the platform further aids in identifying consistent trends and variations in metabolite concentrations over time.

For example, the section "Longitudinal Analysis" showcases raw values of QC, original and remeasured KORA F4 study data, as well as TIGER-normalized subject sample results in the longitudinal analysis of the KORA S4, F4, and FF4 studies (Fig. 5). The boxplots

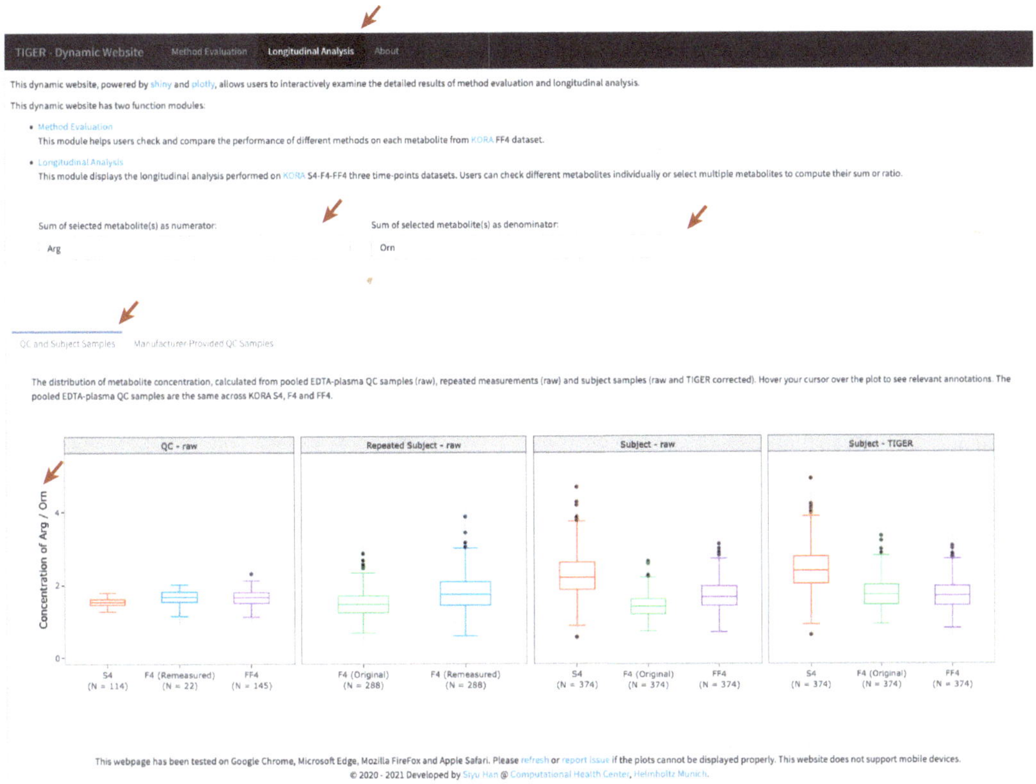

Fig. 6 Longitudinal analysis of the metabolite ratio of Arginine to Ornithine (Arg to Orn) using TIGER. This figure displays how the dropdown menus allow users to select a metabolite ratio from the 103 available metabolites. Boxplots for the ratio of Arginine to Ornithine (Arg to Orn). The plots compare raw values of QC, original and remeasured KORA F4 study data, and TIGER-normalized subject sample results across the KORA S4, F4, and FF4 studies

provide a clear visualization of the metabolite's concentration over time, illustrating how TIGER normalization reduces variability and facilitates the interpretation.

Moreover, this section includes the ratio of metabolites, such as Arginine to Ornithine (Arg to Orn) (Fig. 6) and the ratio of metabolites such as Valine to Glycine + Serine (Val to Gly + Ser) (Fig. 7). These ratios provide additional insights into metabolic changes and demonstrate TIGER normalization's ability to preserve biological signals while reducing technical variation.

Fig. 7 Longitudinal analysis of the metabolite ratio of Valine to Glycine + Serine (Val to Gly + Ser) using TIGER. This figure displays how the dropdown menus allow users to select a metabolite ratio from the 103 available metabolites. Boxplots for the ratio of Valine to Glycine + Serine (Val to Gly + Ser). The plots compare raw values of QC, original and remeasured KORA F4 study data, and TIGER-normalized subject sample results across the KORA S4, F4, and FF4 studies

4 Notes

1. **TIGER requires at least three QCs per plate**

   To effectively correct plate effects, certain basic requirements must be met. A minimum of three identical QC samples per plate is necessary to train multiple base models using cross-validation, where the QCs are split into training and validation folds. This ensures that the models are robust and capable of accounting for variations across different plates.

   Insufficient QC samples:

   - Problem: If fewer than three QC samples per plate are included, the cross-validation process for training base models becomes unreliable, leading to poor correction of plate effects.

- Solution: Ensure a minimum of three QC samples per plate. For enhanced robustness, use five or more QC samples, as demonstrated in the KORA FF4 study.

Alternative QC strategies:

- If identical QC samples are unavailable, consider remeasuring randomly selected subject samples as QC surrogates (*see* **Note 2**).

Sensitivity and timescale:

- Sensitivity to QC number: The accuracy of TIGER's plate effect correction improves significantly with higher QC numbers.
- Timescale: Incorporating additional QC samples may increase preparation time, but the improved reliability justifies this investment.

2. **Validity of variation between subject and QC samples**

TIGER trains its models using QC samples and tests them on subject samples. While both QC and subject samples are measured on the same plates, it is important to ensure that their variation trends are aligned.

Misaligned variation trends:

- Problem: If technical variation in QC and subject samples differ due to measurement inconsistencies, TIGER's predictions can be invalid.
- Solution: Regularly monitor QC and subject sample trends during preprocessing. Remeasure subject samples alongside QCs if significant discrepancies are detected.

Related techniques and variations, such as cross-kit harmonization:

- When working with longitudinal datasets measured using different analytical kits, remeasuring selected subject samples with the newer kit can bridge variation trends (*see* KORA example in **Note 3**).

Sensitivity and timescale:

- Sensitivity to misalignment:TIGER's performance depends heavily on the alignment of variation trends between QC and subject samples. Misalignment can lead to significant errors.
- Timescale: Regular monitoring of variation trends adds minimal time burden but greatly enhances reliability.

3. **TIGER is free of hyperparameter tuning**

TIGER incorporates multiple base models, each trained using a different combination of hyperparameters. In traditional machine learning workflows, hyperparameter tuning is often required to optimize model performance. This process can be time-consuming and computationally intensive, particularly for large datasets.

TIGER eliminates the need for manual hyperparameter tuning by leveraging an ensemble learning approach. By training multiple base models with varied hyperparameter combinations and combining them through a meta-model, TIGER achieves robust performance without requiring explicit tuning, which also reduces the potential for user-induced errors during the tuning process.

Computational complexity for large datasets:

- Problem:While TIGER eliminates hyperparameter tuning, training multiple base models with varied hyperparameters can still be computationally intensive for large datasets.
- Solution: Use TIGER's default settings (e.g., processing smaller subsets of highly correlated metabolites) to reduce computational demands without compromising accuracy. Or use more cores for parallel computing.

Comparison with traditional methods:

- Unlike traditional machine learning approaches, which require extensive hyperparameter tuning, TIGER's ensemble learning approach streamlines the workflow and reduces user intervention.

Sensitivity and timescale:

- Sensitivity to model weights:The meta-model's exponential weighting function ensures robustness by prioritizing well-performing base models.
- Timescale:TIGER's hyperparameter-free design significantly reduces the time required for model development, making it ideal for automated analytical pipelines.

4. **Using an exponential decay function to assign weights**

To build a robust model, TIGER uses an exponential decay function to assign weights to multiple base models.

Imbalanced model contributions:

- Problem: Without an appropriate weighting mechanism, poorly performing base models could disproportionately affect the meta-model's predictions. Meanwhile, under-performed base models can also contribute to the model's robustness by mitigating the overfitting issue.
- Solution: TIGER uses an exponential decay function to assign weights to base models. Models with lower error ratios (better performance during validation) receive higher weights, while those with higher error ratios contribute less to the overall prediction.

Benefits of exponential weighting:

- This approach ensures a balanced contribution of all base models, avoiding over-reliance on any single model. As a result, TIGER achieves robust performance across datasets, even for noisy or heterogeneous data.

5. **Cleaning data before running TIGER**

Metabolites of poor quality can adversely affect the overall training process of TIGER. To obtain reliable performance, it is essential to first clean the dataset, for example, by removing metabolites with high intra- and inter-plate variation or a large amount of missing values.

High variability in metabolites:

- Problem:Metabolites with high intra- or inter-plate variation can introduce noise and reduce the accuracy of model training and prediction.
- Solution:Remove metabolites with high RSD ≥25% or those with inconsistent measurements across plates.

Undetectable metabolites:

- Problem:Metabolites with a high percentage of missing values or those below the LOD may compromise data quality.
- Solution: Retain only metabolites where at least 50% of measured values are above the LOD and where missing values are ≤ 10%.

Practical insights:

- Cleaning the dataset before running TIGER not only improves prediction accuracy but also significantly reduces processing time. By working with reliably measured metabolites, you ensure a more efficient and accurate modeling process.

Timescale considerations:

- Data cleaning is a critical preprocessing step that may take additional time but ultimately saves time during the model training phase.

6. **Training models with highly correlated metabolites**

To further save processing time, TIGER trains models with a group of five to ten highly correlated metabolites and saves up to about five times when compared to using the full dataset of the KORA FF4 study, while comparable results were obtained.

High computational load for large datasets:

- Problem:Training models on the full dataset can be computationally expensive.

- Solution:TIGER groups highly correlated metabolites (five to ten per group by default) to reduce computational demands without sacrificing performance.

  Efficiency gains:

- In the KORA FF4 study, this grouping strategy reduced processing time by approximately fivefold compared to using the full dataset, while delivering comparable results.

  Variations of the technique:

- The grouping size can be adjusted based on dataset characteristics. For smaller datasets, larger groups may be feasible, allowing for acceptable processing time with minimal trade-offs in accuracy.

7. **Evaluating models using both RSD and MAPE**

   To evaluate the models appropriately, TIGER uses both RSD and MAPE to identify overfitting or underfitting the models. Additionally, cross-validation techniques are employed, ensuring that the model generalizes well across datasets. Additionally, hyperparameter-free designs in TIGER help reduce the risk of overfitting by leveraging multiple base models and assigning weights based on validation performance. By evaluating models with both RSD and MAPE, TIGER provides a more comprehensive assessment of model performance, balancing consistency and accuracy while minimizing the risk of overfitting.

   Overfitting or underfitting:

- Problem: A low RSD but high MAPE indicates overfitting, while high values for both metrics suggest underfitting.
- Solution: TIGER uses cross-validation techniques to ensure models generalize well across datasets. Additionally, the hyperparameter-free design minimizes overfitting by leveraging multiple base models and assigning weights based on validation performance.

  Misinterpre.tation of metrics:

- Problem: Solely relying on one metric (e.g., RSD) may lead to incomplete evaluation.
- Solution: Use both RSD (for consistency) and MAPE (for accuracy) to comprehensively evaluate model performance.

  Benefits of combining RSD and MAPE:

- By evaluating models with both metrics, TIGER balances consistency and accuracy, ensuring reliable predictions. This dual-metric approach also facilitates the early detection of overfitting or underfitting, which also mitigates the risk of misinterpretation in subsequent analysis.

  Sensitivity and timescale:

- Sensitivity: The results of evaluating technical variation removal may differ depending on the metric used, as each metric reflects different aspects of the data. Therefore, a combination of RSD and MAPE is essential for robust analysis.
- Timescale: While evaluating both metrics adds minimal additional time, it significantly enhances model reliability.

References

1. Han S et al (2022) TIGER: technical variation elimination for metabolomics data using ensemble learning architecture. Brief Bioinform 23(2):bbab535
2. Martens A et al (2023) Instrumental drift in untargeted metabolomics: optimizing data quality with intrastudy QC samples. Metabolites 13(5):665
3. Holle R et al. (2005) KORA–a research platform for population based health research. Gesundheitswesen 67(Suppl 1):S19–S25
4. Wang-Sattler R et al. (2012) Novel biomarkers for pre-diabetes identified by metabolomics. Mol Syst Biol 8:615
5. Rabel M et al. (2017) The longitudinal association between change in physical activity, weight, and health-related quality of life: results from the population-based KORA S4/F4/FF4 cohort study. PLoS One 12(9):e0185205
6. Han S et al. (2025) LEOPARD: missing view completion for multi-timepoint omics data via representation disentanglement and temporal knowledge transfer. Nat Commun 16(1):3278
7. Fan S et al. (2019) Systematic error removal using random forest for normalizing large-scale untargeted lipidomics data. Anal Chem 91(5):3590–3596
8. Deng K et al. (2019) WaveICA: a novel algorithm to remove batch effects for large-scale untargeted metabolomics data based on wavelet analysis. Anal Chim Acta 1061:60–69
9. Haid M et al. (2018) Long-term stability of human plasma metabolites during storage at -80 °C. J Proteome Res 17(1):203–211
10. Mittelstrass K et al. (2011) Discovery of sexual dimorphisms in metabolic and genetic biomarkers. PLoS Genet 7(8):e1002215
11. Yu Z et al. (2012) Human serum metabolic profiles are age dependent. Aging Cell 11(6):960–967
12. Shi M et al. (2023) Identification of candidate metabolite biomarkers for metabolic syndrome and its five components in population-based human cohorts. Cardiovasc Diabetol 22(1):141
13. Wang W et al. (2003) Quantification of proteins and metabolites by mass spectrometry without isotopic labeling or spiked standards. Anal Chem 75(18):4818–4826
14. Huang J et al. (2020) Machine learning approaches reveal metabolic signatures of incident chronic kidney disease in individuals with prediabetes and type 2 diabetes. Diabetes 69(12):2756–2765
15. Li B et al. (2016) Performance evaluation and online realization of data-driven normalization methods used in LC/MS based untargeted metabolomics analysis. Sci Rep 6: 38881

Check for updates

Chapter 16

Metabolomics Spectral Processing, Data Analysis, and Multiomics Integration Using MetaboAnalyst 6.0

Zhiqiang Pang, Fiona Hui and Jianguo Xia

Abstract

MetaboAnalyst is a comprehensive, user-friendly web-based platform for metabolomics data analysis and interpretation. It offers a unified analytical workflow with an intuitive interface designed to streamline data analysis for both targeted and untargeted metabolomics, while supporting integration with other omics datasets. MetaboAnalyst 6.0 represents a significant advancement toward this vision by supporting LC–MS raw spectral processing and MS2 spectral annotation. This chapter builds on our 2019 protocols, which presented 12 step-by-step procedures for statistical and functional analysis using MetaboAnalyst 4.0, with emphasis on targeted metabolomics. Here, we extend that foundation with eight new or substantially revised protocols for MetaboAnalyst 6.0, centered mainly on LC–MS untargeted metabolomics, advanced statistics, and functional analysis. The main topics covered include:

- **Basic Protocol 1:** LC–MS raw spectral processing with or without MS2 spectra
- **Basic Protocol 2:** MS2 spectral annotation
- **Basic Protocol 3:** Functional analysis of untargeted LC–MS metabolomics data
- **Basic Protocol 4:** Advanced statistical analysis for studies with multiple factors
- **Basic Protocol 5:** Dose–response analysis
- **Basic Protocol 6:** Pathway and joint-pathway analysis
- **Basic Protocol 7:** Causal analysis
- **Basic Protocol 8:** Using MetaboAnalystR for batch processing

Keywords LC–MS, MS/MS, Spectral processing, Compound identification, Covariate adjustment, Pathway analysis, Dose response analysis, Causal analysis, Mendelian randomization

---

These authors contributed equally Zhiqiang Pang and Fiona Hui

Michael Witting (ed.), *Computational Methods and Data Analysis for Metabolomics*, Methods in Molecular Biology, vol. 3063, https://doi.org/10.1007/978-1-0716-5452-1_16,

1 Introduction

MetaboAnalyst (metaboanalyst.ca) is a comprehensive, web-based platform used by 100,000s of researchers worldwide for metabolomicsdata analysis. Through its well-designed user interface, MetaboAnalyst offers streamlined workflows ranging from raw spectral processing to biological interpretation, supporting various common data formats generated in metabolomics studies. The first three versions are centered on statistical and functional analysis for targeted metabolomics [1–3]. Since version 4.0, the focus has been shifted to address key bottlenecks in global or untargeted metabolomics [4–6]. The latest version, MetaboAnalyst 6.0, contains 18 modules to guide users to perform various common tasks in metabolomics. In 2019, we published a book chapter that provides step-by-step protocols on statistical and functional analysis using MetaboAnalyst 4.0 [7]. Based on our 2019 work, this new book chapter offers comprehensive protocols introducing the eight new or significantly updated modules in the last two versions (5.0 and 6.0), including:

- **Basic Protocol 1:**LC–MS raw spectral processing with or without MS2 spectra
- **Basic Protocol 2:** MS2 spectral annotation
- **Basic Protocol 3:** Functional analysis of untargeted LC–MS metabolomics data
- **Basic Protocol 4:** Advanced statistical analysis for studies with multiple factors
- **Basic Protocol 5:** Dose–response analysis
- **Basic Protocol 6:**Pathway and joint-pathway analysis
- **Basic Protocol 7:**Causal analysis
- **Basic Protocol 8:** Using MetaboAnalystR for batch processing

Users can follow the first seven protocols by directly visiting the MetaboAnalyst website. Open a modern web browser (e.g., Chrome, Firefox, or Microsoft Edge) and navigate to the MetaboAnalyst home page (https://www.metaboanalyst.ca). Once the website loads, click the "Click here to start" button located at the center of the page to access the Module Overview page (Fig. 1). The subsequent analyses start with selecting the appropriate module corresponding to your data type and objectives.

2 Materials

Metabolomic data analysis was conducted using the MetaboAnalyst web platform and the MetaboAnalystR R package. For web-based analysis, a desktop or laptop computer running a modern operating

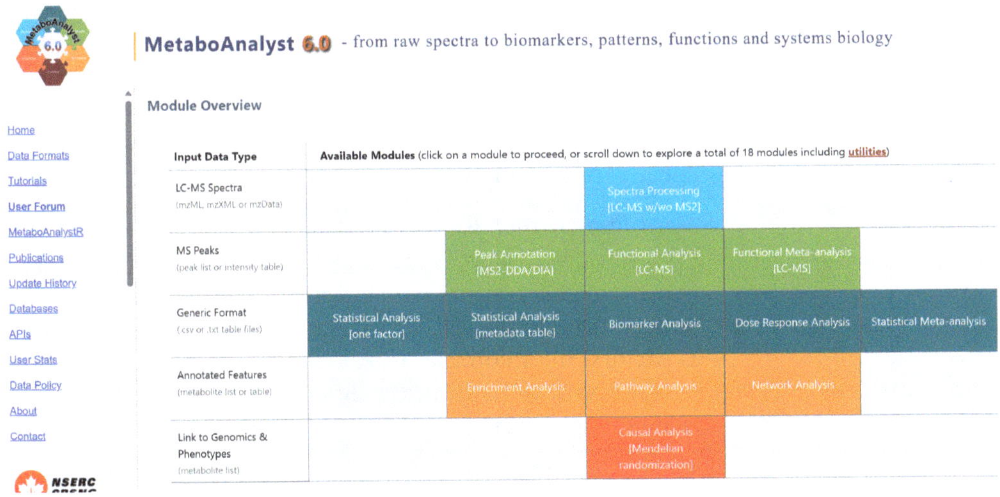

Fig. 1 MetaboAnalyst 6.0 Module Overview page

system (Windows, macOS, or Linux) with an up-to-date web browser (e.g., Chrome, Firefox, or Edge) is required. A minimum screen resolution of 1920 × 1080 pixels is recommended to ensure proper display of interactive plots and tables. For local analyses using MetaboAnalystR, an R environment (version 4.1 or later) with RStudio is recommended, along with at least 8 GB of RAM (16 GB recommended for large datasets). Adequate disk space (≥10 GB free) is required for package installation and intermediate files. A stable broadband Internet connection (≥30 Mbps) is necessary for accessing the web server, uploading data files, and downloading results. For interactive 3D PCA visualization, a GPU or integrated graphics processor that supports WebGL (e.g., OpenGL ES 2.0 or later) is recommended to ensure smooth rendering and user interaction.

3 Methods

3.1 Basic Protocol 1: LC–MS Raw Spectral Processing With or Without MS2 spectra

Global or untargeted metabolomics aims to study all potential metabolic features in biological samples. Due to its high sensitivity and broad coverage, liquid chromatography-mass spectrometry (LC–MS) is the most widely used analytical instrument for untargeted metabolomics. In addition to biological replicates (at least three per group), the experimental design of untargeted metabolomics often includes quality control (QC) and/or blank samples to ensure data reliability and reproducibility [8]. QC samples are prepared by pooling equal aliquots from all study samples and are injected regularly throughout the analytical sequence to monitor instrument stability, retention time drift, and signal intensity variation. Blank samples are used to identify background noises or contaminations from solvents and sample preparation. In addition, pooled QC samples are ideal for tandem MS (MS/MS or MS2) data acquisition, providing high-quality fragmentation spectra to support metabolite identification in untargeted metabolomics. This protocol describes LC–MS raw spectral processing with or without MS2 spectra.

1. *Data Preparation.* Raw spectral data files in vendor-specific formats must first be converted into open formats (e.g., mzML, mzXML, CDF, or mzData) in centroid mode. We recommend using ProteoWizard [9] for the task and setting "peak picking" as the first filter condition[1]. All converted and formatted spectral files should be individually compressed into *.zip files to facilitate uploading.
   - It is highly recommended to include a metadata file indicating the main study groups of each spectral file. The metadata file should contain two tab-separated columns: the first column lists all spectral file names, and the second column contains their group information. Quality Control (QC) and Blank injections should be labeled as "QC" and "BLANK," respectively. MS2 spectra files should be labeled as "MS2" in the metadata table.
2. *Module Selection.* Go to the MetaboAnalyst website and the Module Overview page, click the "Spectral Processing[LC–MS w/wo MS2]" panel to go to the "LC-MS Spectra Upload" page.
3. *Spectral Uploading.* In the center of the page, click the "Select" button to launch the File Selection dialog. Navigate to the folder containing all spectral files and the metadata table (Fig. 2). Once all files have been selected, an "Upload" button will appear next to the "Select" button. Click "Upload" to start uploading the spectral data. This process may take some time, depending on your internet connection speed and the current server load. After all spectral files are successfully uploaded, select the appropriate mode from the "Select Mode" option based on your data, and then click "Proceed."
4. *Data Integrity Check.* On this page, all spectral files are listed in the "Data Integrity Check" table, which summarizes the name, centroid status, file size, and (optionally) MS level and metadata group information for each spectral file.
   - Only centroided spectra can be included in the subsequent data processing steps. If any spectral files are marked as "False" in the "Centroid Status" column, you can try to use our built-in approach by clicking the "Convert" button to centroid the spectra. If this does not fix the issue, you can also exclude those spectra by unchecking the box in the

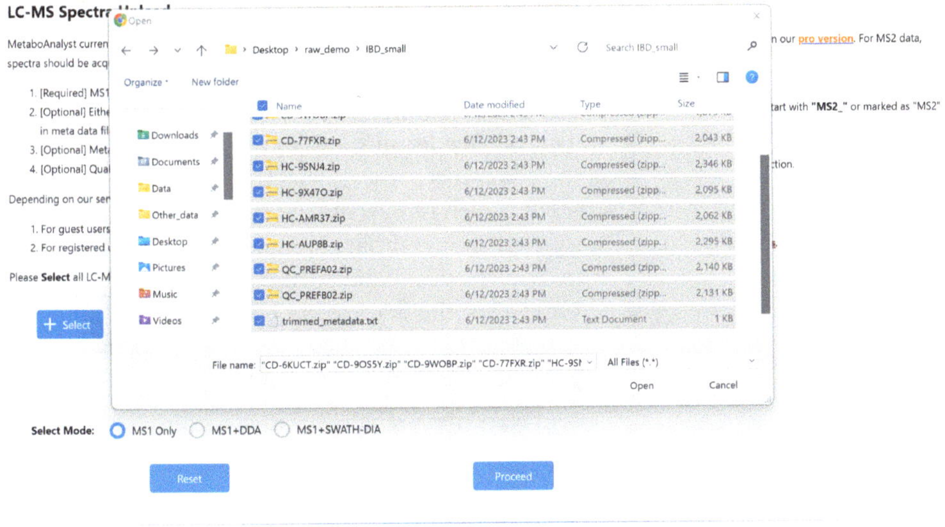

Fig. 2 Spectra uploading by selecting all spectra files together with a metadata table

"Include" column. Once ready, click the "Next" button at the bottom of the page to continue.

5. *(Optional) SWATH Design Check.* The SWATH Design Check is performed only if the dataset contains MS2 spectra acquired using SWATH-DIA (Sequential Window Acquisition of All Theoretical Mass Spectra–Data-Independent Acquisition). In SWATH-DIA, the m/z range is divided into sequential windows, and all ions within each window are fragmented and recorded[2]. A spectra deconvolution is required to correctly assign MS2 spectra to their corresponding precursors.

6. *Parameters Setting.* This is the most critical step in raw spectral data processing. MetaboAnalyst provides several pre-defined parameter sets for different platforms. Select the appropriate platform from the "LC–MS Platform" option to load the default parameters. You can then customize "Peak Picking," "Peak Alignment," "Peak Annotation," "MS2 Processing," and other settings. The key parameters are described below. Note that some parameters may not show depending on your choice for peak picking.

   - Peak Picking: This section includes several parameters. The first is "Algorithms."
     – centWave [10] is a well-established and widely used algorithm for detecting peaks based on the signal-to-noise ratio and peak shape characteristics within each sample at a time. The default option, "centWave-auto,"

automatically optimizes all parameters for peak picking [11].

- Asari [12] is a recent algorithm that leverages the robust, high resolution, and high mass accuracy of modern MS instruments. It works by first creating "Mass Tracks" across all samples for a specific *m/z* value. It then aggregates these mass tracks into a composite mass track, which represents the cumulative signal pattern of a potential metabolite across the entire dataset. Elution peak detection is performed on this composite map. Only after the feature is defined globally are the individual peak areas looked up in the respective samples to generate the final feature table.
- centWave-manual allows advanced users to customize parameters manually.
- MatchedFilter [13] and Massifquant [14] are suitable for low-resolution mass spectrometers or spectra with noisy backgrounds, respectively.

- Peak Alignment: This section contains two key parameters.
  - minFraction specifies the fraction of samples required to form a group.
  - Retention time correction method includes LOESS [15] and Obiwarp [16]. LOESS uses a regression model based on selected features, while Obiwarp uses a dynamic programming algorithm that aligns the entire chromatogram by locally stretching and compressing.
- Peak Annotation: Multiple parameters are available to annotate adducts, isotopes, and isomers.
  - Polarity refers to the ionization mode of ESI, including positive or negative.
  - Various adducts can be customized based on experimental conditions.
  - Other parameters define the annotation types, such as maximum adduct charges and the maximum number of isotopes.
- MS2 Processing: This section is enabled when the uploaded dataset contains valid MS2 spectra. Parameters differ between DDA and SWATH-DIA modes.
  - For DDA, users can define the ppm for MS2 data processing and specify mass spectrometer acquisition parameters.
  - For SWATH-DIA, the filtering value sets the threshold for peak selection during spectra deconvolution.

  - Other customizable parameters include similarity method (dot-product [17] or spectra entropy [18]), target peaks, MS2 databases, and omics type, depending on the study requirements.

Once all parameters are set, click "Submit" and then click "Confirm" to submit the job.

7. *Job Submission and Resuming*. Once the job is submitted, MetaboAnalyst will redirect users to the "Job Status View" page. A job status box displays the detailed job progress. Additional information and logs are shown in the Text Output box at the center of the page (Fig. 3). Since raw spectral processing can take anywhere from minutes to hours, users can click "Create Job URL" and save the job link. This allows users to return later to check the job status. Once the job progress reaches 100% and the status changes from "Running" to "Finished," click the "Proceed" button at the bottom of the page.
   - The submitted job can be canceled at any time by clicking the "Cancel" button. Users can then update the parameters and submit it as a new job.
8. *MS1 Results Visualization*. This page summarizes raw spectral processing results. It contains two main panels. The upper panel includes interactive 3D PCA, intensity summaries, Base Peak Ion (BPI) plot, or Total Ion Chromatograms (TICs) of all samples, metabolome classification, and exposome classification (applicable for exposomics data). By default, the interactive

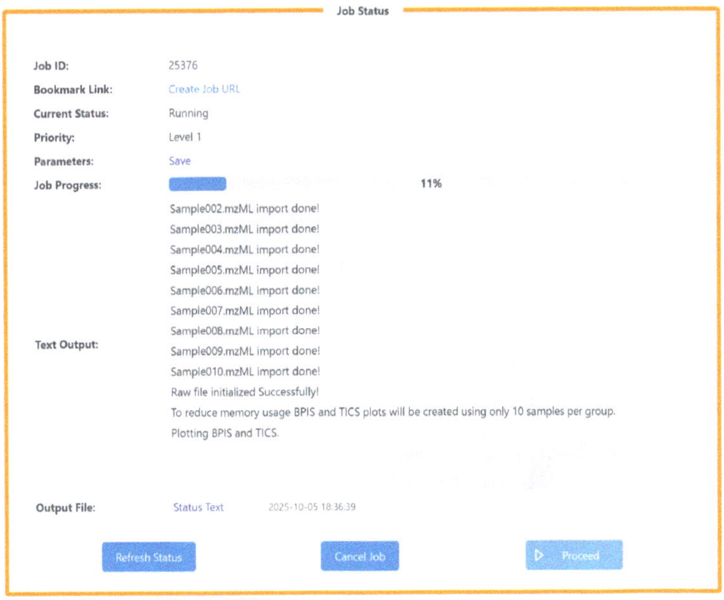

Fig. 3 Raw spectra job processing status page

PCA plot shows the score plot (samples) at the center, while the loading plot (features) as the inset. Double-click a sample in the score plot to view its individual TIC. Users can click the "Switch" icon (double arrows at the top left) to move the loading plot to the center (Fig. 4). Double-clicking a feature in the loading plot will launch an interactive box plot. Users can visually compare its intensities across different groups. Double-clicking a point in the box plot generates a stacked Extracted Ion Chromatogram (EIC) (Fig. 5). Click the other tabs to view the corresponding figures.

9. *Samples and Features Summary.* The lower panel contains multiple tables summarizing all detected MS1 features and identified compounds. The first tab, "Result Summary," provides a general overview of all detected MS1 features, annotated adducts, and isotopes. The second tab, "Spectra/ Sample Table," summarizes detected MS1 features from the perspective of individual samples. It includes detailed information such as the spectral file name, group, peak number, percentage of missing values, retention time range, *m/z* range, etc. The "View" button allows users to view the TIC of the

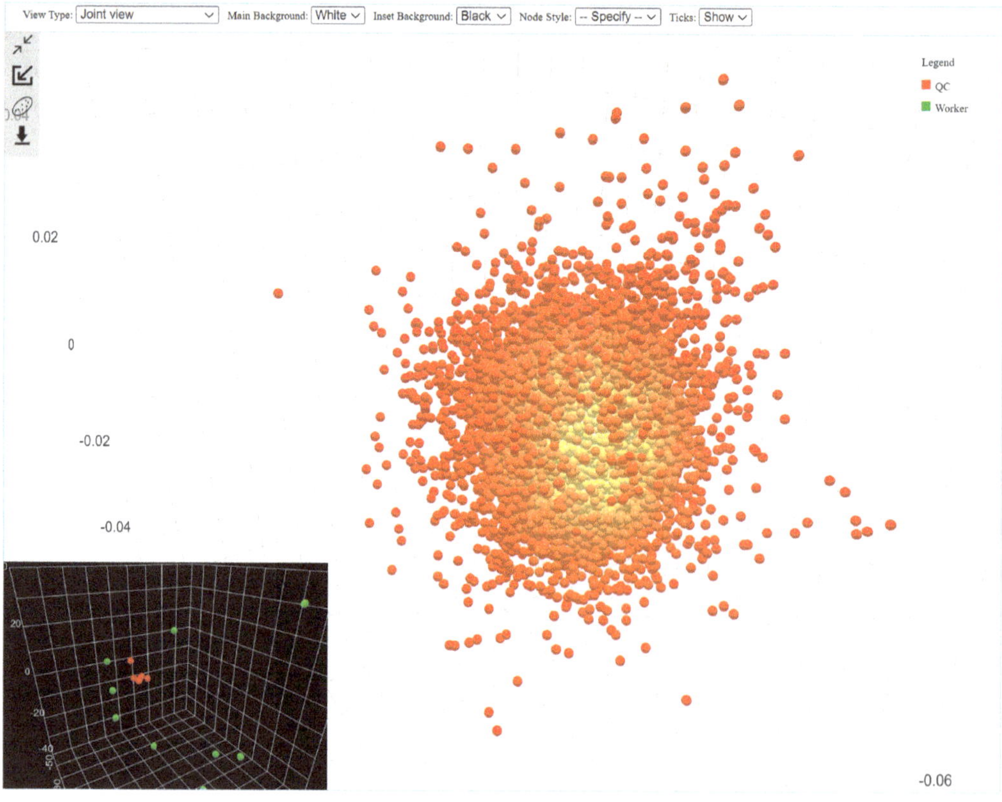

Fig. 4 Visualization of raw LC–MS spectra features in an interactive 3D-PCA from the spectra result page

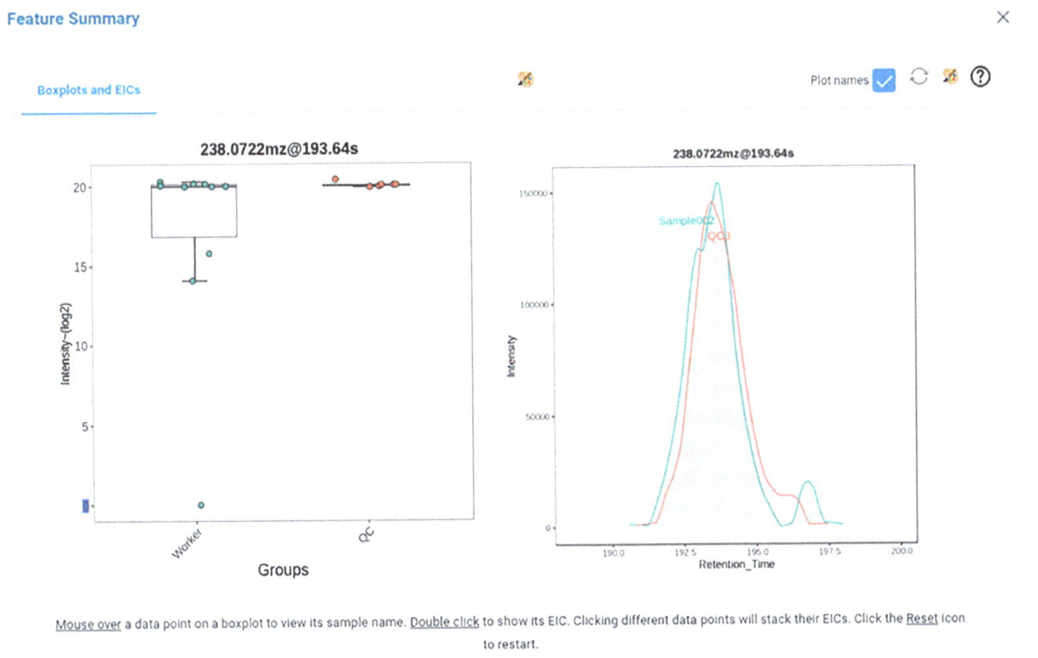

Fig. 5 Statistics of features across samples in a box plot and EICs generated by clicking specific dots from the box plot

corresponding sample. The third tab, "Feature/Peak Table," presents a comprehensive table of all detected MS1 features along with their corresponding information, including *m/z*, retention time, average intensity, coefficient of variation, statistical *p* values, FDR values, and annotation results. This tab also includes a button to generate stacked EIC plots, as described in Step 8.

10. *MS2 Results Exploration*. If the raw spectra include MS2 spectra, the fourth tab, "MS/MS Results," in the lower panel provides detailed compound identification results (Fig. 6). MS2 spectra are organized based on their parent ions. In this table, each row represents an MS1 feature. Click a row to expand it and display all candidate compounds for that MS1 feature, along with their chemical information and similarity matching scores (ranging from 0 to 100) [5, 19]. A score of 100 indicates a perfect match, while lower scores correspond to lower confidence in the identification. Click the "View" button for each chemical candidate to open an interactive mirror plot, which visualizes the matching pattern and fragment information (Fig. 6, inset).

Fig. 6 MS2 results summary table and an example mirror plot showing the matching pattern of MS2 spectra against the reference spectral database

- If the job includes MS2 data, click the "Metabolome" tab in the upper panel to view the metabolome classification pie chart (Fig. 7). By default, the pie chart summarizes all detected compounds across all groups. You can select a specific group using the Groups dropdown menu. You can also change the chemical taxonomic level to view classification information at different levels [20]. Double-click the piece of the metabolome pie chart to view the details of compounds belonging to the class.

11. *(Optional) Exposomics Result Exploration.* If the Omics type was set to "Exposomics" on the previous "Parameters Setting" page, the Processing Results page displays two extra results specific to exposomics data analysis. The first appears on the upper panel under the "Exposome" tab, where exposome classification [21] is visualized as a pie chart (Fig. 7). Use the "Groups" option and click "Update" to view the exposome classification for the selected group. Double-click the piece of the exposome pie chart to view the details of compounds belonging to the class. Additionally, in the "MS/MS Results" tab on the lower panel, an extra column labeled "Classes" will appear in the

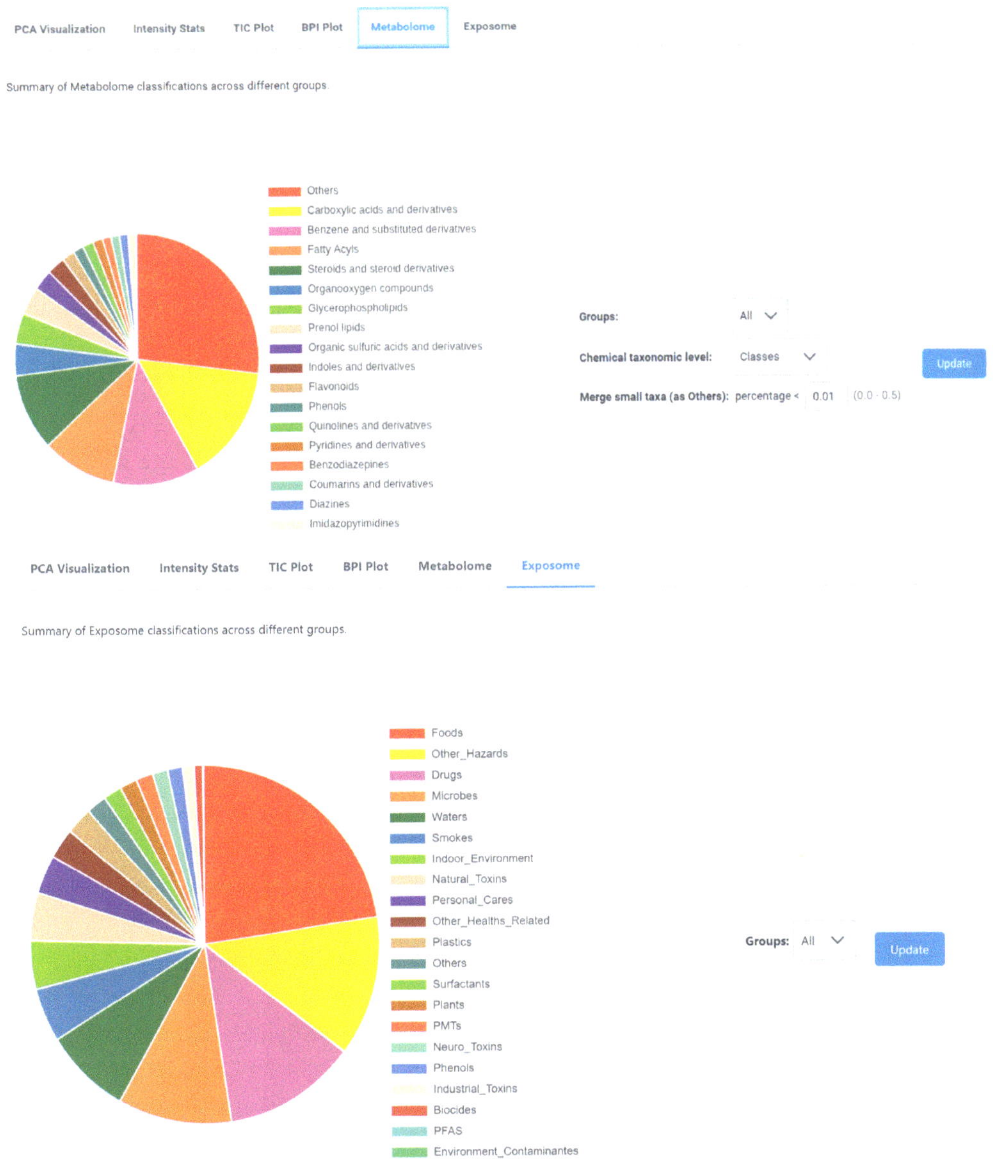

Fig. 7 Metabolome and exposome classification pie charts

expanded row, showing detailed exposomics classification for each compound. Click the hyperlink in this column to view the results in a pop-up dialog.

12. *Results Exporting and Downloading*. Click the "Download Page" button to complete the analysis and go to the results download page. All images and tables generated during raw spectral processing can be downloaded directly from this page. Click "Exit" in the left navigation tree to leave the module.

3.2 Basic Protocol 2: MS2 Spectral Annotations

The module accepts three types of MS2 data: a single MS2 spectrum or multiple MS2 spectra organized in standard msp or mgf formats from DDA or SWATH-DIA. A single MS2 spectrum contains two tab-delimited columns: the first column represents m/z values, and the second column contains intensity or relative intensity values. DDA MS2 data can be generated from raw files in mgf format using ProteoWizard, while SWATH-DIA MS2 data must be preprocessed to deconvolve the spectra. This module is designed for quick MS2 spectral searching. It is highly recommended to follow the workflow described in the previous section for SWATH-DIA analysis rather than using this module directly.

1. *MS2 Spectral Uploading*. Select the "Peak Annotation [MS2-DDA/DIA]" module from the Module Overview page to access the MS2 spectra upload page. The page contains two tabs (Fig. 8). The "Single Tandem Spectrum" tab allows you to copy and paste a single MS2 spectrum into the box on the left side of the page. Similarly, the "Multiple Tandem Spectra"

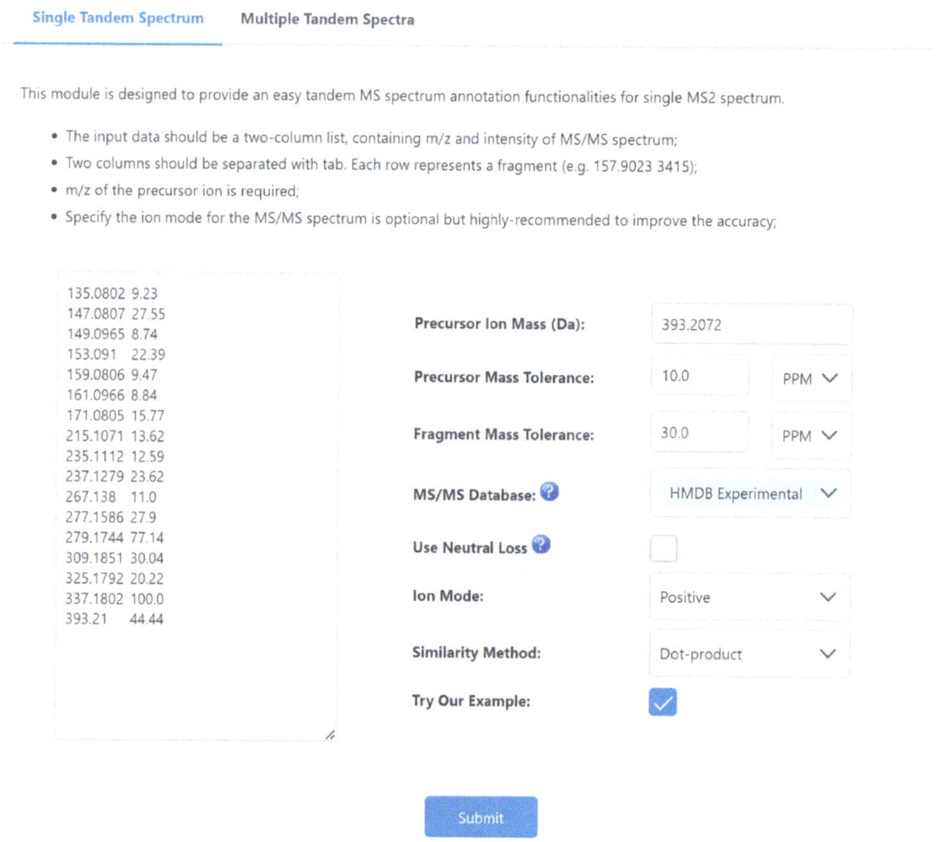

Fig. 8 MS2 spectra uploading interface for annotation. It includes two tabs: Single Tandem Spectrum or Multiple Tandem Spectra

tab allows you to upload a file containing multiple spectra in either mgf or msp format[3].

2. *Parameter Setting*. Multiple parameters must be customized on the upload page. Mass tolerances for precursors and fragments should be set according to the MS instrument. MetaboAnalyst integrates various public MS/MSdatabases for searching, with "HMDB Experimental" as the default. Use the dropdown menu to choose other databases if needed. MS/MS searching can be performed by calculating and comparing neutral losses [22]. Check the "Use Neutral Loss" box to enable this option. Ion Mode must also be specified by the user. Two similarity assessment methods are available: "Dot-product" and "Spectral Entropy." For Single Tandem Spectrum, the mass of the precursor ion must also be defined. Click the "Submit" button to upload the data and parameter settings.
3. *Integrity Check of MS2 Spectra*. An integrity check is performed for multiple MS2 spectra (Fig. 9). To accommodate large users, only 20 spectra can be processed in a single batch search. For a large spectra file, the first 20 are selected by default. Users can click the "Edit Data" button at the bottom to customize which ions to include in the search. In the "MS/MS Spectral Inclusion List Editor," users can manually include or exclude specific ions for the subsequent search (Fig. 9). Click "Submit" to start the MS/MS database search.
4. *Results Exploration*. Once the search is complete, users are redirected to the "MS/MS Annotation Results" page. The MS2 spectra search results are summarized in a table. For a single spectrum, the results are displayed directly in the table.

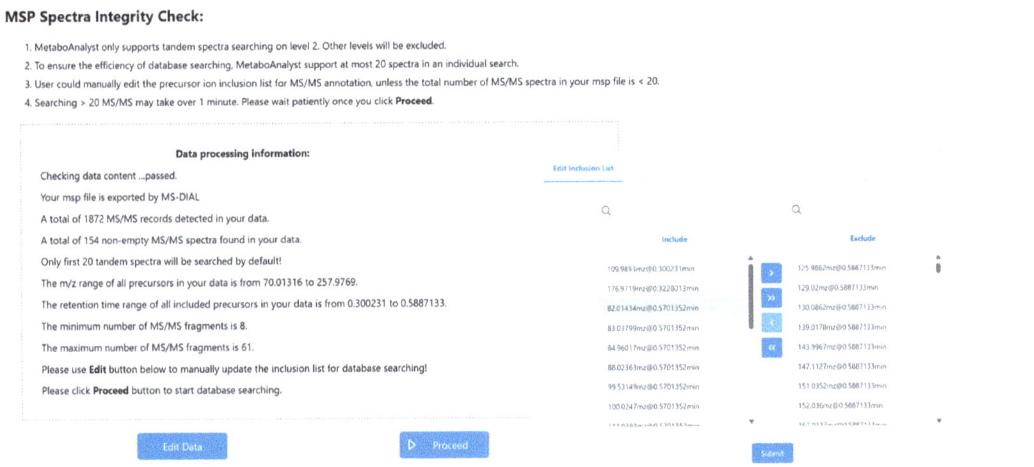

Fig. 9 MS2 spectra integrity check page, summarizing the spectra uploaded, and a dialog allowing customization of the inclusion list MS2 spectra

Fig. 10 MS2 spectra matching results page. A mirror plot is used to show the matching pattern

Click a row to expand it and view the matching pattern using the interactive mirror plot (Fig. 10). For multiple spectra, all results are summarized in the same table, with each row representing a precursor. Click a row to expand it and view all chemical candidates. The matching pattern or mirror plot can be visualized by clicking the "View" button in the last column of the table.

5. *Result Export.* Click the "Download Page" button at the bottom of the table to go to the download page. All images and results can be downloaded directly from the page.

3.3 Basic Protocol 3: Functional Analysis of Global LC–MS Metabolomics Data

Functional analysis for untargeted metabolomics requires a complete set of peaks (not significant peaks) or a peak intensity table generated from high-resolution LC–MS instruments such as Orbitrap or Time-of-Flight (TOF). This is because (1) the algorithms (Mummichog [23] or GSEA [24]) use permutations to correct the instrument bias in coverage; and (2) the algorithms use putative annotations that are "approximately" correct to detect enriched functions. High-resolution MS peaks can significantly reduce false positives. An empirical rule is at least 3000 peaks for common biofluid samples from humans for robust performance.

1. *Data Preparation and Uploading.* Users can upload a complete peak list, a peak table, or a peak list combined with an

annotated compound list. Below is a guide for formatting data prior to uploading:

- A complete peak list: Ideally, a peak list should contain at least four columns: "m.z," "r.t," "t.score," and "p.value," representing the mass-to-charge ratio, retention time, *t* scores, and *p* values of LC–MS features, respectively. The m.z column is always required, while r.t is required for Mummichog version 2. If t.score or p.value columns are not included, all m/z values must be ordered from the most significant to the least significant. For datasets containing mixed ion modes, a "mode" column should also be included in the peak list.
- Peak table: A generic LC–MS peak table format is acceptable for functional analysis. For enrichment analysis, it is recommended to include only two groups for the ease of interpretation.
- Peak list with annotated compound list: Users can upload an MS1 peak list together with a corresponding annotated compound list. The input can be prepared in one of the three ways:
  - If the raw spectra are processed using the LC–MS raw spectral module of MetaboAnalyst, users can directly upload the result file ("compound_annotation_list.csv") obtained from the download page.
  - Prepare a list with the same number of rows as the peak list, where each row corresponds to a row in the MS1 peak list. If compound information is missing for any peaks, fill those rows with NA.
  - Prepare a list containing only compounds. Each row must correspond to a specific row in the MS1 peak list, with the corresponding index (row number) included as the first column.

All compounds should be identified using InChIKeys [25]. Refer to the two examples in the "MS Peaks and Annotations" tab on the Functional Analysis [LC–MS] module upload page to prepare your data.

2. *Data Integrity Check.* The "Data Integrity Check" page summarizes all information about the uploaded peaks. If the uploaded data is formatted as a peak list, review the information and click "Proceed" to continue to the next parameter setting page. If the uploaded data is a peak table, users will be guided through the data filtering and normalization steps.
3. *(Optional) Data filtering and normalization.* On the Data Filtering page, customize the filtering conditions and click the "Submit" button to perform data filtering. Then click

"Proceed" to go to the Normalization Overview page. Adjust the data transformation, normalization, and scaling parameters to normalize the data. Review the resulting distribution using the "View Result" button at the bottom of the page and click "Proceed."

4. *Parameter Setting and Library Selection.* This page consists of two panels. The upper panel allows users to customize algorithms, visual analytics methods, and specify currency metabolites and adducts. Two methods are available for functional prediction: Mummichog [23] and GSEA [24]. For Mummichog, the most critical parameter is the p value cutoff, which defaults to the top 10%. Users can customize this threshold. There are two versions of Mummichog: Version 1 considers only m/z values for empirical compounds; Version 2 further considers retention time when generating the empirical compound list [11]. Note: V2 will be disabled if retention time is not available. GSEA uses the overall rank based on p values or t scores for functional enrichment. Users can visualize results using either a scatter plot or a heatmap. Note that heatmap visualization is only available when the uploaded data is a peak table. Currency metabolites and adducts can also be customized according to the experimental settings. The bottom panel allows selection of different pathway libraries. MetaboAnalyst currently supports over 130 species-specific metabolic pathways based on KEGG [26]. Click the "Proceed" button.

5. *Pathway Result Visualization.* If the visual analytics method is set to "Scatter Plot" in the previous step, the enriched pathways are displayed in a scatter plot (Fig. 11). Each dot

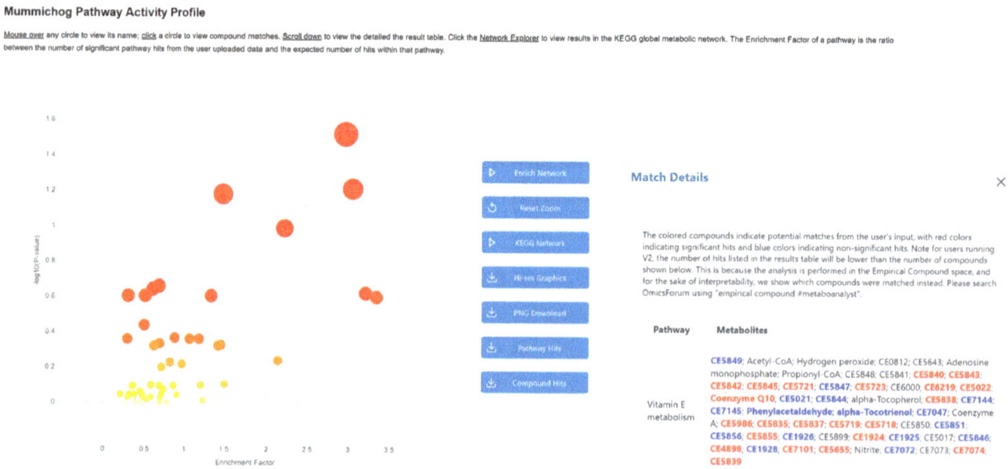

Fig. 11 Data integrity check page on the uploaded feature table. The RSD distribution in different groups is summarized as a violin plot

represents a pathway. The x-axis represents the enrichment factor, while the y-axis shows −log10(p) of the enrichment pathways. The enrichment factor is the ratio between the number of significant hits detected from the user uploaded data and the expected number of hits based on the current model. Pathways clustered in the top-right corner are most likely to be activated/enriched. Click a dot to view detailed matched results in the pop-up dialog (Fig. 11). A summary of all pathway information is provided in the table at the bottom panel of the page.

6. *Enrichment Network.* In addition to a scatter plot, the enriched pathways can be visualized as networks. This visualization is useful to find out if some pathways are driven by shared metabolites. Click the "Enrich Network" button next to the scatter plot. In the enrichment network, pathways are connected based on shared compounds. Double-click a pathway node to expand it and view the detailed compounds (Fig. 12). Use the top-left panel to locate a pathway of interest and view its complete list of compounds in the bottom-left panel. Click the "View Result" hyperlink in the top menu to return to the results page.
7. *KEGG Global Network.* Click the "KEGG Network" button next to the scatter plot. All enriched pathways are mapped onto a global metabolic network (Fig. 13). Pathways are listed in the top-left panel, and all hits for the selected pathway are shown in the bottom-left panel. Click a pathway to highlight it in the

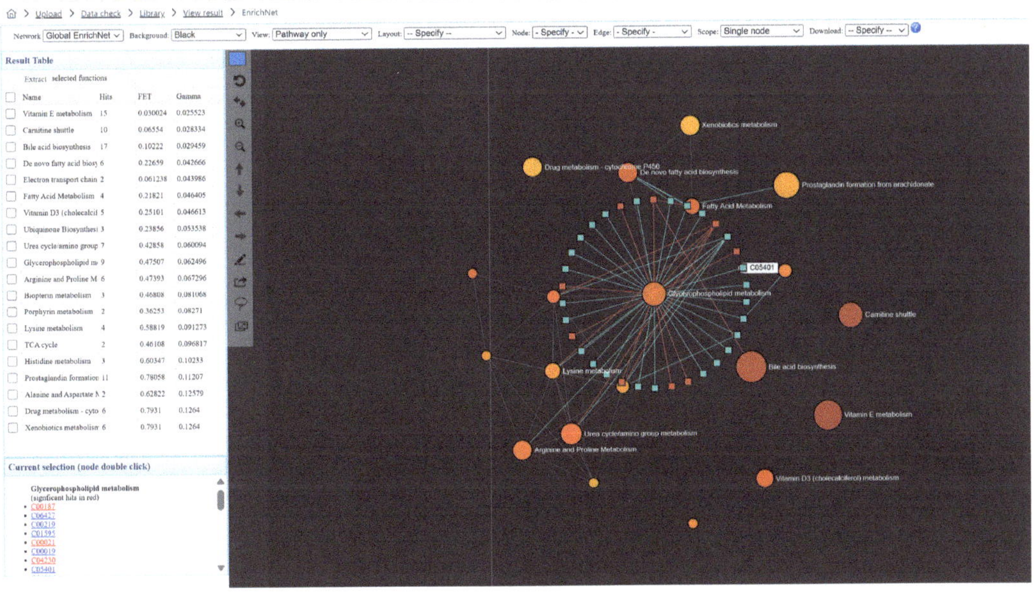

Fig. 12 Metadata editor to adjust the order of metadata factors

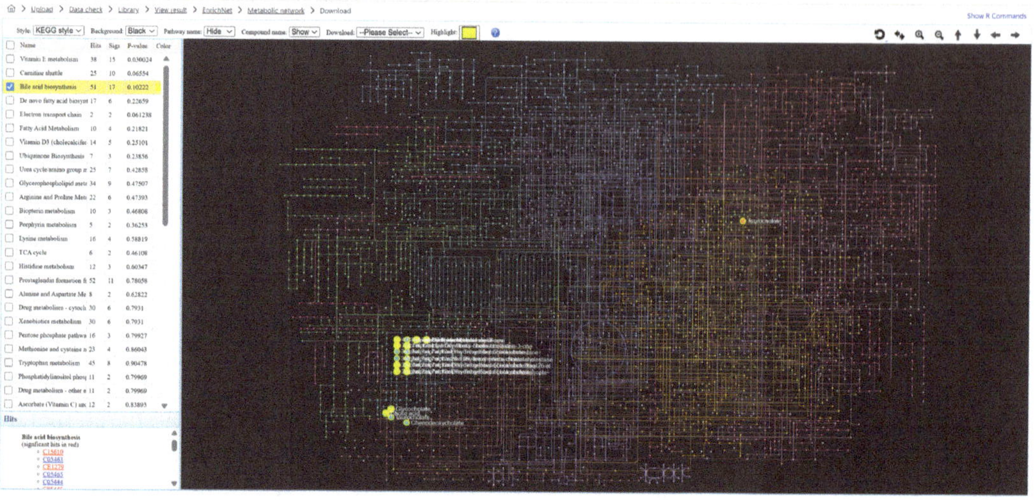

Fig. 13 Missing value estimation panel showing missing value percentage and average abundance in a box plot. Missing values across samples are shown in a heatmap

global network view. It complements scatter plot and enrichment network, by providing an intuitive overview of functional changes within the context of the global metabolism.

8. *Heatmap-Based Functional Analysis.* When a peak intensity table is uploaded, and the visual analytics method is set to

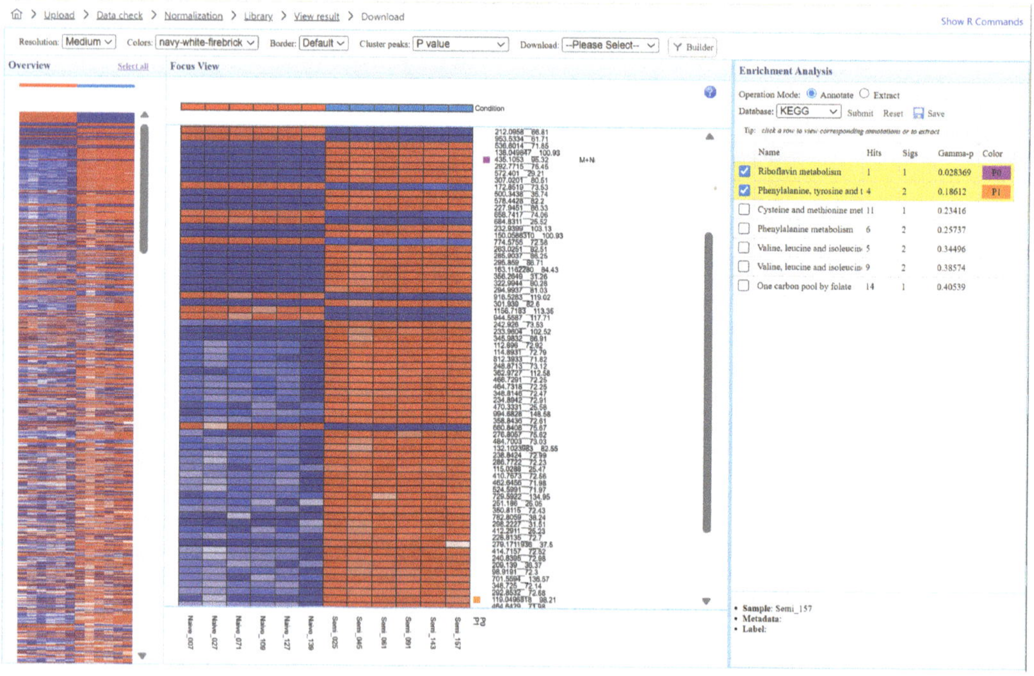

Fig. 14 Visualization of data normalization results

"Heatmap," clicking the "Proceed" button on the Parameters and Library View page will enter the Heatmap View page. As shown in Fig. 14, the left panel (Overview) shows heatmaps for all features, and the center panel (Focus view) shows the currently selected features of interest (default to all significant features). The enrichment analysis is performed for the features in the Focus view. A unique feature of the heatmap-based visualization is to allow users to perform functional analysis based on any patterns of interest. Locate the top menu bar and change the "Cluster Peaks" option from the default "*p* value" to "Ward's Method" to facilitate pattern detection. To focus on a specific region, click and drag on the left panel to select a pattern of interest, which will appear in the center focus view.

- Users can also "stitch" multiple regions of interest. Click the "Builder" icon in the top menu bar. A "heatmap editor" area (stitching area) will show at the bottom of the page. Continue selecting areas from the focus view to add them to the stitching area. Repeat this process until all regions of interest are selected and stitched. Click "To Focus View" to move the stitched patterns to the focus view. Finally, click the "Submit" button in the right panel to perform functional analysis. All enriched pathways are then listed in the right-side panel. Click a pathway name to display detailed matching information in the focus view heatmap (Fig. 14).

9. *Result Export.* Click the "Download" button at the bottom of the page (Scatter plot view) or click the "Download" hyperlink from the top navigation tree (other result view pages). On the download page, save all the images and tables generated during the analysis session.

3.4 Basic Protocol 4: Advanced Statistical Analysis for Studies with Multiple Factors

Metabolomics is increasingly applied to clinical and population studies. Compared to designed experiments where there is only one or two study factors (e.g., control vs. disease), clinical or observational studies are characterized by many metadata factors or descriptors associated with each subject (e.g., sex, age, BMI). In addition to the primary experimental factors, these metadata factors should be considered during data analysis. MetaboAnalyst offers multiple advanced statistical and visualization approaches to address this need.

1. *Metadata and Data Preparation.* From the Module Overview page, select the "Statistical Analysis [Metadata Table]" module for analysis. A metadata table is required for statistical analysis involving multiple factors. The metadata table typically consists of multiple columns, with the first column containing sample names and the remaining columns containing various descriptors of the samples, which can be either discrete or continuous. The table should be saved as a CSV file. The data file contains

all samples, features, and their corresponding compound concentrations or peak intensities. Multiple example datasets are provided on the data upload page. Select the data and metadata files, specify the Study Design and Data Format based on your dataset, and click the "Submit" button to submit your data. We will use the "E-waste" data in the following analysis steps.

2. *Data Integrity Check*. The "Data Integrity Check" page summarizes all features and samples for the uploaded data (Fig. 15), including details on missing value distributions (when detected). Click the "Proceed" button to continue.
   - When QC samples are present, MetaboAnalyst will automatically generate a relative standard deviation (RSD) plot. Click the "View RSD plot" hyperlink to compare the RSD statistics for biological samples with QC replicates.
3. *Metadata Integrity Check*. Metadata factor information is summarized in the gray box on the Metadata Check page. All metadata factors can be edited and updated using the lower panel of the page. Verify the type of each factor to ensure it is correctly identified as either Categorical or Continuous. To update a group label, click the "Edit" hyperlink. This dialog contains three tabs. If the order of a categorical factor matters,

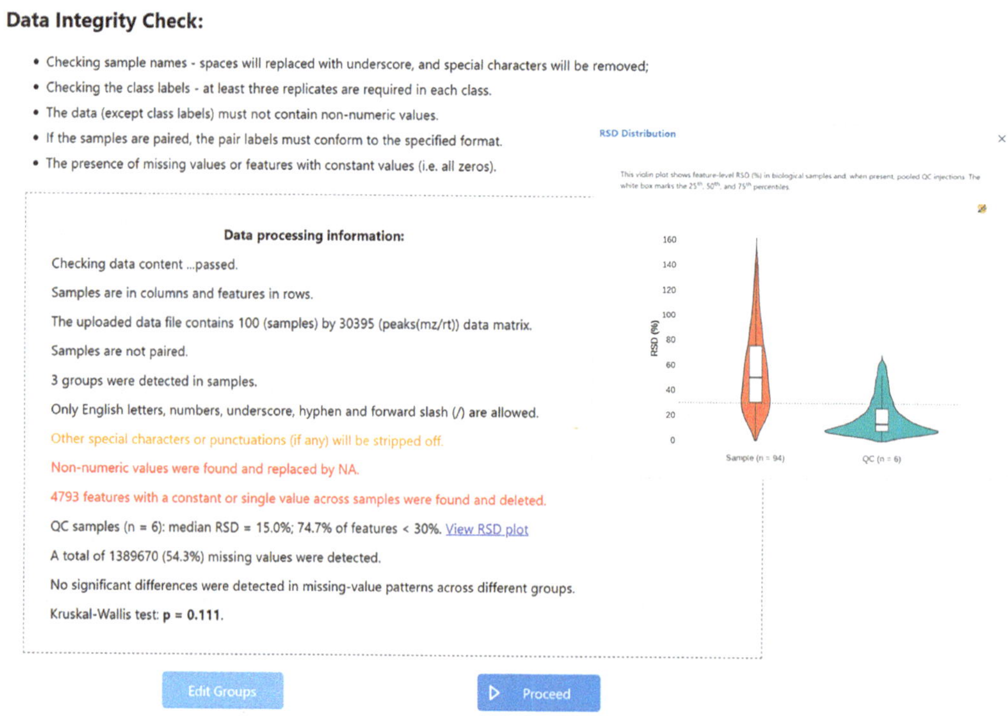

Fig. 15 Metadata visualization in clustering and correlation heatmaps

make sure the order in the second tab, Order (factor-level), is correct. You can manually adjust the order by dragging the labels or using the direction arrows (Fig. 16). Click "Update" to confirm the order, then click "Proceed."

4. *Data Filtering.* Data filtering is important for removing variables that are unlikely to be useful in downstream data analysis. Four different types of filtering methods are available, based on data quality, repeatability, variance, and abundance. More details for each category are provided on the web page. Users can update the default settings for each type and click "Submit" to perform the filtering. The results will be summarized in a message at the top right of the page. Click "Proceed" to continue.

5. *Missing Value Estimation.* When missing values are detected, users will be directed to the Missing Value Estimation page after data filtering. Users can visualize missing value distributions across groups as bar plots or across samples as heatmaps (Fig. 17). The lower panel of the page allows users to specify a method for missing value estimation. There are three categories: left-censored data estimation, univariate statistical methods, and multivariate statistical methods. By default, MetaboAnalyst uses the left-censored data estimation method by replacing missing values with one-fifth of the lowest positive values as their proxies of detection limits. Choose a method and click "Submit" to perform the estimation. Once the estimation is complete, click "Proceed."

6. *Data Normalization.* The Normalization Overview page offers multiple methods organized into three main categories: sample normalization, data transformation, and data scaling. Users

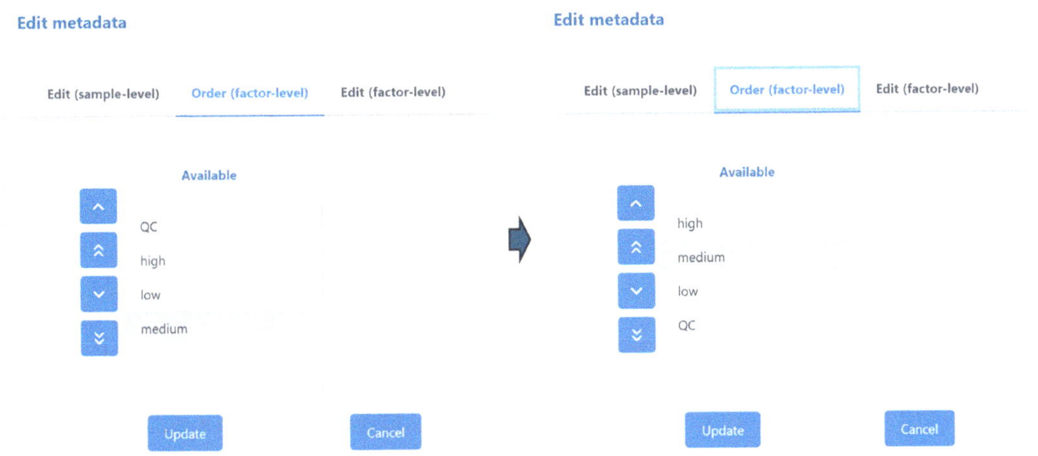

Fig. 16 A pairwise score plot and interactive 3D-PCA visualization of data based on different metadata factors

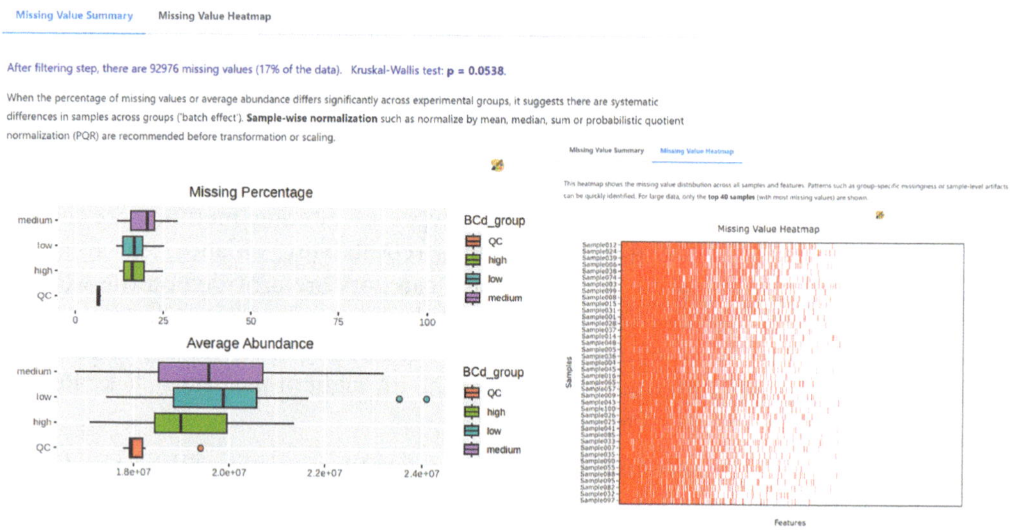

Fig. 17 Covariate adjustment results and feature view dialog of a specific feature in different groups after covariate adjustment

can use one or more methods based on their experiments and data. For example, select "Normalization by median" and "Log transformation (base 10)" to normalize the data, then click "View Result" to examine the distribution. A bell-shaped distribution typically indicates good-quality data normalization (Fig. 18). Click "Proceed."

7. *Metadata and Data Overview*. The Data and Metadata Overview section includes "Metadata Visualization," "Interactive PCA Visualization," and "Hierarchical Clustering and Heatmap Visualization." Click the corresponding hyperlink on the Analysis Overview page to visualize the results.
   - Metadata factors are converted to numerical values and visualized as heatmaps (Fig. 19). The hierarchical clustering heatmaps (left) show the detailed patterns of each metadata factor and their similarity. The correlation heatmaps (right) summarize the associations among metadata factors based on different correlation measures. Since highly correlated metadata factors can cause instability in statistical modeling (i.e., multicollinearity), we should choose only one of such factors for further analysis. Six correlation measures are available to evaluate these relationships, and multiple parameters can be adjusted to optimize the heatmaps.
   - The data can be visualized as interactive 3D PCA plots (Fig. 20), with the color and shape of nodes customized according to different metadata factors. This interactive 3D PCA provides insight into overall data patterns in relation

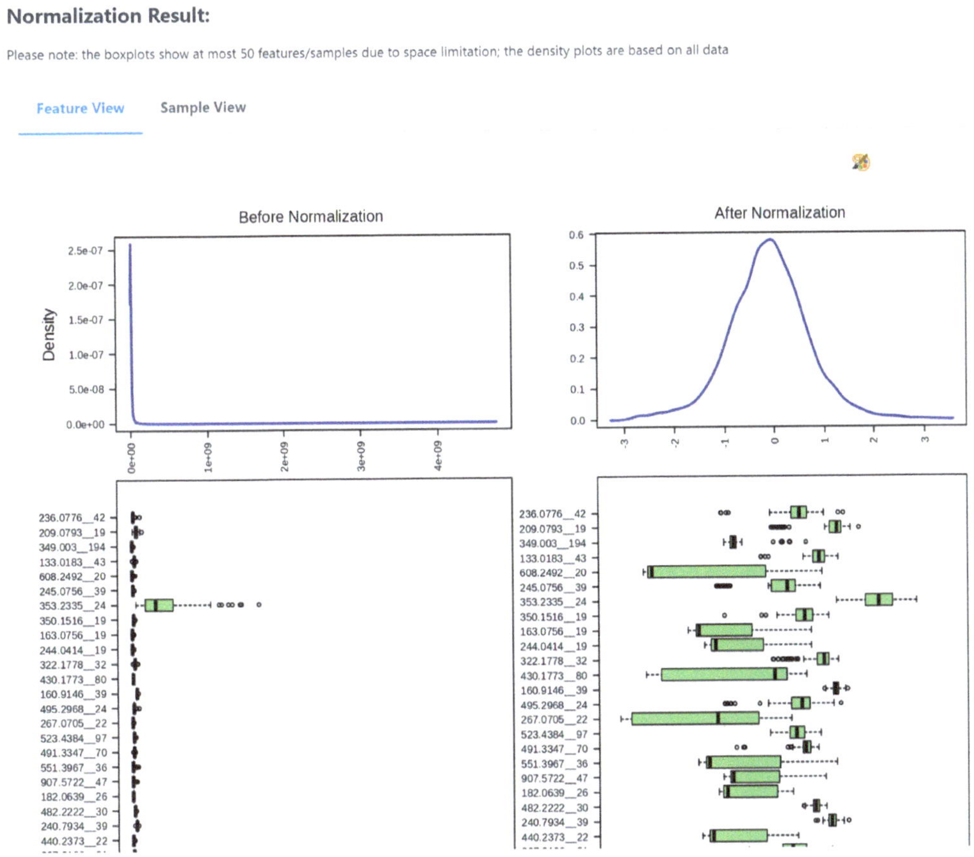

Fig. 18 Scatter plot of functional analysis, and the match details of a specific enriched pathway

to metadata. Similarly, an interactive hierarchical clustering heatmap allows direct visualization of feature abundances across samples and all metadata factors.

8. *Univariate and Multivariate Analysis.* Univariate analysis includes "Linear Models with Covariate Adjustment," "Correlation and Partial Correlation Analysis," and "Multifactor ANOVA." Linear model-based covariate adjustment is designed to perform differential analysis while accounting for covariates. The primary metadata, reference group, and contrast objective can be customized according to the study goals, and one or more covariates can be defined for adjustment. Click "Submit" to perform the analysis. The results are summarized in an interactive scatter plot (Fig. 21), where features that become significant after adjustment are labeled in green, and those that become non-significant are colored red. Double-click a feature node to view its distribution across all groups.

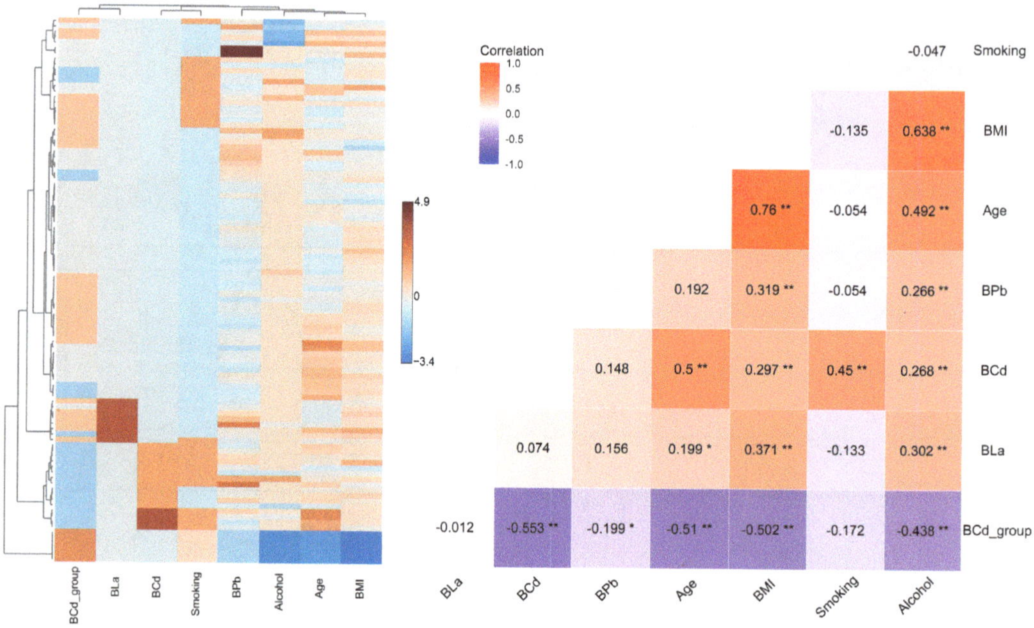

Fig. 19 Enriched pathway network view and exploration

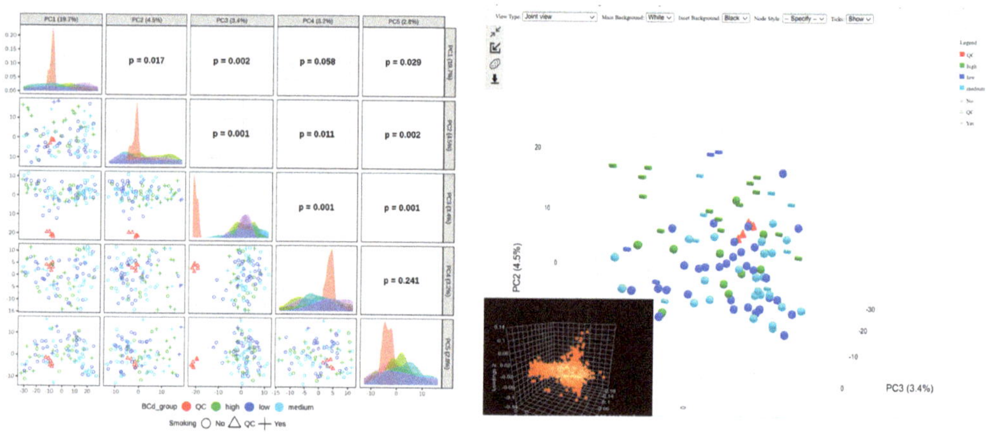

Fig. 20 KEGG global network mapping of enriched pathways

- Users can also perform conventional Correlation Analysis (including Partial Correlation Analysis), two-way ANOVA, and ANOVA-Simultaneous Component Analysis (ASCA) [27] to identify features that are significantly related to different experimental factors and potential interactions. Compared to linear modeling, those methods usually cannot deal with more than two metadata factors at the same time.

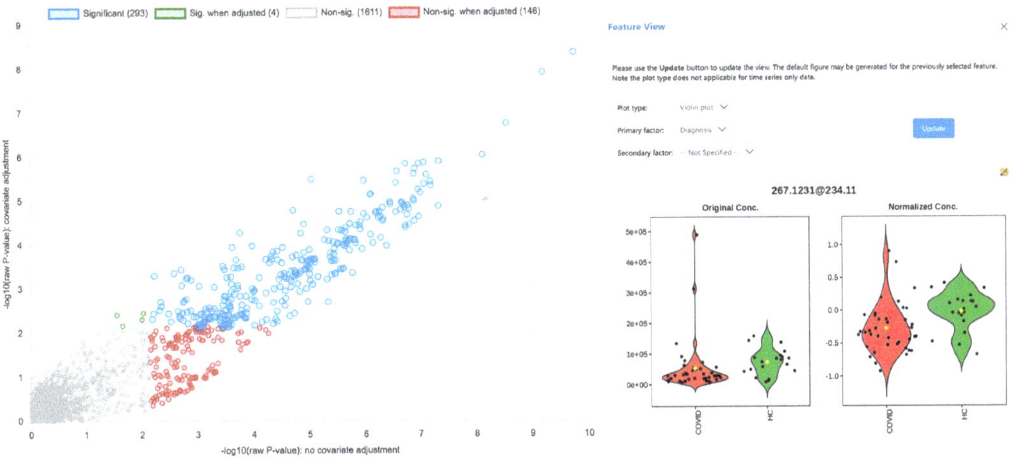

Fig. 21 Heatmap-based metabolic pattern customization and functional enrichment

9. *Random Forest.* The random forest is a powerful and versatile machine learning method that works by constructing a multitude of decision trees at training time and outputting the class based on majority vote (for classification) or the mean prediction (for regression) of the individual trees in the ensemble [28]. The important features are identified based on their contributions to the classification performance when they are permuted. Select the primary metadata factor of interest and other metadata factors as predictors from the dropdown menus. The randomness can be kept "on" or set to a fixed value to ensure reproducibility of the results. The classification curve of the Random Forest model is displayed at the bottom of the page. The "OOB error" (Out-of-Bag error) provides an estimate of how well the model will perform on a new dataset, with lower OOB error indicating a more robust model. The top-ranked features that contribute most to classification error are displayed in the dot plot under the "Var. Importance" tab.
10. *Results Export.* Click the "Download" node from the navigation tree of the webpage. On the download page, download the result files, including images and all generated tables.

3.5 Basic Protocol 5: Dose–Response Analysis

Dose–response analysis is designed to evaluate concentration-dependent metabolic changes across varying exposure levels. This analysis is widely used to characterize toxicological effects, pharmacological responses, or environmental exposures. By modeling metabolite abundance as a function of dose, users can identify compounds exhibiting monotonic or non-monotonic responses in accordance with levels of exposure. A key output is the benchmark dose (BMD), the minimum concentration associated with a measurable biological change relative to controls or the baseline [29, 30]. Proper experimental design is critical to ensure robust curve fitting and statistical inference. Typical study design falls into one of the two categories: (1) Controlled experiments involve at least three to five dosing levels with biological replicates at each level. Such designs enable the controlled evaluation of concentration-dependent metabolic perturbations under defined exposure conditions. This is particularly useful for characterizing chemical potency, identifying threshold effects, and linking exposure levels to specific biochemical mechanisms or adverse outcome pathways. (2) Observational studies, on the other hand, consist of continuous exposure data collected across individuals, with each subject contributing paired exposure–metabolite measurements. This type of data allows associations to be modeled as continuous functions. Observational analyses aim to reveal metabolic features associated with real-world exposure gradients and to infer potential mechanisms or pathways underlying observed exposure–response relationships.

1. *Data upload*: To begin, select the "Dose Response Analysis" module from the Module Overview page. Two types of experimental design are accepted: Repeated dosing involves controlled experiments with a control group (dose = 0) and at least three different treatment groups at discrete dosing levels, each with an equal number of replicates; continuous exposure is usually based on observational data (e.g., exposomics), where individuals have varying, measured exposure levels without predefined dose groups. Examples of repeated dosing and continuous exposure are provided on this page to help users align their data format prior to uploading. Here, we select the example data with repeated dosing from a breast cancer study [31] to demonstrate this module. Click "Submit" to proceed to the next page.
2. *Data integrity check*: Verify the sample and group information and click the "Proceed" button.
3. *Data filtering*: Perform data filtering by choosing from the available options (low-quality, low-repeatability, low-variance, and low-abundance) and set a percentage to filter out. In this example, we set "Low-variance filter: Inter quantile range (IQR)" at 40% to filter out the near- constant metabolic features[4]. Click "Submit," then "Proceed."
4. *Data normalization*: Normalize data by selecting "Log transformation (base 10)," click "Normalize." Click "View Results" to inspect the normalization result.
5. *Adjusting order for dose group*: By default, MetaboAnalyst interprets the dose groups as nominal or factor variables, resulting in alphanumeric sorting (lexicographical ordering) rather than

numerical ordering. For instance, a dose of "100" is ranked before "5." To fix this issue, click the "Data editor" option in the left workflow panel to enter the Data Editor page. Click the "Edit Groups" tab to verify the order of the dosing groups. Use the arrow button to adjust the order when necessary. Ensure the "Group order matters" box is checked and click "Submit." Users will be redirected to the Data Normalization page and need to re-perform normalization. Normalize data once again by log transformation (base 10). Click "Proceed."

6. *Significance analysis*: This step uses statistical tests to identify metabolites that show meaningful variations across dose levels. Set "Dose" as primary metadata, and control for any covariate when necessary. Set the thresholds for fold change and p value. In this example, we will use the default p value cut-off at 0.05, with the "Adjusted p-value (FDR)" box checked. Click "Submit." A volcano plot (Fig. 22) will appear below, showing the statistically significant features (up-regulated features in red and down-regulated features in blue) to be used for curve fitting. Click "Proceed."

7. *Dose–response curve fitting*: MetaboAnalyst provides ten different models, including exponential models (Exp2, Exp3, Exp4, Exp5), linear model (Linear), polynomial models (Poly2, Poly3, Poly4), sigmoidal model (Hill), and Power law model. Select Exp2, Exp3, Exp4, Linear, Poly2, and Power, as Exp5 and Hill models are not suitable for this example data. Set Lack-of-fit p value to 0.10 (i.e., there is no significant lack-of-fit when >0.10), BMR factor to 1.00 (i.e., BMD is the dose that produces a change ≥1 standard deviation from the control), and Control abundance to "Evaluate model at 0." Click the "Submit" button. Once the computation is complete, the Result Summary section below shows the number of features with fitted models and the number of features with BMD, as well as a bar plot summarizing each model's performance

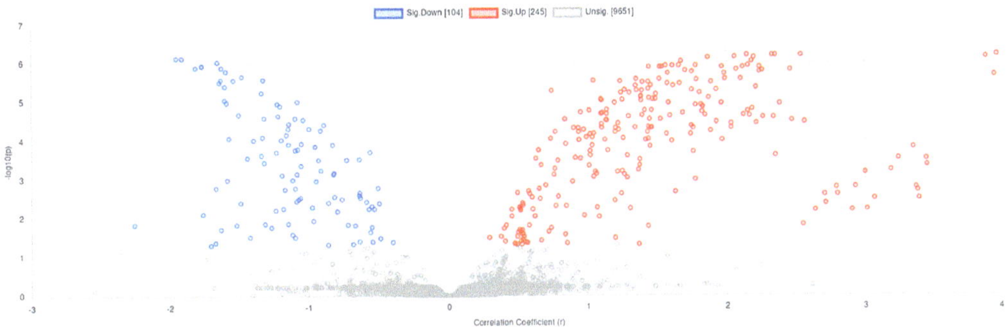

Fig. 22 Pre-filtered statistically significant metabolites for dose–response analysis

(Fig. 23). Click "Proceed" to view the detailed curve-fitting results for each feature.

8. *Feature-level BMD results*: All features with BMDs are shown in the results table, detailing their BMDs, lower and upper bounds, and *p* values. To view the dose–response curve of an individual feature, click on the image icon in the table, and a pop-up window will appear illustrating the dose–response curve with its BMD (Fig. 23).

9. *Results Download*. Click the "Download" node from the left navigation tree. On the download page, download the result files, including images and all generated tables. Click "Exit" to complete the analysis session.

3.6 Basic Protocol 6: Pathway Analysis and Joint Pathway Analysis

Pathways are naturally integrative, as metabolites, genes, or proteins can be mapped to reference pathways to reveal coordinated biological processes. This module supports pathway analysis for data input from metabolomics only, or joint pathway analysis for metabolomics data coupled with data from other omics. MetaboAnalyst combines quantitative enrichment analysis with pathway topology analysis [32]. Enrichment analysis evaluates whether certain pathways contain more significant metabolites than expected by chance, while topology analysis considers the relative importance of each compound within the pathway based on its position and connectivity. Results from these two orthogonal methods can be visualized through a scatter plot (bubble charts) coupled with interactive pathway visualization. A major enhancement in MetaboAnalyst 6.0 is the substantial expansion of its pathway libraries to support over 130 organisms—encompassing a broad taxonomic range of mammals, plants, insects, fungi, and

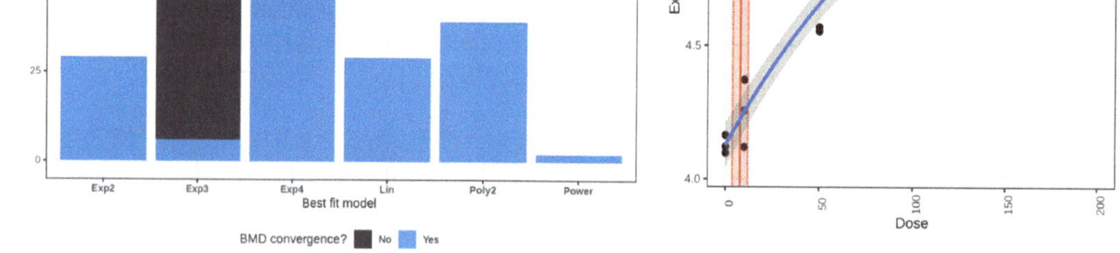

Fig. 23 Dose–response curves and best-fit model with benchmark dose for each metabolite

bacteria—to support organism-specific analyses across a wider range of model and non-model species.

1. *Data upload*: To begin, access the Pathway Analysis module from the MetaboAnalyst Module Overview page. The data upload interface includes several tabs corresponding to different input types. Accepted input formats for metabolomics data include:
   - A compound list identified by common name, HMDB ID, or KEGG ID;
   - A compound concentration table;
   - A study identifier from the *Metabolomics Workbench* repository;
   - A .xlsx file directly exported from *Metabolon.*

   The last tab on the upload menu, Joint Pathway Analysis, supports multiomics integration by combining metabolomics data with transcriptomics or proteomics datasets. Metabolomics inputs can be provided as a compound list with optional fold changes (for targeted data) or as a tab-delimited table containing m/z, p values, t-scores, and optionally retention time (rt) (for untargeted data). For transcriptomics or proteomics input, users must upload a list of significant genes or proteins, identified by Official Gene Symbol or UniProt ID, with optional fold change information. In this example, we will demonstrate the Joint Pathway Analysisworkflow. Select the appropriate tab, then click "Try our example" to view the available demonstration datasets. For illustration, we will use a built-in example dataset derived from a COVID-19 study [33] containing proteins (Official Gene Symbol) and compounds (HMDB ID) lists from human subjects. Select the example dataset and click "Submit" to continue.

2. *Pathway library selection*: On the next page, KEGG pathway libraries from 130 species are available for users to choose from. Select the pathway library corresponding to the species—in this example, "Homo sapiens." Click "Proceed."
3. *Name/ID standardization*: In the following page, the platform performs automated compound and gene mapping, providing users with a summary of matched and unmatched entries prior to pathway mapping. Inspect both compound name mapping and gene name mapping. Proper identifier matching is essential, as it ensures that only accurately annotated molecules contribute to downstream enrichment and topology calculations. If necessary, an individual entry may be removed by clicking on the delete option at this stage. Click "Proceed" to continue to the next step.
4. *Selecting pathway libraries*: Within the Analysis Parameters page, users can select the type of pathways to be examined:

integrated (i.e., gene–metabolite), metabolite only, or gene only, and whether to include gene-only pathways in addition to metabolic pathways. Use the default Metabolic pathways (integrated) in this example.

5. *Algorithm parameters for targeted metabolomics*: Under Algorithm Selection, specify options for enrichment analysis, pathway topology assessment, and integration strategy.

   - The enrichment analysis assesses whether a set of features (e.g., genes/proteins or metabolites) is overrepresented in a pathway, and users may choose to use the Hypergeometric Test or Fisher's Exact Test. Despite the difference in terminology, they give the same results and can be used almost interchangeably. Here, we will keep the Hypergeometric Test (default).
   - Pathway topology measures help identify important metabolites based on their positions within a given pathway. Degree centrality measures how many direct connections a node has, betweenness centrality captures how often a node lies on the shortest paths between others, and closeness centrality reflects how quickly a node can reach all other nodes in the network. Select "Degree Centrality" in this example.
   - For integration, genes/proteins and metabolites may be integrated at the query level, or they may undergo enrichment analyses separately, and their *p* values will subsequently be combined with different weighing methods (equal weights, weights based on the database, or weights based on individual pathways). By default, the Joint Pathway Analysis module merges metabolites and genes into a unified molecular list for simultaneous enrichment against composite KEGG pathways containing both molecular types. Keep the default choice.

6. *(Optional) Parameter settings for untargeted metabolomics*: If an untargeted metabolomics peak list is submitted as input, the Analysis Parameters page will have a slightly different layout compared to targeted metabolomics. The Mummichog algorithm [23] will be applied to predict pathway activity from the peak list data. Users may optionally modify the list of currency metabolites or potential adducts for peak annotation. The resulting metabolic pathways with *p* values will be integrated with the pathway p-values computated based on gene/protein list[5]. Users may define whether to query only "Metabolic Pathways" or integrated pathways ("All Pathways") in the pathway library option before proceeding to view pathway results.

7. *Pathway result visualization*: The resulting output (Fig. 24) shows an interactive overview plot summarizing all matched

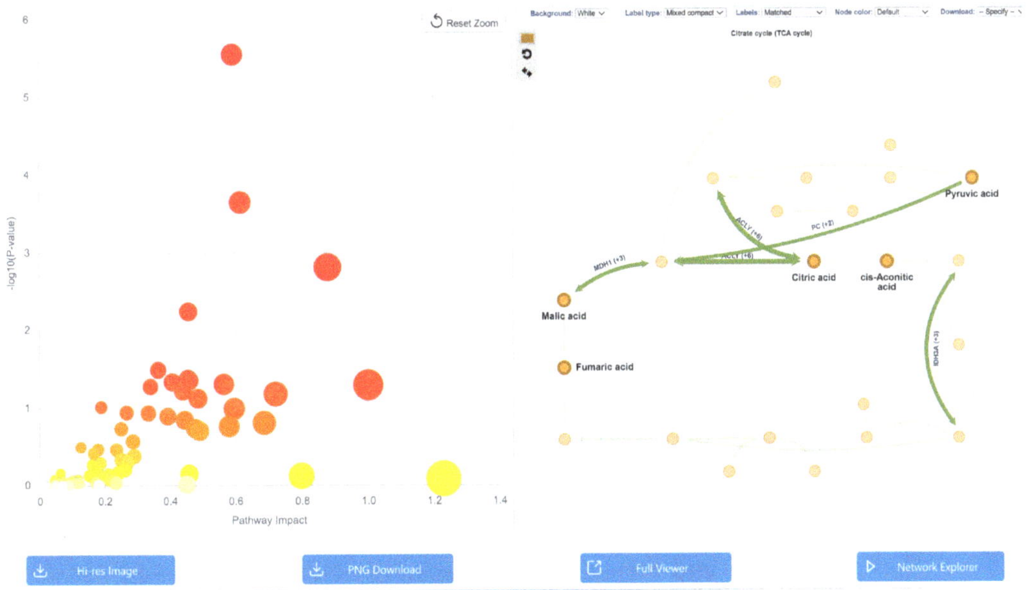

Click **Pathway Name** to highlight the corresponding point on the scatter; click **Match Status** to view pathway members (with matched ones highlighted).

Pathway Name	Match Status	p	-log(p)	Holm p	FDR	Impact	Details
Glycerophospholipid metabolism	19/87	2.8321E-6	5.5479	2.294E-4	2.294E-4	0.59302	KEGG
Ether lipid metabolism	10/40	2.2578E-4	3.6463	0.018062	0.0091441	0.61538	KEGG
Citrate cycle (TCA cycle)	9/42	0.0015599	2.8069	0.12323	0.042117	0.87805	KEGG
Inositol phosphate metabolism	11/69	0.0058073	2.236	0.45297	0.1176	0.45588	KEGG
Glyoxylate and dicarboxylate metabolism	8/56	0.032763	1.4846	1.0	0.39524	0.36364	KEGG

Fig. 24 Joint pathway analysis visualization of biologically relevant pathways

pathways as nodes whose color intensity reflects the statistical significance (p value) and whose size corresponds to pathway impact. Users may click individual pathways to visualize detailed maps, where matched genes and metabolites are highlighted. Circular nodes represent metabolites, while rectangular nodes denote genes/proteins, allowing for immediate visual interpretation of molecular roles within pathways. This visualization provides a biologically meaningful context for interpreting correlated alterations, such as co-regulated enzymes and metabolite intermediates.

3.7 Basic Protocol 7: Causal Analysis

MetaboAnalyst 6.0 offers two-sample Mendelian Randomization (MR) [34, 35] to help provide causal analysis between a metabolite and phenotype of interest by leveraging the SNP–metabolite association summary statistics curated from 65 published metabolomics genome-wide association studies (mGWAS) [36] and SNP–phenotype association summary statistics from the IEU OpenGWAS database [37]. The general workflow for conducting causal analysis based on two-sample MR in MetaboAnalyst consists of three tasks: (1) identify candidate SNPs; (2) perform SNP filtering and harmonization; and (3) conduct MR tests and interpret results.

1. *Input selection*: To start, click "Casal Analysis [Mendelian randomization]." Select an exposure (i.e., metabolite) and an outcome (i.e., disease) of interest from the available options. As an example, choose *L-carnitine* as the metabolite of interest, type and select *Dementia | finn-b-F5_DEMENTIA* as outcome[6].
2. *Parameter setting*: In the SNP filtering & harmonization step, it is crucial that the instrumental variables (IV) are independent, have strong effects, and have no horizontal pleiotropy. Several methods that support proper IV selection are provided: LD-based harmonization ensures IV independence through LD clumping and proxy SNP identification; allele harmonization aligns SNP effects across exposure and outcome; pleiotropic SNPs associated with multiple metabolites are removed; and Steiger filtering excludes SNPs more strongly associated with the outcome than the exposure to avoid reverse causality. In this example, select "1. LD-based Harmonization: Use clumping to prune SNPs for LD," "2. Allele Harmonization: Try to infer the forward strand alleles using allele frequency information," and "3. Remove Pleiotropic SNPs: Exclude SNPs associated with multiple metabolites." Click "Update," then click "Proceed."
   - The system automatically identifies SNPs satisfying the predefined IV selection criteria. Users may review the SNP candidates and manually include or exclude SNPs.
3. *MR method selection*: For MR analysis, a suite of 14 methods is available. Detailed explanations for each method are provided on the webpage. Users can view descriptions by hovering over the question mark icon next to each method. We recommend using a core set of robust and interpretable methods to estimate the causal effect reliably and assess the robustness of your findings. Here, we select the following: Wald ratio, MR Egger, Weighted median, Inverse variance weighted (MRE), Inverse variance weighted (FE), Simple mode, and Weighted mode. Click "Proceed" to perform MR analysis.
4. *View MR result*: In the results page, the causal effect estimates are reported with standard errors and *p* values for each selected MR method. In this example, across most methods, the estimated effect is positive, suggesting a possible association, but the results are not statistically significant in any method. Additionally, there was no evidence of horizontal pleiotropy or heterogeneity among the SNPs, supporting the validity of the instruments. Overall, this result suggests a potential association, but there is no strong evidence for a causal relationship. To visualize the MR results, choose from the following tabs:

- *Forest Plot* (Fig. 25a) shows the causal estimates from each of the 18 SNPs used as instrumental variables and the pooled estimates.
- *Scatter Plot* (Fig. 25b) shows the trend between SNP associations with L-carnitine and dementia across the 18 instruments. The positive slopes suggest a potential positive association between exposure and outcome; a flat slope of the MR Egger line indicates no strong causal effect.
- *Funnel Plot* (Fig. 25c) shows symmetry of SNP-specific estimates with a symmetrical plot indicating no directional pleiotropy.
- *Sensitivity Analysis* (Fig. 25d) shows the leave-one-out plot, where in this case, the removal of any individual SNP did not substantially alter the causal estimate.
- Literature Evidence provides external links to published studies that have investigated the exposure and outcome pair.

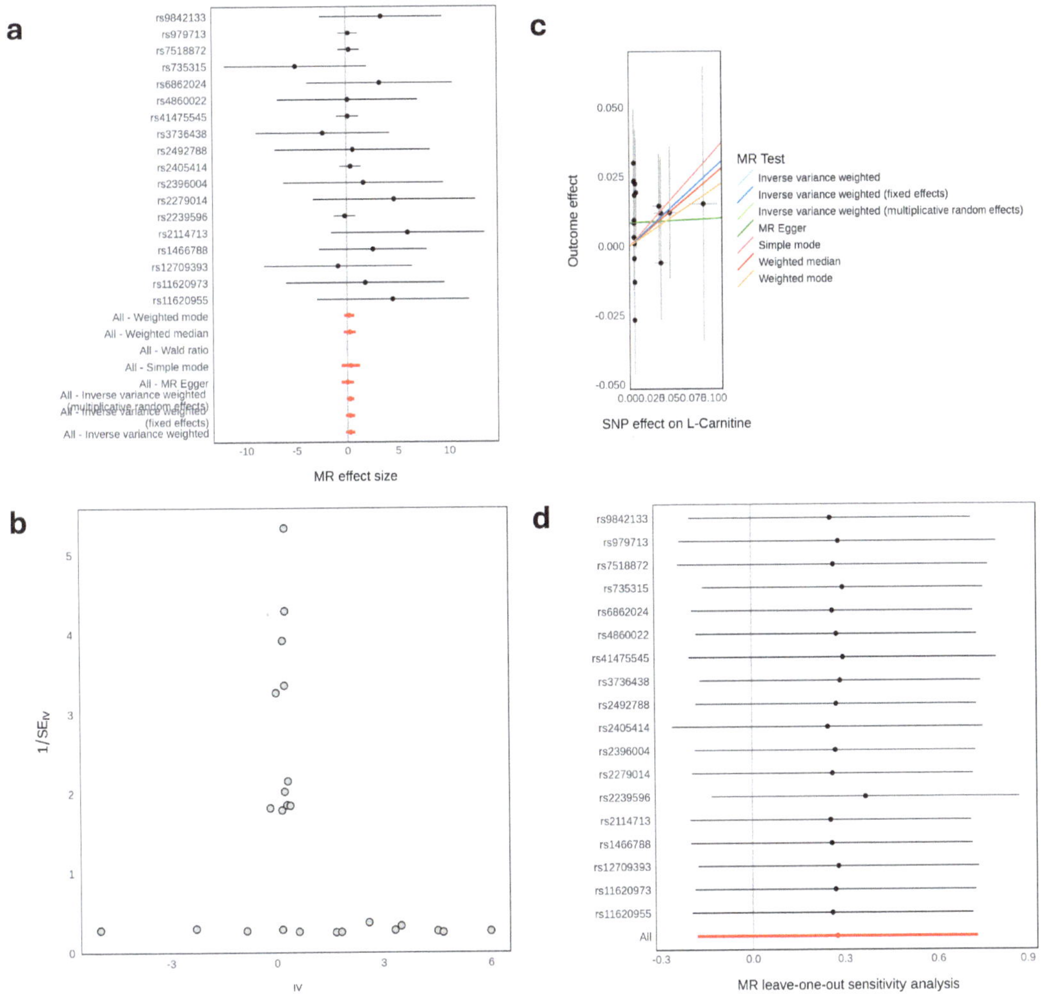

Fig. 25 Mendelian randomization graphical outputs for causal analysis

3.8 Basic Protocol 8: MetaboAnalystR for Local Batch Processing and Results Reproducibility

The MetaboAnalyst web platform is designed for general users, offering an intuitive and user-friendly interface. However, its flexibility in data analysis capacity to process large or multibatch datasets and compatibility with external tools are inherently limited by the constraints of web-based applications and server resources. To address these limitations, we release the underlying MetaboAnalystR package. This package promotes transparency, reproducibility, and flexibility, allowing integration with other analytical pipelines or customized workflows. As the companion package to the MetaboAnalyst web platform, MetaboAnalystR can reproduce all tables and figures generated online and can be further extended by advanced users to tailor their data analyses.

1. *Hardware*: A modern computer with internet access (at least 10 Mbps) is required. The system can be either a desktop or a laptop equipped with a minimum of 8 GB of RAM (32 GB or more is recommended for optimal performance). At least 10 GB of available hard drive space is necessary for raw spectral data processing, and a solid-state drive (SSD) is preferred to enhance data handling speed. A multi-core CPU with at least eight cores is required, with a recommended clock speed of 3.5 GHz. The display should support a minimum screen resolution of 1280 × 720.
2. *Software*: MetaboAnalystR is an R package that can be installed across major operating systems (OS), including Windows, Linux, and macOS. For Windows users, Rtools must be installed to enable package compilation. On Linux and macOS systems, several OS-level dependency packages are required prior to installation. The specific dependencies may vary depending on the Linux distribution and version; however, the following packages are generally needed for Linux: libcairo2-dev, libnetcdf-dev, libfribidi-dev, libfreetype6-dev, libpng-dev, libtiff5-dev, and libjpeg-dev. In addition, R (version newer than 4.2.0) and RStudio (version newer than 2023.06.0+421) are required to ensure proper execution and visualization of graphical outputs.
3. *Tutorials and manuals*: Comprehensive tutorials for MetaboAnalystR are available at the MetaboAnalyst website (https://www.metaboanalyst.ca/docs/RTutorial.xhtml). These tutorials provide a detailed introduction to the package, information on dependency requirements, installation guidance, and detailed instructions for various data processing and analysis modules (Fig. 26). Users can navigate to specific sections using the content panel at the top of the page and follow the step-by-step instructions for their selected analysis. To return to the main content panel, click the "Back to Content" hyperlink located at the bottom of each section.

MetaboAnalystR offers detailed manuals for individual functions that guide users in customizing parameters and graphical outputs (Fig. 26).

4. *Installation guidance*: The MetaboAnalystR package and its source code have been released in a GitHub Repository (https://github.com/xia-lab/MetaboAnalystR). To ease the installation and application, MetaboAnalystR has split the raw spectra processing functions into a companion package (OptiLCMS, https://github.com/xia-lab/OptiLCMS) to simply and speed up the installation. The installation of OptiLCMS[7] is required for raw spectral processing only. There are three ways to install MetaboAnalystR and OptiLCMS.

   a. Install the latest version of the R package from GitHub[8]. Click to start the RStudio interface, and execute the following R command line from the RStudio console panel to install the latest version of MetaboAnalystR and OptiLCMS by using devtools. This approach is applicable for Windows, Linux, and macOS.

      - *(Optionally)* set multi-core parallel compiling for package source code: *Sys.setenv(MAKE="make -j8")*
      - MetaboAnalystR: *devtools::install_github("xia-lab/MetaboAnalystR", build = TRUE)*
      - *OptiLCMS: devtools::install_github("xia-lab/OptiLCMS", build = TRUE)*

   b. Pull the latest source codes from the repositories by using Git and execute the compilation directly on the code. This approach is applicable for Unix-like OS, e.g., Linux. Open a terminal and execute the following command lines.

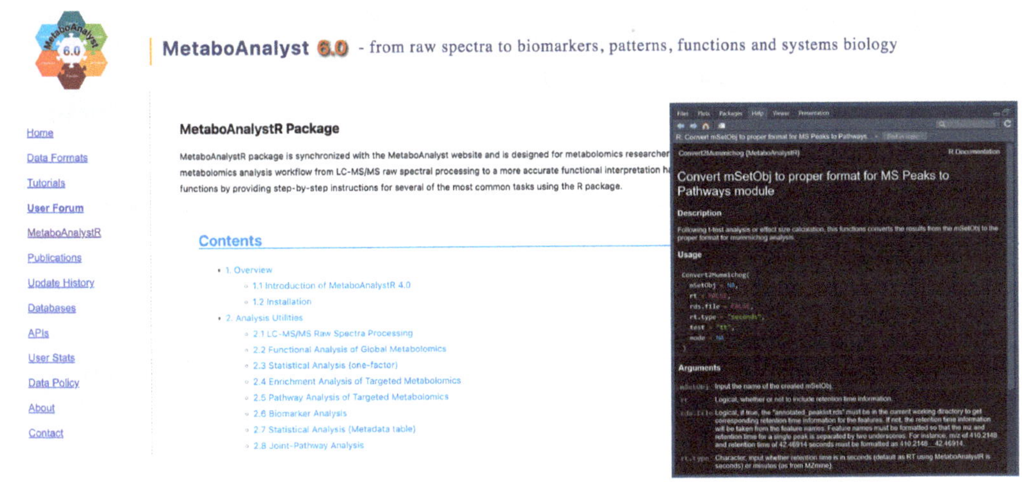

Fig. 26 MetaboAnalystR tutorial page and an example function manual

- *git clone*https://github.com/xia-lab/MetaboAnalystR.git
- *R CMD build MetaboAnalystR*
- *R CMD INSTALL MetaboAnalystR_4.2.0.tar.gz*
- *git clone*https://github.com/xia-lab/OptiLCMS.git
- *R CMD build OptiLCMS*
- *R CMD INSTALL OptiLCMS_1.2.0.tar.gz*

c. Install a stable release. MetaboAnalystR releases a stable release roughly semi-annually. Start RStudio interface and execute the following R commands from the RStudio console panel to install the latest version of MetaboAnalystR and OptiLCMS by using *devtools*. This option should only be used to reproduce the previous results, but it is not recommended for regular implementation, as the latest updates are not included and reflected in the stable version. For example, install MetaboAnalystR 4.2.0 and OptiLCMS 1.2.0 by using the following commands from the RStudio console panel.

- *install.packages(*"https://github.com/xia-lab/MetaboAnalystR/releases/download/v4.2.0/MetaboAnalystR_4.2.0.tar.gz", *repos = NULL)*
- *install.packages(*"https://github.com/xia-lab/OptiLCMS/releases/download/v1.2.0/OptiLCMS_1.2.0.tar.gz", *repos = NULL)*

5. *Local processing for large datasets*: The MetaboAnalyst website has a file size limit of 200 spectra files and 200 MB per file for online processing. In this case, MetaboAnalystR together with OptiLCMS can be applied. To demonstrate the application of MetaboAnalystR on large-scale LC–MS raw spectra datasets processing, MetaboAnalystR offers a detailed explanation on functionalities here.

a. Loading the libraries (MetaboAnalyst and OptiLCMS).

- *library(MetaboAnalystR); library(OptiLCMS)*

b. (Optional) Parameter optimization is required for feature detection using the *centWave* algorithm. We recommend performing automatic parameter optimization by using the functions *PerformROIExtraction* and *PerformParamsOptimization*. The optimal parameters will be returned as a list object upon completion of this step.

c. LC–MS1 Spectra Import: Users need to locate their MS^1 spectra files, which must be unzipped and centroided prior to processing. There are no intrinsic limitations on file size or number; however, hardware capacity may become a bottleneck when handling large datasets. In practice, a minimum of 32 GB RAM is recommended for processing more than 200 spectra files. Use the function *ImportRawMSData* to import the raw spectra. Optionally, TIC and BPI plots for all spectra files can be generated by setting the argument *plotSettings = SetPlotParam(Plot = TRUE)*.

d. LC–MS1 Feature Profiling: The *PerformPeakProfiling* function serves as a wrapper for raw spectral processing, performing peak detection,alignment, and grouping in an automated workflow. The first argument of this function is the *mSetObj* object generated in the previous step, while the second argument, Params, specifies the parameter settings. If parameters were optimized in step (b), this argument should be assigned the optimal parameter list from that step; otherwise, parameters must be defined manually. Depending on the dataset size and computational resources, this step may take several minutes to several hours (or even days) to complete.

e. Peak Annotation: Following MS1 feature profiling, peak annotation is performed to identify all potential adducts, isotopes, and isomers. Use *SetAnnotationParam* to configure the annotation parameters and then execute *PerformPeakAnnotation* to perform the feature annotation. This step is critical for reducing feature redundancy. It typically completes within a few minutes.

f. Feature Formatting and Export. Detected features can be formatted into a standardized feature table using the function *FormatPeakList*. The formatted table can then be used for subsequent statistical, functional, or biomarker analyses in MetaboAnalystR. Several export functions—*Export.Annotation*, *Export.PeakTable*, and *Export.PeakSummary*—are available to save all feature detection and annotation results. In addition, the *mSetObj* object generated by FormatPeakList can serve as the input for subsequent MS2 spectra processing.

6. *Reproducibility and flexibility:* Another major advantage of MetaboAnalystR is its ability to reproduce results generated on the MetaboAnalyst web server directly on a local computer. As shown in Fig. 27, users can easily obtain the corresponding R code from the left panel of the web page by clicking the "Show R Commands" hyperlink. All R commands used for data analysis and graphical generation are displayed in a newly opened panel.

   a. Click the "Save" button to download these commands as an R script file ("Rhistory.R").

b. Locate this file in your download folder and open it in RStudio. All command lines used in the web session will be loaded automatically. Modify the input file path based on your local environment.

c. To reproduce the results or extend the workflow for batch processing, load the MetaboAnalystR package first and then execute the provided commands.

Note that the input file path must be adjusted according to the location of your local data files (as illustrated in Fig. 27). All result files are automatically stored within the *mSetObj* object in the R session environment. This object follows the S3 class structure in R and can be easily accessed or exported as needed, ensuring high flexibility in data processing. By simply modifying the input file paths, users can efficiently perform analyses on multiple datasets in batch mode—eliminating the need for repetitive manual operations on the web interface. This feature represents a significant advantage for advanced R users seeking reproducible, scalable, and automated workflows.

7. *Interfacing with other tools.* MetaboAnalyst and MetaboAnalystR support a generic data format applicable to multiple data processing modules, including statistical analysis, biomarker discovery, dose–response analysis, and functional

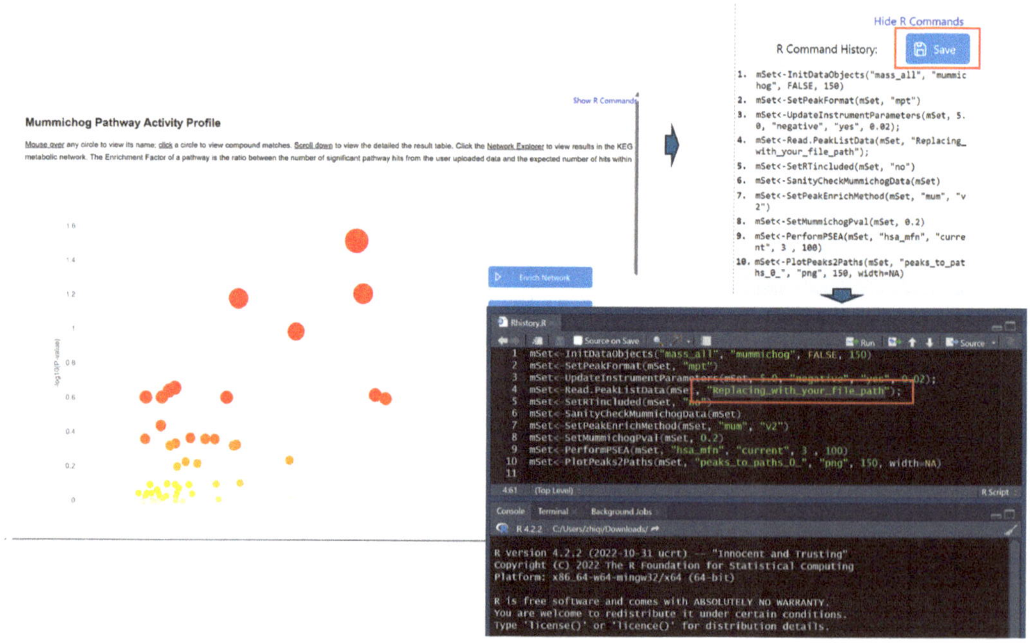

Fig. 27 Demonstration of reproducibility and flexibility of MetaboAnalystR for data processing, results reproduction, and extended analysis

enrichment. However, many commonly used raw spectral processing tools generate pre-processed data in proprietary formats that cannot be directly used within MetaboAnalyst. Converting these results into a compatible format can often be labor-intensive and may require programming expertise. To enhance interoperability with other widely used platforms, MetaboAnalystR accepts LC–MS1 feature detection results from MS-DIAL,MZmine,Asari, and XCMS. These outputs can be automatically reformatted into MetaboAnalystR's standardized input structure for seamless downstream processing. In addition, MetaboAnalystR supports pre-processed LC–MS2 data in MSP format generated by MS-DIAL and MZmine. Detailed guidance on adapting outputs from external tools to the MetaboAnalyst data processing and analysis pipeline is provided in our online tutorial section.

8. *Export and saving results.* All intermediate results generated during analysis in MetaboAnalystR are stored within the *mSetObj* objects. Graphical outputs and result tables are automatically saved in the current working directory. To prevent overwriting previous analysis outputs, it is recommended to create and use a dedicated working directory for each individual project.

4 Notes

1. For raw spectra data preparation: When MS1 and MS2 spectra are mixed, users should use the "Subset" filter to extract (or subset) MS1 and MS2 spectra separately. In the settings for this filter, you can set the "MS levels" to include, for example, 1–1 for MS1 or 2–2 for MS2. To extract MS1 scans, set "MS levels" to 1 and click Add. Click Start to convert the file, which will contain only MS1 scans. After the conversion is complete, remove the MS level 1 filter by selecting it and clicking Remove. Set "MS levels" to 2 and click Add. Repeat the process to extract MS2 scans.
2. On the SWATH-Design Check page: By default, the designed windows are extracted directly from the spectral data. If this information is missing, users must manually enter the "Lower m/z" and "Upper m/z" values for each window to ensure correct spectral deconvolution in the subsequent steps.
3. For MSP and MGF file uploading for MS2 spectral annotation, ensure the files are organized into a standard format. Check the examples from the uploading page to get details. Non-standard files may lead to failure of spectral recognition.

4. In the Data filtering step of Dose–response analysis: Only a small percentage of features are expected to respond to exposures. The IQR filter will remove those features unlikely to be useful in BMD modeling.
5. For joint pathway analysis of untargeted metabolomics data, only pathway-level *p* value integration is supported. As peaks cannot be directly mixed with gene/protein IDs for pathway mapping, query-level integration cannot be used in this case.
6. On the Causal analysis input selection page: MetaboAnalyst will automatically update the available options based on user input text. The finn-b-F5_DEMENTIA is the dataset name in the OpenGWAS data (https://opengwas.io/datasets/finn-b-F5_DEMENTIA).
7. If you are using *OptiLCMS* from Windows, make sure the software, RTools, has been installed to compile the source code correctly.
8. MetaboAnalyst and MetaboAnalystR are in constant development to further expand their knowledge library, functionalities, and stability. The source code of all functionalities is available from https://github.com/xia-lab/MetaboAnalystR. If you encounter any issues or have any advice, feel free to contact us via Omics Forum (https://omicsforum.ca/).

MetaboAnalyst is one of the most widely used bioinformatics platforms for metabolomicsdata processing,analysis, and interpretation. This chapter introduced the major updates and newly implemented features in MetaboAnalyst 6.0, extending the functionalities covered in the previous protocols. MetaboAnalyst 6.0 now supports LC–MS raw spectral processing and MS2 spectral annotation, allowing users to begin analysis directly from raw spectra. The platform also expands analytical support for exposomics analysis using *asari* for improved detection of trace-level signals and chemical classification [21]. Several modules for downstream data interpretation have also been enhanced. The dose–response analysis module supports quantitative modeling of concentration-dependent effects, while the cause analysis module allows evaluation of potential causal links between a metabolite or metabolic exposure and a phenotype of interest. Together, these developments strengthen the capacity of MetaboAnalyst to identify and evaluate biologically and clinically relevant metabolic signatures. These capabilities provide an important step toward the mechanistic interpretation of complex biological systems. As with other omics studies, careful experimental design and data collection remain essential for ensuring data quality and reliable interpretation. For advanced users, the MetaboAnalystR package can be coupled with the web platform to create flexible and reproducible workflows, enabling seamless transition from raw spectral data to functional and causal biological insights.

Acknowledgments

This work is supported by the Canadian Foundation for Innovation (CFI), the Natural Sciences and Engineering Research Council of Canada (NSERC), and the Canada Research Chairs (CRC).

References

1. Xia J, Psychogios N, Young N, Wishart DS (2009) MetaboAnalyst: a web server for metabolomic data analysis and interpretation. Nucleic Acids Res 37(Web Server issue):W652–W660. https://doi.org/10.1093/nar/gkp356
2. Xia J, Mandal R, Sinelnikov IV, Broadhurst D, Wishart DS (2012) MetaboAnalyst 2.0–a comprehensive server for metabolomic data analysis. Nucleic Acids Res 40 (Web Server issue):W127–W133. https://doi.org/10.1093/nar/gks374
3. Xia J, Sinelnikov IV, Han B, Wishart DS (2015) MetaboAnalyst 3.0–making metabolomics more meaningful. Nucleic Acids Res 43(W1):W251–W257. https://doi.org/10.1093/nar/gkv380
4. Chong J, Soufan O, Li C, Caraus I, Li S, Bourque G, Wishart DS, Xia J (2018) MetaboAnalyst 4.0: towards more transparent and integrative metabolomics analysis. Nucleic Acids Res 46(W1):W486–W494. https://doi.org/10.1093/nar/gky310
5. Pang Z, Lu Y, Zhou G, Hui F, Xu L, Viau C, Spigelman AF, MacDonald PE, Wishart DS, Li S, Xia J (2024) MetaboAnalyst 6.0: towards a unified platform for metabolomics data processing, analysis and interpretation. Nucleic Acids Res 52(W1):W398–W406. https://doi.org/10.1093/nar/gkae253
6. Pang Z, Chong J, Zhou G, de Lima Morais DA, Chang L, Barrette M, Gauthier C, Jacques PE, Li S, Xia J (2021) MetaboAnalyst 5.0: narrowing the gap between raw spectra and functional insights. Nucleic Acids Res 49(W1):W388–W396. https://doi.org/10.1093/nar/gkab382
7. Chong J, Xia J (2020) Using MetaboAnalyst 4.0 for metabolomics data analysis, interpretation, and integration with other omics data. Methods Mol Biol 2104:337–360. https://doi.org/10.1007/978-1-0716-0239-3_17
8. Dunn WB, Broadhurst D, Begley P, Zelena E, Francis-McIntyre S, Anderson N, Brown M, Knowles JD, Halsall A, Haselden JN, Nicholls AW, Wilson ID, Kell DB, Goodacre R, Human Serum Metabolome C (2011) Procedures for large-scale metabolic profiling of serum and plasma using gas chromatography and liquid chromatography coupled to mass spectrometry. Nat Protoc 6(7):1060–1083. https://doi.org/10.1038/nprot.2011.335
9. Kessner D, Chambers M, Burke R, Agus D, Mallick P (2008) ProteoWizard: open source software for rapid proteomics tools development. Bioinformatics 24(21):2534–2536. https://doi.org/10.1093/bioinformatics/btn323
10. Tautenhahn R, Böttcher C, Neumann S (2008) Highly sensitive feature detection for high resolution LC/MS. BMC Bioinformatics 9: 504. https://doi.org/10.1186/1471-2105-9-504
11. Pang Z, Chong J, Li S, Xia J (2020) MetaboAnalystR 3.0: toward an optimized workflow for global metabolomics. Metabolites 10(5):186. https://doi.org/10.3390/metabo10050186
12. Li S, Siddiqa A, Thapa M, Chi Y, Zheng S (2023) Trackable and scalable LC-MS metabolomics data processing using asari. Nat Commun 14(1):4113. https://doi.org/10.1038/s41467-023-39889-1
13. Smith CA, Want EJ, O'Maille G, Abagyan R, Siuzdak G (2006) XCMS: processing mass spectrometry data for metabolite profiling using nonlinear peak alignment, matching, and identification. Anal Chem 78(3):779–787. https://doi.org/10.1021/ac051437y
14. Conley CJ, Smith R, Torgrip RJ, Taylor RM, Tautenhahn R, Prince JT (2014) Massifquant: open-source Kalman filter-based XC-MS isotope trace feature detection. Bioinformatics 30(18):2636–2643. https://doi.org/10.1093/bioinformatics/btu359
15. Domingo-Almenara X, Siuzdak G (2020) Metabolomics data processing using XCMS. Methods Mol Biol 2104:11–24. https://doi.org/10.1007/978-1-0716-0239-3_2
16. Prince JT, Marcotte EM (2006) Chromatographic alignment of ESI-LC-MS proteomics data sets by ordered bijective interpolated warping. Anal Chem 78(17):6140–6152. https://doi.org/10.1021/ac0605344
17. Tsugawa H, Cajka T, Kind T, Ma Y, Higgins B, Ikeda K, Kanazawa M, VanderGheynst J, Fiehn O, Arita M (2015) MS-DIAL: data-independent MS/MS deconvolution for comprehensive metabolome analysis. Nat Methods 12(6):523–526. https://doi.org/10.1038/nmeth.3393
18. Li Y, Kind T, Folz J, Vaniya A, Mehta SS, Fiehn O (2021) Spectral entropy outperforms

MS/MS dot product similarity for small-molecule compound identification. Nat Methods 18(12):1524–1531.https://doi.org/10.1038/s41592-021-01331-z
19. Pang Z, Xu L, Viau C, Lu Y, Salavati R, Basu N, Xia J (2024) MetaboAnalystR 4.0: a unified LC-MS workflow for global metabolomics. Nat Commun 15(1):3675. https://doi.org/10.1038/s41467-024-48009-6
20. Djoumbou Feunang Y, Eisner R, Knox C, Chepelev L, Hastings J, Owen G, Fahy E, Steinbeck C, Subramanian S, Bolton E, Greiner R, Wishart DS (2016) ClassyFire: automated chemical classification with a comprehensive, computable taxonomy. J Cheminform 8:61. https://doi.org/10.1186/s13321-016-0174-y
21. Pang Z, Viau C, Fobil JN, Basu N, Xia J (2024) Comprehensive blood metabolome and exposome analysis, annotation, and interpretation in e-waste workers. Metabolites 14(12):671. https://doi.org/10.3390/metabo14120671
22. Aisporna A, Benton HP, Chen A, Derks RJE, Galano JM, Giera M, Siuzdak G (2022) Neutral loss mass spectral data enhances molecular similarity analysis in METLIN. J Am Soc Mass Spectrom 33(3):530–534. https://doi.org/10.1021/jasms.1c00343
23. Li S, Park Y, Duraisingham S, Strobel FH, Khan N, Soltow QA, Jones DP, Pulendran B (2013) Predicting network activity from high throughput metabolomics. PLoS Comput Biol 9(7):e1003123. https://doi.org/10.1371/journal.pcbi.1003123
24. Subramanian A, Tamayo P, Mootha VK, Mukherjee S, Ebert BL, Gillette MA, Paulovich A, Pomeroy SL, Golub TR, Lander ES, Mesirov JP (2005) Gene set enrichment analysis: a knowledge-based approach for interpreting genome-wide expression profiles. Proc Natl Acad Sci U S A 102(43):15545–15550. https://doi.org/10.1073/pnas.0506580102
25. Heller SR, McNaught A, Pletnev I, Stein S, Tchekhovskoi D (2015) InChI, the IUPAC International Chemical Identifier. J Cheminform 7(1):23. https://doi.org/10.1186/s13321-015-0068-4
26. Kanehisa M, Furumichi M, Sato Y, Matsuura Y, Ishiguro-Watanabe M (2025) KEGG: biological systems database as a model of the real world. Nucleic Acids Res 53(D1):D672–D677. https://doi.org/10.1093/nar/gkae909
27. Bertinetto C, Engel J, Jansen J (2020) ANOVA simultaneous component analysis: a tutorial review. Anal Chim Acta X 6:100061. https://doi.org/10.1016/j.acax.2020.100061
28. Couronne R, Probst P, Boulesteix AL (2018) Random forest versus logistic regression: a large-scale benchmark experiment. BMC Bioinformatics 19(1):270. https://doi.org/10.1186/s12859-018-2264-5
29. Committee ES, More SJ, Bampidis V, Benford D, Bragard C, Halldorsson TI, Hernández-Jerez AF, Bennekou SH, Koutsoumanis K, Lambré C (2022) Guidance on the use of the benchmark dose approach in risk assessment. EFSA Jl 20(10):e07584
30. Ewald J, Soufan O, Xia J, Basu N (2021) FastBMD: an online tool for rapid benchmark dose–response analysis of transcriptomics data. Bioinformatics 37(7):1035–1036
31. Yao CH, Wang L, Stancliffe E, Sindelar M, Cho K, Yin W, Wang Y, Patti GJ (2020) Dose-response metabolomics to understand biochemical mechanisms and off-target drug effects with the TOXcms software. Anal Chem 92(2):1856–1864. https://doi.org/10.1021/acs.analchem.9b03811
32. Xia J, Wishart DS (2010) MetPA: a web-based metabolomics tool for pathway analysis and visualization. Bioinformatics 26(18):2342–2344
33. Su Y, Chen D, Yuan D, Lausted C, Choi J, Dai CL, Voillet V, Duvvuri VR, Scherler K, Troisch P, Baloni P, Qin G, Smith B, Kornilov SA, Rostomily C, Xu A, Li J, Dong S, Rothchild A, Zhou J, Murray K, Edmark R, Hong S, Heath JE, Earls J, Zhang R, Xie J, Li S, Roper R, Jones L, Zhou Y, Rowen L, Liu R, Mackay S, O'Mahony DS, Dale CR, Wallick JA, Algren HA, Zager MA, Wei W, Price ND, Huang S, Subramanian N, Wang K, Magis AT, Hadlock JJ, Hood L, Aderem A, Bluestone JA, Lanier LL, Greenberg PD, Gottardo R, Davis MM, Goldman JD, Heath JR (2020) Multi-omics resolves a sharp disease-state shift between mild and moderate COVID-19. Cell 183(6):1479–1495.e1420. https://doi.org/10.1016/j.cell.2020.10.037
34. Davey Smith G, Hemani G (2014) Mendelian randomization: genetic anchors for causal inference in epidemiological studies. Hum Mol Genet 23(R1):R89–98. https://doi.org/10.1093/hmg/ddu328
35. Hartwig FP, Davies NM, Hemani G, Davey Smith G (2016) Two-sample Mendelian randomization: avoiding the downsides of a powerful, widely applicable but potentially fallible technique, vol 45. Oxford University Press
36. Chang L, Zhou G, Xia J (2023) mGWAS-explorer 2.0: causal analysis and interpretation of metabolite–phenotype associations. Metabolites 13(7):826
37. Elsworth B, Lyon M, Alexander T, Liu Y, Matthews P, Hallett J, Bates P, Palmer T, Haberland V, Smith GD, Zheng J, Haycock P, Gaunt TR, Hemani G (2020) The MRC IEU OpenGWAS data infrastructure. bioRxiv: 2020.08.10.244293. https://doi.org/10.1101/2020.08.10.244293

Check for updates

Chapter 17

Multi-Omics Guided Pathway and Network Analysis of Clinical Metabolomics and Proteomics Data

Christina Schmidt and Thomas Naake

Abstract

Metabolomics, the study of small molecules in biological systems, is a powerful tool for understanding biochemical pathways, discovering biomarkers, and elucidating disease mechanisms. This chapter provides a guide to performing metabolomics data analysis in *R*, focusing on enrichment analysis and network-based approaches. It covers essential steps in data processing, quality control (QC), differential expression analysis, integration with proteomics using multi-omics factor analysis (MOFA), and statistical network analysis, as well as enrichment analysis using prior knowledge. The methods outlined provide a framework for biomarker discovery and advancing systems-level understanding of disease processes using metabolomics data in combination with prior knowledge and proteomics data.

Keywords Metabolomics, Proteomics, Processing, Quality control, Network analysis, Pathway, Knowledge graphs

1 Introduction

With the increasing number of metabolomics studies, standardized and reproducible analysis strategies are essential to ensure accurate biological interpretation. To support this, high-quality prior knowledge (PK) resources, such as curated metabolite sets and gene-metabolite associations, combined with the application of metabolic networks, are effective to facilitate data interpretation. Networks provide a powerful framework to study metabolomics and enable the integration with proteomics data by representing metabolites and proteins as vertices and their relationships (e.g., correlations, biosynthetic enzymes) as edges. This approach enables insights into system-level changes and interactions between metabolites, proteomics, and other omics layers.

In this chapter, we will showcase the statistical programming language R's extensive ecosystem for metabolomics analysis and integration with proteomics PK using network analysis among other

Michael Witting (ed.), *Computational Methods and Data Analysis for Metabolomics*, Methods in Molecular Biology, vol. 3063, https://doi.org/10.1007/978-1-0716-5452-1_17,

strategies, including packages like *igraph* [1], *MOFA* [2], *decoupleR* [3], and *MetaProViz* [4]. As an example of multiple data modalities integration, we integrate metabolomics with proteomics that share the same sample set using Multi-omics factor analysis (MOFA), a statistical technique to identify latent factors of shared variation in multi-omics datasets. We use the results from the MOFA model to guide downstream analysis in pathway and network analysis. *MetaProViz* provides access and quality control plots for the integration of metabolomics with prior knowledge, while *decoupleR* uses those biological signatures to calculate scores based on prior knowledge, offering the interpretation of the datasets by known biological networks and pathways.

In this chapter, we distinguish between knowledge and experimental networks [5]. Knowledge networks are constructed using existing biochemical and biological information, helping to interpret metabolomics and proteomics data within the framework of known pathways. For example, a metabolic reaction network represents a knowledge network, where metabolites act as vertices, and their biochemical conversions are the edges. In contrast, experimental networks are directly derived from metabolomics and proteomics data, linking metabolites and proteins based on observed relationships like correlation or, in the case of MS/MS metabolomics data, spectral similarity. Both types of networks can be explored using advanced statistical methods, graph analysis, and data-driven techniques to reveal meaningful patterns and connections within the dataset.

2 Materials

We will use the dataset from [6] to showcase functionality in *R* to analyze metabolomics and proteomics using network and pathway analysis tools. We present here an end-to-end pipeline starting with data import, data processing quality assessment and control, multi-omics integration, and downstream analysis to interpret the latent factors of the model. The datasets are proteomics and metabolomics measurements from fresh-frozen tumor and non-tumorous (adjacent) tissue of patients with lung adenocarcinoma. Metabolomics measurements were acquired using the MxP® Quant 500 kit (Biocrates) using LC–MS/MS and FIA-MS/MS measurements run on a UPLC I-class PLUS (Waters) system coupled to a SCIEX QTRAP® 6500 + mass spectrometry system in electrospray ionization (ESI) mode. Proteomics measurements were acquired using an Easy-nLC™ 1200 system (Thermo) coupled to a timsTOF Pro mass spectrometer (Bruker Daltonics). For further details on the analytical methods, refer to [6].

Due to space constraints, the code for all analyses will not be displayed in this chapter. Interested readers are referred to the complete and reproducible protocol available at www.github.com/tnaake/MiMB_networks/R/chapter_network_multiomics.qmd.

3 Methods

3.1 Data Processing and Quality Assessment and Quality Control (QA/QC)

Processing and quality assessment and quality control (QA/QC) are critical to ensure the integrity of metabolomics data. Steps typically include detection control of outliers, removal of batch effects, data normalization and transformation, and missing value imputation (optional). Each step requires careful parameter selection based on the dataset and experimental conditions. In the case of multi-omics analysis using MOFA, as presented here, it is paramount to ensure that all modalities share the same set of samples, with identical sample names and order, to facilitate accurate integration and analysis.

3.2 Load the Datasets, Data Wrangling

1. Load the proteomics and metabolomics datasets via import functions from *MatrixQCvisUtils* (available via www.github.com/tnaake/MatrixQCvisUtils). Importing the XLSX sheets into *R* via the *MatrixQCvisUtils* package, creates *SummarizedExperiment* ([7]) objects. The *SummarizedExperiment* class stores the measured data (available via *assay*), next to the metadata associated to the samples (available via *colData*) and features (available via *rowData*).
2. After importing the datasets, check the following properties:
   - Correct representation of missing values (*NA*) in dataset,
   - Correct data dimension and structure,
   - Consistent number of samples and matching sample names between datasets,
   - Verify that the sample order is identical across the modalities,
   - Correct formatting of feature names, complete number of features for each dataset,
   - Complete and correctly formatted metadata associated to each dataset
3. (Optional) Depending on 2., identify the nature of missing features, harmonize sample names, correct feature names, harmonize metadata and check for completeness.
4. (Optional) Identify additional metadata, for example, clinical parameters or translated IDs for metabolites (e.g., HMDB, KEGG, ChEBI ids) and proteins (e.g., Entrez, SYMBOL, Uniprot ids), from other sources. Add the metadata to the respective slots of the *SummarizedExperiment* object: *rowData* for feature-related and *colData* for sample-related metadata).

3.3 Filtering of Features in the Datasets

Each dataset typically includes information on the reliability of features, which can be used to remove non-reliable features. This step improves the quality of the datasets. Filtering and removal of the features should be tailored to the study design and the characteristics of the dataset. While stricter filtering can reduce false positives, it may also remove true low-abundance features.

1. Remove low or high abundance features. Low intensity features may be due to noise or below the limit of detection or quantification. High intensity may be above the limit of quantification, the linear dynamic range, or the calibration range. Removing these features prevents artifacts in downstream statistical analyses. For the targeted metabolomics dataset, ensure that the dataset only includes high-quality metabolite measurements. The MetIDQ output provides assessment of the quality of each measurement, encoded by a color code of each cell (dark blue: < LOD (Limit of Detection; 3x signal to noise); light blue: < LLOQ (Lower Limit of Quantification; 10x signal to noise) or > ULOQ (Upper Limit of Quantification); green: valid; Yellow: Internal Standard out of range). Remove the low-quality features that do not meet the defined quantification criteria (*choice of user*). In this case, we will retain only those metabolites where at least 50% of the measurements fall within the quantification limits (valid). This filtering step improves data reliability and minimizing the impact of imprecise data points. Other filtering steps can also be applied, e.g., the 80% filtering rule or adapted-80% filtering rule [8]
2. Verify that all measurements fall within expected ranges (e.g., non-negative values for intensities). For measurements that fall outside the expected range, correct them, e.g., by setting these values to *NA* (missing data).
3. (Optional) Remove those features that are only detected in a subset of samples (*choice of user*). High levels of missing data (*NA*) values can reduce statistical power. While imputation techniques can be applied, imputation of excessive numbers of missing values per feature can lead to unreliable results. Additionally, features identified in only a few biological replicates may lack reproducibility, impacting downstream analyses. Filtering criteria can be set on a global level, such as removing features absent in more than 50% of the samples, or based on metadata, such as removing features detected in fewer than two biological replicates per condition. More stringent thresholds, such as requiring detection in at least 50% of replicates within each condition, can further improve dataset reliability. The choice of filtering thresholds should be guided by domain-specific knowledge and the characteristics of the datasets. Overly strict filtering may remove biologically relevant proteins that are present only in specific subsets of samples, such as particular cell lines or patient groups. Filtering strategies should be

carefully considered to maintain a balance between data quality and the retention of meaningful biological information.

4. Remove the features with high coefficient of variation (CV, *choice of user*). High variability in technical replicates or QC pools suggests inaccurate quantification. For example, features with CV > 30-50% of raw intensities across technical replicates may be removed.
5. (Optional) Remove features with a standard deviation of 0 as these features do not vary across samples and do not contain information for biological interpretation.
6. For proteomics data, ensure that the dataset only includes high-quality protein measurements. In the proteomics dataset generated by MaxQuant [9], peptide counts serve as an indicator of the reliability of protein identification. To ensure data quality, we will filter out proteins identified with low confidence (*choice of user*). In this case, we will remove any protein that has fewer than two peptides supporting its identification. This approach helps retain only well-supported protein identifications, enhancing the robustness of downstream analyses.
7. For proteomics data, remove common contaminants by using a contaminant database (e.g., MaxQuant's contaminant list). Contaminants like keratins (from human skin), trypsin (from digestion), or bovine serum albumin (BSA) can interfere with interpretation of results.
8. For the proteomics dataset, remove the features with reverse sequences as they do not correspond to real biological proteins. Reverse sequences refer to decoy sequences generated by reversing the amino acid sequence of real proteins. Reverse sequences are commonly used as negative controls in target-decoy strategies for false discovery rate estimation in mass-spectrometry-based proteomics.

3.4 Quality Assessment and Quality Control (QA/QC)

The next step in the processingpipeline is to assess the quality of the data, including both the measurements and the metadata values, and to control for it.

The *MatrixQCvis* package [10] performs QA/QC on *SummarizedExperiment* objects and offers users to perform these steps in an interactive *shiny* application.

1. Launch the application via *shinyQC(metabolomics)* or *shinyQC(proteomics)* to begin the quality assessment for the respective datasets.
2. Check the number of measured and missing values per sample and per feature. The metric informs about the dynamic range of the acquisition. Differences between samples of an

experiment may indicate differences in the dynamic range and/or in the sample content.

3. Check the distribution of measurements per sample. The metric informs about the dynamic range of the acquisition. Differences between samples of an experiment may indicate differences in the dynamic range and/or in the sample content.
4. Examine the mean-sd plot (Fig. 1A and B). The metric informs about the level of homoskedasticity of the dataset, which may be important for parametric tests. Homoscedasticity is an assumption of parametric tests, e.g., *t*-tests. In the case of sd-mean independence, the running median should be approximately horizontal.
5. Examine the MA plots and Hoeffding's D statistic plot. The metric informs about systematic biases and variability in the data by taking into account the log2-fold change between two conditions ($M = log2(I\_i) - log2(I\_j)$) and the mean expression of two conditions ($A = 1/2\ (log2(I\_i) + log2(I\_j))$), where I_i can be the intensity from one sample i and I_j the averaged intensities from a set of samples excluding sample i. Hoeffding's D statistic measures the dependency between A and M. It is a measure of the distance between $F(A, M)$ and $G(A)H(M)$, where $F(A, M)$ is the joint cumulative distribution function (CDF) of A and M, and G and H are marginal CDFs. The higher the value of D, the more dependent are A and M.
6. Examine the empirical cumulative distribution function (ECDF) plots. The metric informs about the distribution, the skewness of a sample and helps to identify outliers.
7. Examine the distance matrix between samples (Fig. 1C and D). The metric informs how similar or different samples are from one another based on the measured values. It informs about relationships and consistency between samples of the same group (e.g., treatment vs. control). The metric also helps to detect outliers and to identify batch effects.
8. Examine dimension reduction, e.g., principal component analysis (PCA, Fig. 1E and F) and loadings plot. PCA preserves the variance between data points and aids in exploring relationships between samples and features. PCA is helpful in identifying structure in the dataset and in detecting outliers. Make sure that pool samples cluster at the zero-zero axes of a PC1–PC2 plot.
9. Identify and remove any outliers based on the steps above. Reassess the data after the removal by repeating the Steps 1–8.

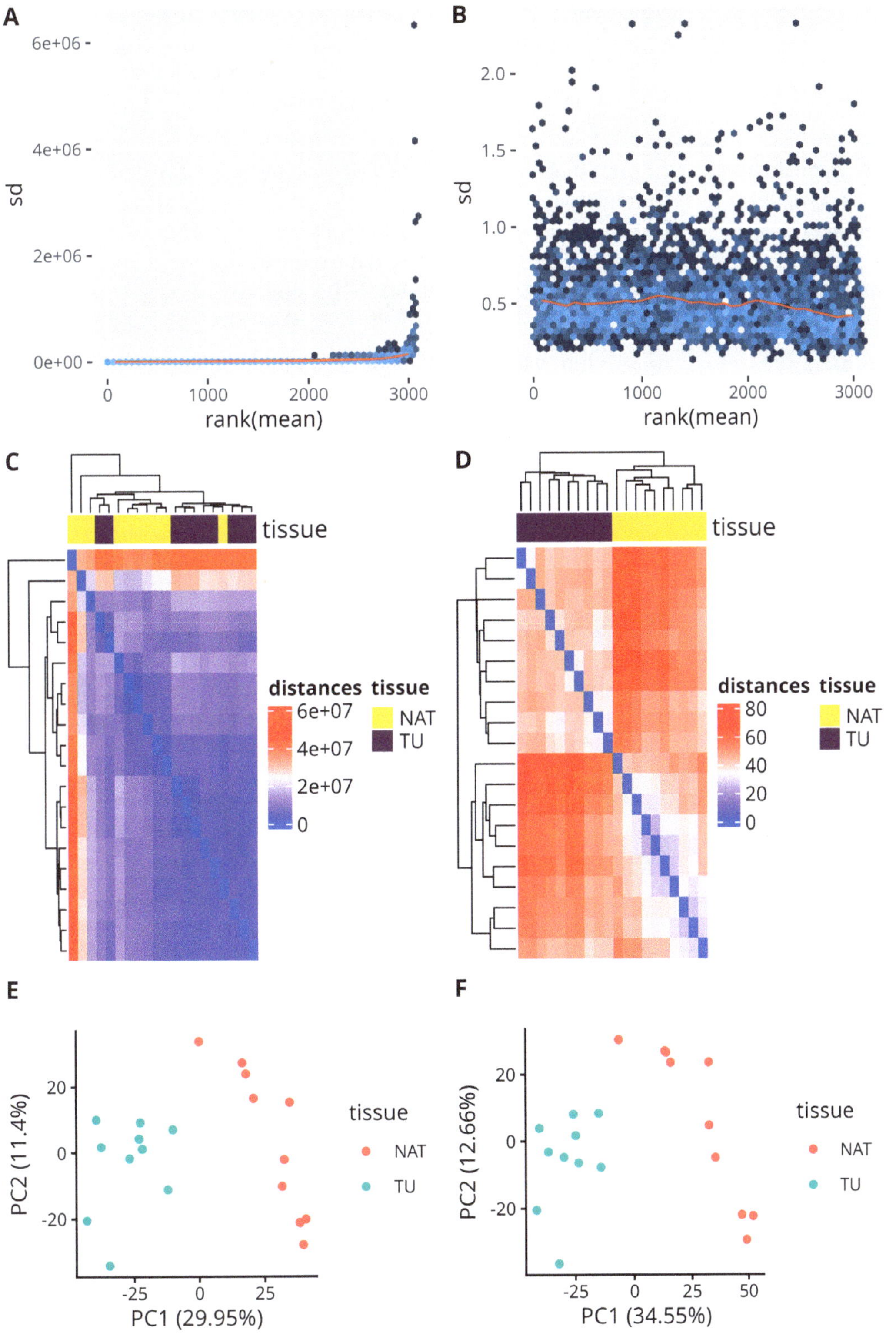
A
6e+06
4e+06
2e+06
0e+00
sd
0
1000
2000
3000
rank(mean)
B
2.0
1.5
1.0
0.5
sd
0
1000
2000
3000
rank(mean)
C
tissue
distances
6e+07
4e+07
2e+07
0
tissue
NAT
TU
D
tissue
distances
80
60
40
20
0
tissue
NAT
TU
E
PC2 (11.4%)
20
0
-20
-25
0
25
PC1 (29.95%)
tissue
NAT
TU
F
PC2 (12.66%)
20
0
-20
-25
0
25
50
PC1 (34.55%)
tissue
NAT
TU

10. After reviewing the metrics 3–9, identify the most appropriate methods for normalization, batch correction, transformation, and imputation for your datasets to ensure consistency and quality across samples and features.

3.5 Normalization, Batch Correction, Transformation, Missing Value Imputation

Tools, such as *MatrixQCvis* [10], can be used to explore data quality interactively and to optimize the choice of batch correction, normalization, or transformation methods. The *MatrixQCvis* package offers utility functions to perform normalization, batch reduction, transformation, and imputation using a variety of methods (*normalizeAssay*, *batchCorrectionAssay*, *transformAssay*, *imputeAssay*).

1. (Optional) Normalize the dataset [11]. Normalization adjusts for technical variability and ensures comparability of metabolite and protein levels across samples. The choice of method depends on the dataset's variability and experimental design. Common methods are median normalization, quantile division normalization (division of measurements of each sample by the given quantile, e.g., 50% for median division), sum normalization (division of measurements of each sample by the sum of intensities column for that sample).
2. (Optional) Correct for batches. Batch effect correction is typically essential in multi-batch experiments acquired over a longer time, but may not be needed in smaller-scale experiments. *ComBat*, for instance, requires specifying batch labels and optionally including covariates to retain biological variability. In the case of linear effects of batches, the batch effect can be removed via *limma*'s *removeBatchEffect* function and specifying the batch labels.
3. Transform the dataset. For example, *log* transformation is commonly used to reduce skewness and heteroskedasticity, other methods, such as *log2*, *log10*, or variance stabilizing normalization (vsn) can be applied as well. After transformation, check the mean-sd plot to assess homoskedasticity and determine if the transformation has balanced the variance.
4. (Optional) Scale the features. Scaling standardizes the magnitude of intensities across features, making them comparable by putting them on a common scale (e.g., z-score).

Fig. 1 Assessment of log-transformation of proteomics dataset on quality metrics. (**A**) Mean-sd plot of raw data. (**B**) Mean-sd plot of transformed data. Log-transformation controls heteroskedasticity. (**C**) Matrix of Euclidean distances between samples of raw data. (**D**) Matrix of Euclidean distances between samples of transformed data. (**E**) PCA of raw data. (**F**) PCA of transformed data. Log-transformation increases variation on principal components 1 and 2. The percentages indicate the explained variance per principal component. PC principal component, sd standard deviation

5. (Optional) Impute the missing values (*NA*) of the datasets. Metabolomics and proteomics data acquired by mass spectrometry typically contain missing values. The choice of the imputation method depends on the mechanism of missingness (e.g., missing completely at random, missing at random, missing not at random). Various imputation techniques are available in *MatrixQCvis*, such as k-nearest neighbors imputation or imputation by minimum values. For a more in-depth discussion on imputation methods in metabolomics, refer to [8], and for proteomics, see [12]. Generally, missing values should be imputed with caution, as doing so requires assuming a model of missingness and can introduce bias. In many cases, downstream analyses can tolerate a certain level of missing values without a significant impact on results. Tools such as *limma* [13] and *MOFA* are designed to handle datasets with incomplete observations, making imputation unnecessary for many applications.

3.6 Multi-Omics Integration Using MOFA

MOFA [2] integrates omics modalities using a factor analysis to uncover latent factors that explain the shared variation across different datasets. Proper data wrangling and processing as discussed in previous sections (e.g., normalization transformation) are critical steps in ensuring the reliability and interpretability of the results.

3.7 Model Building

1. Define the modalities (omics data types) to be included in the model. Make sure that the modalities are properly normalized and transformed. (Optional) Define group information. Group information refers to the different subsets or categories within the data that can be used to model variation across multiple omics layers. This could represent different experimental conditions, sample types (e.g., tumor vs. non-tumor), or other biological categories that might drive the variation in the data.
2. Define the parameters for training of the MOFA model (*choice of user*). Parameters control how data is preprocessed before training, such as whether views or groups are scaled to the same variance or centered (with a mean of zero). Parameters can define the structure of the MOFA model, such as the number of latent factors to infer, which can be guided by prior biological knowledge or model selection techniques. The user can also specify if a spike-and-slab prior on factor loadings is applied, which promotes sparsity in factor loadings by ensuring that only a few features have strong contributions while others are near zero. The user may apply the automatic relevance determination (ARD) priors on factor loadings, allowing the model to automatically determine which factors are relevant for each view. In addition, parameters can specify the training process of the model, including the mode of convergence

of the algorithm, the use of stochastic inference, the seed for reproducibility, and the maximum number of iterations before the model training is interrupted.

3. Add the model parameters to the untrained MOFA model and initiate the training of the model. The model will learn latent factors that explain the shared variation across the different omics datasets, based on the specified configurations and data characteristics.
4. Inspect the sanity of the model output. Plot data overview and verify that the dimensions for each dataset are as expected. Calculate the correlation scores between the factors. The correlation coefficients between the factors should be low, indicating that each factor captures distinct variation in the data. If factors are found to be highly correlated, it may suggest that the model is not effectively separating the variation across the datasets. In such cases, consider reducing the number of trained factors. This will not only help in improving the interpretability of the model but also enhance the biological relevance of the factors by focusing on more distinct sources of variation.
5. Examine the explained variance of the factors and per modality (Fig. 2A). The analysis shows the proportion of variance explained by each latent factor across different views (modalities) and groups (if applicable). Higher variance explained per factor means that the factor captures a significant portion of the variation in that particular view. If the first few factors explain most of the variance, it suggests that only a subset of factors drives the structure in the data. If certain factors explain more variance in one view compared to another, it suggests that the factor is more specific to that modality. Covariance in most factors indicates a strong relationship between the data types. If plotted separately, differences in variance explained between groups may indicate distinct biological patterns or variation in data structure across conditions.

3.8 Model Interpretation: Variance Decomposition and Analysis of Factors

1. Obtain sample covariates and cofactors, and add them to metadata. Metadata can be retrieved from the *colData* slot of the *SummarizedExperiment* objects.
2. Analyze the association between latent factors and metadata (e.g., clinical metadata, information on demographics, experimental condition, Fig. 2B). Compute correlations between the learned MOFA factors and the user-specific covariates (continuous or categorical). Visualize and examine the correlation coefficients and p-values. The intensity and sign of the color indicate the strength and direction of the correlation. Positive correlations suggest that an increase in the covariate is associated with an increase in the factor values. Negative correlations suggest an inverse relationship.

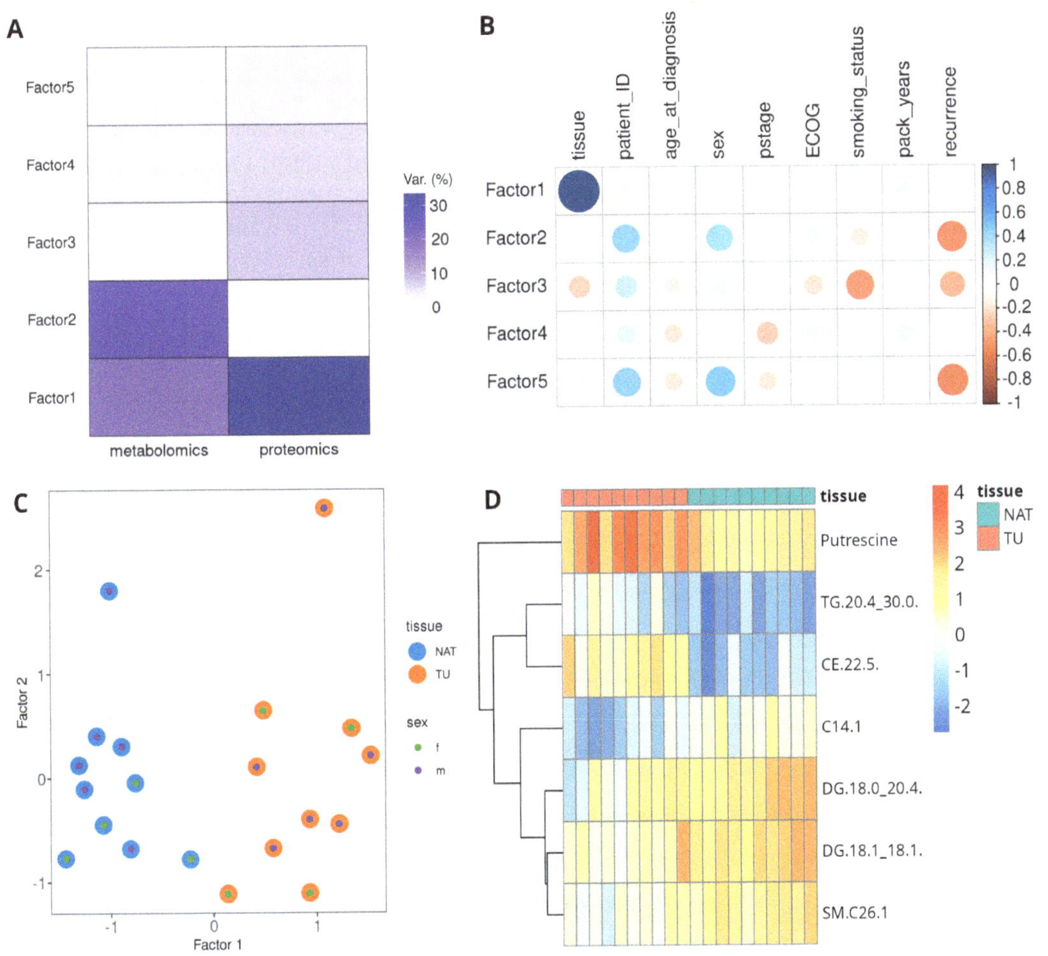

Fig. 2 Multi-omics factor analysis, diagnostic, and loading plots. (**A**) Variance explained by Factors 1–5 in the metabolomics and proteomics layers. (**B**) Sample-based Pearson correlation coefficients between MOFA factors and metadata variables. (**C**) Scatterplot of Factor values for each sample for Factor 1 and Factor 2. Factor 1 separates the samples along the NAT-TU axis. (**D**): Heatmap of transformed metabolite values for Factor 1. Shown are the top seven metabolites with the highest absolute loadings

3. Plot a scatterplot of the samples by factors (Fig. 2C). Each factor arranges the samples along a one-dimensional axis centered at zero, where the absolute value is not important; only the relative positioning of the samples matters. Samples with different signs on the axis indicate opposite "effects" along the inferred variation axis, with larger absolute values suggesting stronger effects. The interpretation of these factors is similar to that of principal components in PCA, where each factor represents a direction of variation in the data, and the separation of samples along these factors provides insight into the underlying patterns.

4. Analyze the weights of the factors. The weights represent how strongly each feature is associated with each factor. Features with little or no association to a factor will have values near zero, while those with a strong association will exhibit large absolute values. The sign of the weight indicates the direction of the effect: a positive weight means the feature has higher levels in samples with positive factor values, while a negative weight indicates higher levels in samples with negative factor values.
5. Analyze the feature intensities across samples and relationships between features using the information from the MOFA model (Fig. 2D). The visualizations are helpful for identifying patterns clusters, assessing data quality and batch effects, detecting potential outliers, and understanding how different features, factors, or sample groups contribute to the overall variation in the dataset. The visualizations should be created for each factor of interest and for the most important (e.g., top 10 or 20) features per factor (*choice of user*).

3.9 Differential Expression Analysis

Differential analysis identifies metabolites with significant abundance changes between conditions [14]. Here, selecting appropriate statistical tests depends on the data distribution and experimental design.

3.10 Statistical Testing

To systematically define which statistical tests to consider, one can follow these steps:

1. To define which comparison(s) should be performed, it is important to understand the meta information in the data and define the biological question. For ***one_vs_one (single comparison)*** analysis, where a single condition is compared against another (e.g., numerator = "Condition1" and denominator = "Condition2"), statistical tests such as the Student's *t*-test [15], Wilcoxon rank-sum test [16], or linear regression modeling [13] can be used. For ***all_vs_one (multiple comparison against a reference)***, where multiple conditions are compared individually against a single reference condition (e.g., numerator = ["Condition2", "Condition3", ..., "ConditionN"], denominator = "Condition1") or ***all_vs_all (multiple comparison)***, where all conditions are compared against each other (e.g., numerator = ["Condition1", "Condition2", ..., "ConditionN"], denominator = ["Condition1", "Condition2", ..., "ConditionN"]), multiple testing has to be performed for which statistical tests such as ANOVA (analysis of variance), Welch's ANOVA [17], Kruskal–Wallis test [15], or linear modeling (e.g., using limma's linear model fitting [13]) are appropriate.
2. Check the assumption of normality to understand if the data follow a normal- or not-normal distribution by, for example,

performing a Shapiro test [15]. For normally distributed data, *t*-tests or ANOVA can be used, while for not-normally distributed data, non-parametric tests like the Wilcoxon rank-sum test or Kruskal–Wallis test are appropriate [15]. The R package *MetaProViz* can be used to perform the Shapiro test for normality. and the Bartlett's test for homoscedasticity

3. Check whether the data are homoscedastic, meaning whether they have equal variance, using Levene's test or Bartlett's test [18]. If the data are not homoscedastic, consider using a test like the Wilcoxon rank-sum test since tests like Kruskal–Wallis assume equal variance [15]. *MetaProViz* can be used to perform Bartlett's test for homoscedasticity.
4. Perform differential expression analysis. If there are missing values (*NA*s) in the data linear modeling can be used, such as limma, which fits a linear model to the data and can handle missing values [13]. Yet, the limma method assumes that the residuals of the linear model are approximately normally distributed [13]. Other options are to either perform missing value imputation, which comes with its own pitfalls in metabolomics, or to assume that the missing values are true zeros, which in turn can lead to zero inflation [14]. *MetaProViz* can be used to perform differential analysis. It gives the choice between different statistical tests. Importantly, *MetaProViz* also adjusts for multiple testing using the Benjamini–Hochberg, False Discovery Rate (FDR), Bonferroni, or Holm method [4].

3.11 Visualization of Differential Analysis Results

1. Check the results of the differential analysis by a volcano plot. Check the proportion of the metabolites that are significantly up- or down-regulated, e.g., between the two conditions for both, the *t*-test and Wilcoxon rank-sum test (Fig. 3C and D). The classical volcano plot of the differential analysis results(-Fig. 3B) displays the log2 fold change on the x-axis, while the y-axis represents the negative logarithm of the adjusted p-value. The volcano plot is a scatter plot that displays the relationship between the magnitude of change (log2 fold change) and the significance (adjusted p-value) of each metabolite. Points above a certain threshold are considered statistically significant, and hence, it is important to set thresholds for significance (e.g., adjusted p-value < 0.05) and log2 fold change (e.g., $|\log2FC| > 0.5$).
2. (Optional) Add additional information to the volcano plot. For example, with the *MetaProViz* package, add additional relevant information to the plot [4], e.g., the metabolite name and color code for metabolite classes as provided by Biocrates (Fig. 3E).

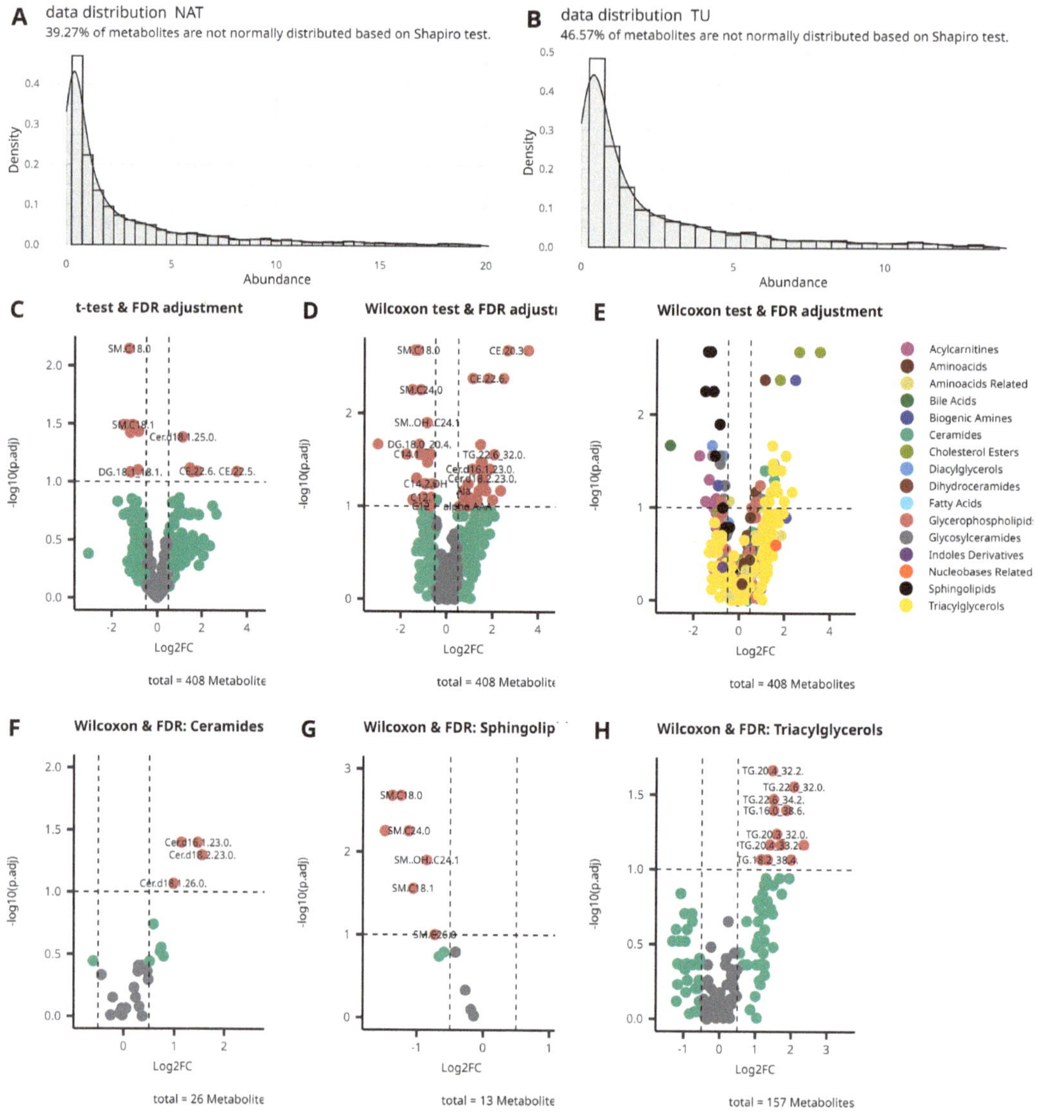

Fig. 3 One-versus-one differential expression analysis between tumor (TU) versus normal adjacent tumor (NAT). (**A**) Data distribution for each metabolite abundance of NAT. (**B**) Data distribution for each metabolite abundance of TU. The Shapiro test results show that the majority of the metabolites follow mostly a normal distribution, which means it is appropriate to perform a *t*-test. It is important to note that almost half of the metabolites are not normally distributed, which means that the *t*-test may not be appropriate for these metabolites. In this case, it might be better to use a non-parametric test such as the Wilcoxon rank-sum test or Kruskal–Wallis test. (**C–H**) Volcano plots of differential analysis comparing TU versus NAT using either the *t*-test and FDR adjustment (**C**) or the Wilcoxon rank-sum test and FDR adjustment (**D–H**). (**E**) is color-coded based on the metabolite class as defined by Biocrates, while (**F–H**) are selected metabolite panels based on the Biocrates class

3. (Optional) Examine individual plots for each metabolite class for better interpretability (Fig. 3F–H).

3.12 Analysis Using Biological Prior Knowledge

Prior Knowledge (PK) is a powerful tool in omics data analysis, allowing researchers to leverage existing biological information to gain mechanistic insights into their data.

1. Choose the PK that should be applied. The type of PK will dictate the biological question that can be answered and how the results can be interpreted.
   - Collections of biological pathways or processes, i.e., sets of genes/metabolites with a common characteristic (e.g., GO-term, KEGG, etc.), are commonly also referred to as pathway analysis. Pathway analysis is used to display that the features of interest are statistically more enriched in specific pathways [18]. The dysregulation of a pathway is inferred from measurements of its own components (e.g., metabolite abundance).
   - Collections of enzymes/molecules and their targets, for example, kinases and target phosphopeptides, which are usable for footprint analysis. Footprint analysis estimates the activities from molecular readouts of the targets (e.g., phosphosites) downstream of the enzyme (e.g., kinase) [19].
   - Collections of enzymes/molecules, their targets, and the interaction between them, which will depict a PK network (PKN), for which network-based approaches are needed to systematically integrate the data and extract coherent patterns [20].
2. Choose a statistical method to be applied on the PK, taking into account the biological perspective and method performance [3]. In case of classical mapping and footprint analysis, statistical methods such as Gene Set Enrichment Analysis (GSEA), Over-Representation Analysis (ORA), multivariate linear models (MLM), and univariate linear models (ULM), can be used to obtain a score for the activity of a kinase (footprinting) or the dysregulation of a pathway (mapping) [3]. The choice of statistical method will influence the results and the biological interpretation. For instance, GSEA uses a feature list ranked by their expression changes between conditions and hence finds pathways enriched at the top or bottom of the ranked gene list, while ORA is based on Fisher's exact test and checks if a selection of upregulated features (e.g., log2FC >1, p.adj < 0.05) are over-represented in a specific pathway [3]. In a benchmark study evaluated on transcriptomic and phospho-proteomic perturbation, linear models (e.g., ULM, MLM, etc.) outperformed GSEA, while ORA performs similarly well [3]. Here, it is also important to note that the chosen thresholds for e.g., up-regulated metabolites will affect the ORA results as it affects the basket size. The choice of ranking by log2FC versus ranking by t-value, which is a combination

of change and statistics, will alter the GSEA results. Dependent on other available analyses, such as MOFA or PCA, the ranking of features can also be done using those outputs. Similarly, using the MOFA results created above, one can also define the top and bottom 20% of altered features, which in turn can be used to perform ORA.

Yet, it is relevant to decide if performing enrichment analysis will be able to answer the biological question at hand. Metabolites are not genes, which means if a pathway is up-regulated in the context of genes, all the genes involved in that pathway produce more copies, while metabolite levels show lower coherence during regulation since the up-regulation may increase the abundance of the end-product but not of the metabolite intermediates [21]. Hence, pathway analysis in metabolomics can produce misleading or nonsensical results due to the unique characteristics and constraints of metabolites compared to genes [21]. Moreover, pathway enrichment results from metabolite measurements in urine or blood plasma can not be directly interpreted as altered pathway biosynthesis, since the enzymes required for many pathways are not present, and it is rather to be thought that these metabolites were released into those fluids from organs or blood cells [21]. Here, it is also relevant to consider that the origin of the pathways is based on well-established characterization, which has been discussed to hinder novel biological insights [22]. Hence, we would recommend to always perform literature searches for the top/bottom altered metabolites, especially if they are not well studied. It is also important to take into account that beyond metabolic pathways, metabolites play an active role in gene-regulation by, for example, binding to receptors and elucidating signaling cascades [23].

3.13 Annotated Metabolites and Coverage in the Prior Knowledge

There are different points to consider when integrating metabolomics data with PK, which can be divided into several steps:

1. Choose a database for metabolite IDs. Different from gene names, there are no standardized names for metabolites [24], but different databases that collect metabolite information and assign a metabolite ID to each entry, as for example, the Human Metabolome Database (HMDB) with HMDB IDs [25] or the Kyoto Encyclopedia of Genes and Genomes (KEGG) with KEGG IDs [26].
2. Investigate metabolite ID mappings. The database often includes multiple entries for the same metabolite with different degrees of ambiguity [27]. Experimentally, different degrees of ambiguity are needed due to the machine sensitivity, where, for example, stereoisomers are not distinguished [28]. Yet, standard metabolite-sets used as PK often contain one specific metabolite ID, which can hinder mapping to the detected data [4]. If metabolite detection is unspecific due to machine sensitivity, assign all possible metabolite IDs (Table 1) to ultimately

Table 1

Preview of selected metabolites to showcase ChEBI and HMDB information provided by Biocrates. Biocrates results provide all possible identifiers for each trivial metabolite name. For some metabolites, multiple ChEBI and HMDB IDs are available, while for other metabolites those databases have no entry

Trivial name	Class	ChEBI	HMDB
Ala	Aminoacids	15570, 16449, 16977, 76050	HMDB0000161, HMDB0001310
C0	Acylcarnitines	11060, 16347, 17126	HMDB0000062
Cer d16:1/18:0	Ceramides	None	None
DG 16:0_20:3	Diacylglycerols	84402	HMDB0007111

increase the overlap with metabolite-sets of interest, translate metabolite IDs, and/or quantify ambiguities, for example, by the *MetaProViz* package [4].

3. Visualize the available metabolite IDs for the detected metabolites using an upset plot and check available metabolite ID types for metabolite classes (Fig. 4A). For instance, for Reactome [29], which uses ChEBI IDs for pathway enrichment analysis, only a small subset of the detected metabolites will be included, which needs to be taken into consideration when interpreting the results biologically. In order to ensure the maximum overlap between measured metabolites and prior knowledge of choice, we recommend assignment of multiple metabolite IDs, dependent on the level of ambiguity of the detected metabolites.
4. Given the incomplete coverage in mass spectrometry-based metabolomics and hence the potential low coverage of the PK, adjust the choice of statistical tests. For metabolomics, methods such as ORA should generally perform well since the method offers the ability to restrict the background to the detected metabolites (assay-specific background) [30].

3.14 Available Prior Knowledge For Metabolites

1. Choose a suitable database containing PK:
   - For classical mappings with metabolite sets, obtain PK from databases such as *KEGG*, *Reactome*, or *WikiPathways*, *RaMP-DB*, *Omnipath*, or *MetSigDB* (*choice of user*):
     a) *RaMP-DB* [31] has combined many classical resources, trying to account for mismappings between the original resources by using molecular weight. Apart from classical pathway-metabolite sets such as wikipathways, *RaMP-DB* [31] has used *ClassyFire* [32] to assign chemical classes to the metabolites based on their

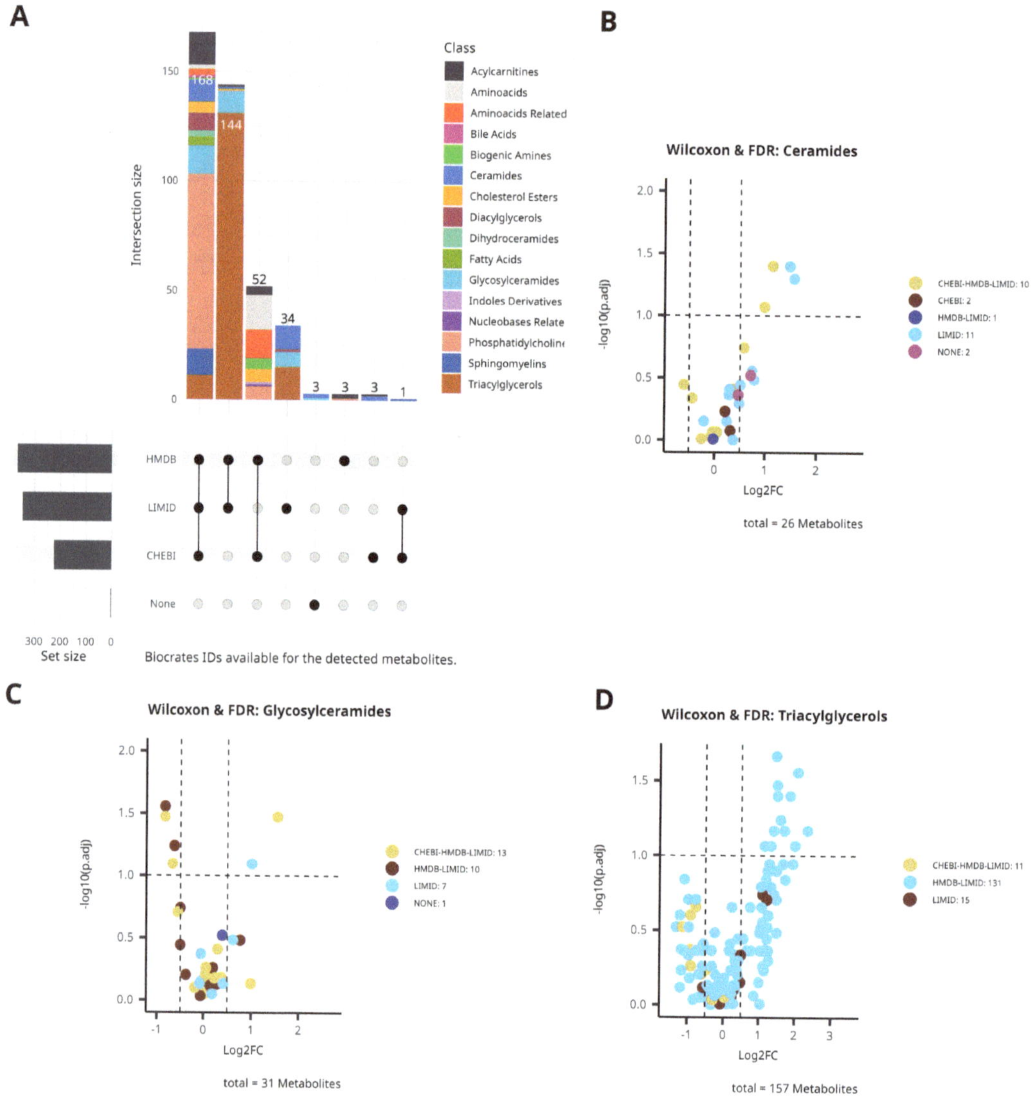

Fig. 4 (**A**) Upset plot of the metabolites IDs of the detected metabolites in the Biocrates example dataset color-coded for the Biocrates-defined metabolite classes. (**B–D**) Volcano plots of differential analysis comparing TU versus NAT using the Wilcoxon rank-sum test and FDR adjustment, selecting for metabolite panels based on Biocrates class and color coding for the available metabolite ID types

structural components. Chemical class-metabolite sets can be used to understand if certain compound classes are altered in the comparison of interest [31].

b) *Omnipath* [33] [34], a database of molecular biology, PK combines data from more than 1000 resources and incorporates many database clients containing metabolites and utilities for metabolite identifier translation.

Apart from pathway-metabolite sets, *Omnipath* also offers many other ontology information.

c) To explore signaling function of metabolites by binding to receptors, use *MetalinksDB* [23]. *MetalinksDB* contains information about metabolite-receptor and metabolite-transporter interaction, as well as whether the binding of the metabolite to a receptor is activating or inhibiting [23]. Using *MetalinksDB* not only enables combined pathway enrichment analysis on metabolite-gene sets or building a network of those interactions, but also adds biological value by highlighting the signaling potential of the metabolites [23].

d) *MetaProViz* includes a collection of annotated metabolite sets usable for classical mapping (Metabolism Signature Database, *MetSigDB*), which includes classical pathway-metabolite sets, chemical class-metabolite sets, and *MetalinksDB* metabolite-receptor and transporter sets, all adapted for classical mapping [4]. Moreover, *MetaProViz* uses the *COSMOS* enzyme-metabolic reaction prior knowledge network [20] to convert any gene set to gene-metabolite sets, which can be used to perform combined enrichment analysis [4].

e) When working with human data, the PKN *ocean* includes a manually curated reduced model of human metabolism based on *redHUMAN* of *Recon3D* and has developed a specific adaptation of the weighted mean, using enzyme distance to weight interaction, to perform enrichment analysis [35]. *ocean* explores coordinated deregulations of metabolite abundances to generate a metabolic footprint for each metabolic enzyme and ultimately define enzymatic bottlenecks [35].

f) If no metabolomics data are available, but the focus is still on metabolic pathways, use the gene-sets generated by [36], which include metabolic enzymes and enable weighting for enzymes that are unique for a pathway compared to enzymes present in multiple pathways.

2. (Recommended) Exclude certain metabolites from the database. The same metabolite can be part of multiple pathways and the database often includes unspecific metabolites like cofactors, ions, H2O or CO2, which can not be detected with a classical mass spectrometry setup. Many metabolic pathways originate from genome-scale metabolic models, which cover the whole reaction in an organism, and those models have not been adapted for the coverage in a metabolomics experiment. We recommend to exclude those metabolites prior to performing enrichment analysis. *MetaProViz* provides pathway-metabolite sets without unspecific metabolites [4].

3.15 Biological Interpretation of the Results

The biological interpretation of the results is crucial and should be done in the context of the biological question and the experimental design. Here, it is important to take into account multiple points:

1. Understand the limitations of the PK used, such as potential biases in the underlying original resources. For example, if the PK is based on a specific database, it may not capture all relevant metabolites or pathways. This can lead to biased results and misinterpretation of the data. Moreover, the choice of PK dictates the biological question one can address, as discussed in the paragraph above for classical pathway-metabolite sets, chemical class-metabolite sets, or receptor-metabolite sets.
2. Understand the coverage of your experimental data in the PK resource. If the PK only includes metabolites that are available in, for example, HMDB, this would in turn exclude many detected ceramides, glycosylceramides, and triglycerols, while using ChEBI IDs would exclude almost all triacylglycerols (Fig. 4A–D). Here, triacylglycerols include 226 metabolites of which only 11 have a ChEBI ID (Fig. 4D). If the results of the enrichment analysis do not show those pathways as significantly altered, it does not mean that they are not altered, but rather that the PK used is not able to capture them. Consequently, it is important not to over-interpret the importance of the most altered pathways, but rather to use them as a starting point for further investigation, not ignoring the detected metabolites that were not covered by the PK resource.
3. (Optional) Combine multiple PK resources to increase the coverage of the detected metabolites.
4. Check how many metabolites of, e.g., a pathway in the PK have been detected. For example, if a pathway only contains 20 metabolites and 13 of them are detected (65%), this is a stronger evidence than detecting 40 metabolites of a pathway containing 110 metabolites (36%).
5. Understand the impact of metabolite set inflation and deflation on the results. Due to the different metabolite ID types in combination with the metabolite ID coverage in metabolite sets, it is sometimes required to translate metabolite IDs from one identifier type to another. For example, in the Biocrates example data, no KEGG IDs are available, yet, if KEGG pathways are used for enrichment analysis, either the KEGG pathways or the metabolite IDs of the data have to be translated, which can lead to one-to-none, one-to-one, one-to-many, and many-to-many mappings between the metabolite ID types. This mapping ambiguity can lead to inflation or deflation of the metabolite set size, which needs to be taken into account when interpreting the results. In case of combined enrichment analysis for multiple modalities, take into account the difference in

feature space between metabolomics and transcriptomics/proteomics, since this can underestimate the altered metabolites.

6. Take into account the statistical method used for enrichment analysis, as it impacts the interpretation of the results (see discussion above).

3.16 Experimental Network Analysis Using Statistical Methods

Statistical analysis uncovers pairwise relationships between metabolites and/or proteins within each modality. The analysis can be done to assess the co-occurrence, association, or regulation of features within a dataset without any PK. The calculation of pairwise coefficients is often done via the *Pearson* or *Spearman correlation* calculation methods, which are useful for exploratory analysis to identify correlated feature pairs, suggesting potential biological relationships or shared pathways (guilt-by-association principle). Other methods may be more suitable depending on the research question, such as *ARACNE* (Algorithm for the Reconstruction of Accurate Cellular Networks), *Bayesian network learning*, *CLR* (Context Likelihood of Relatedness), *GGM* (Gaussian Graphical model), linear regression using *LASSO* (Least Absolute Shrinkage and Selection Operator) regularization, *partial Pearson correlation*, *Random Forest*, or *partial Spearman correlation*. In the resulting network, vertices represent features, such as metabolites or proteins, while edges represent the statistical relationship between them, such as correlation coefficients.

We will showcase the creation of statistical networks based on the implemented functionality of the *MetNet* [37] package.

1. Choose an algorithm based on the data and the biological question:

   - *ARACNE* represents a mutual information (MI)-based approach that removes indirect interactions using data processing inequality. The algorithm should be used when aiming to extract the most relevant direct feature interactions from MI-based networks. It is more accurate than pure MI-based methods by reducing false positives by filtering indirect edges. The algorithm requires large sample sizes to estimate MI accurately.
   - *Bayesian network learning* infers regulatory networks by modeling conditional dependencies between genes using a probabilistic graphical model. The algorithm may be used when prior biological knowledge is available and probabilistic modeling is needed. The algorithm handles noise well, allows for causal inference, and is able to handle missing data. It may be, however, computationally expensive, require strong prior assumptions, and may be sensitive to sample size.
   - *CLR* uses MI to infer regulatory relationships, normalizing interactions for each feature. The algorithm works well when

working with noisy or indirect regulatory influences. It improves MI-based inference by reducing false positives and performs well for global network structures. The algorithm assumes that regulatory interactions follow a Gaussian-like distribution, which may not always be true.

- *GGM* estimates a sparse precision matrix (inverse covariance) to infer direct interactions. The algorithm can be used when assuming a multivariate Gaussian structure in expression data. It identifies direct dependencies and avoids indirect correlations. The algorithm requires large sample sizes to obtain estimates accurately.
- *Linear regression using LASSO regularization* identifies a regulatory network by selecting a sparse set of predictor features for each target feature, thereby reducing the number of spurious edges in a network. The algorithm handles high-dimensional data well, reduces overfitting, and provides an interpretable sparse network. The algorithm assumes a linear relationship, may struggle with correlated predictors (features with similar expression profiles, and may be computationally expensive.
- *Pearson correlation* measures linear relationships between feature expression/abundance profiles. The algorithm should be used when linear dependencies exist between feature expression/abundance levels. The algorithm is simple, computationally efficient, and easily interpretable. It does not capture non-linear relationships and is sensitive to noise.
- *Partial Pearson correlation* identifies direct regulatory interactions by controlling for indirect effects. It can be used when confounding influences in a network must be corrected. The algorithm removes indirect correlations, providing clearer direct relationships. It requires large sample sizes to compute reliably.
- *Random forest* infers a regulatory network using feature importance scores from an ensemble of decision trees. The algorithm works well when dealing with non-linear relationships and heterogeneous data. It captures complex dependencies, is robust to noise, and does not assume a specific distribution. The algorithm can be computationally expensive and may overfit if not tuned properly.
- *Spearman correlation* captures monotonic (rank-based) relationships between feature expression/abundance levels. The algorithm may be used when the data does not follow a normal distribution or when looking for ranked relationships. It is robust to outliers and captures non-linear relationships. The algorithm is less powerful for detecting direct gene interactions.

- *Partial Spearman correlation* is similar to partial Pearson correlation, but for rank-based dependencies. The algorithm controls for confounding variables while dealing with non-linear relationships. It handles non-linearity and indirect interactions. It is less widely used in regulatory network inference and may be computationally demanding.

2. Calculate the adjacency matrix using the *MetNet* package [37]. Specify the algorithms to be run. (Optional) Prior to running the model, check if the model tolerates missing values (*NA*) and impute values if needed. (Optional) Apply the Benjamini–Hochberg method or any other method for adjustment for p-value adjustment for Pearson and Spearman correlation. The result of each algorithm is stored in a respective assay of the *AdjacencyMatrix* object, an extension of the *SummarizedExperiment* class. Note that, depending on the chosen algorithm, the adjacency matrices are not necessarily symmetric.
3. Examine the distribution of similarity coefficients and p-values, if available (Figure 3A and B).
4. (Optional) Retain highly correlating values (*choice of user*). Feature pairs with low coefficients will be regarded as noise and removed from the adjacency matrix.
5. Create a graph from the consensus adjacency matrix. Specify the graph characteristics, e.g., if the graph should be unweighted or weighted or undirected or directed.
6. (Optional) Remove the singleton components or small networks and obtain the induced subgraph.
7. Visualize the network. Results from upstream analyses can be mapped to the network. As an example, the weights of latent factor 1 of the trained MOFA model can be used for coloring of the network vertices (Fig. 5C). Alternatively, we can use logFC of the differential expression analysis.
8. Choose and apply the module determination algorithm (*choice of user*). To detect substructures in the graph, module determination algorithms can be applied to the graph and visualize the results (Fig. 5D). Common algorithms are (*igraph* implementation in brackets, see Table 2 for further details):

   - *Community structure detection based on edge betweenness* (*cluster_edge_betweenness*) detects communities by iteratively removing edges with the highest betweenness centrality, splitting the network into distinct modules. It works well for small graphs, provides a clear hierarchical structure. The algorithm may be computationally expensive for large networks.

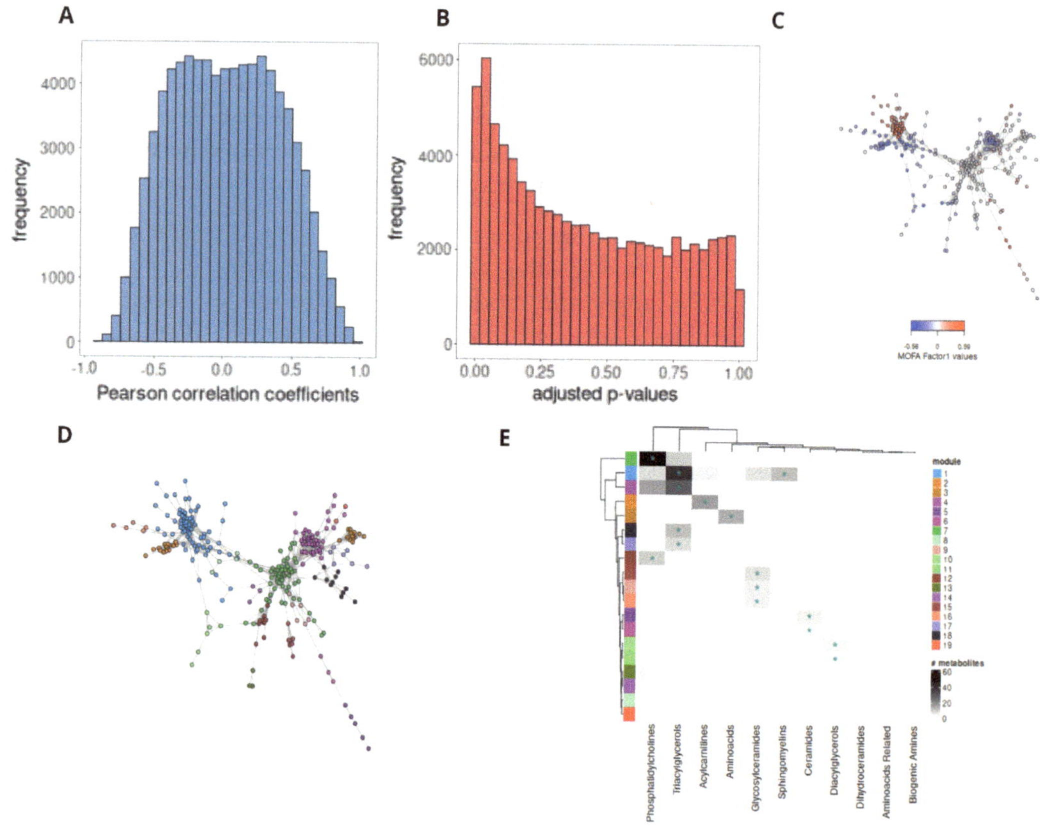

Fig. 5 Statistical network analysis of metabolomics data. (**A**) Histogram of coefficients of Pearson correlation-based network. (**B**) Histogram of corresponding adjusted p-values of Pearson correlation-based network using the Benjamini–Hochberg method). (**C**) Thresholded network based on Pearson and Spearman correlation. Each node represents a metabolite and is colored according to its MOFA Factor 1 loading. Edges were retained if the absolute correlation coefficient was >0.8 (Pearson) and >0.7 (Spearman), with corresponding adjusted p-values < 0.05 using the Benjamini–Hochberg method. (**D**) Thresholded network based on Pearson and Spearman correlation. Nodes represent metabolites and are colored according to community membership based on the label propagation community detection algorithm. For a legend of module membership, refer to panel **E**. (**E**) Heatmap of chemical class distribution across network modules based on label propagation algorithm. The heatmap shows the abundance of chemical classes (columns) within each module (rows), with color intensity reflecting class count. Asterisks (*) indicate statistically significant overrepresentation of a chemical class within a module (Fisher's exact test, FDR-adjusted $p < 0.05$)

- *Community structure detection* via *greedy optimization of modularity* (*cluster_fast_greedy*) uses hierarchical agglomeration to merge vertices into communities, optimizing the modularity at each step. The algorithm works efficiently for large graphs, but is less effective for networks with overlapping communities.
- *Community structure detection algorithm based on interacting fluids* (*cluster_fluid_communities*) uses label propagation with a fixed number of communities, allowing them to

Table 2
Compatibility of community detection algorithms with network properties

Algorithm	Undirected	Directed	Unweighted	Weighted
cluster_edge_betweenness	✓	✓	✓	✓
cluster_fast_greedy	✓	×	✓	✓
cluster_fluid_communities	✓	×	✓	×
cluster_infomap	✓	✓	✓	✓
cluster_label_prop	✓	✓	✓	✓
cluster_leading_eigen	✓	×	✓	✓
cluster_leiden	✓	×	✓	✓
cluster_louvain	✓	×	✓	✓
cluster_optimal	✓	✓	✓	✓
cluster_spinglass	✓	×	✓	✓
cluster_walktrap	✓	×	✓	✓

expand dynamically based on node density. The algorithm is fast. The algorithm requires to predefine the number of communities and may yield inconsistent results across runs.

- *Community structure detection based on Infomap* (*cluster_infomap*) works by simulating random walks on the network and compressing the path description to reveal the best modular structure. The idea is that if a network has well-defined communities, a random walker will spend more time inside a community before jumping to another one. The algorithm captures this by trying to minimize the description length of the walk, effectively grouping vertices that are frequently visited together. The algorithm may be computationally demanding and is sensitive to parameter choices.
- *Community structure detection based on the propagating labels* (*cluster_label_prop*) assigns labels to vertices, which spread iteratively based on majority voting until convergence. The algorithm scales linearly with the number of vertices and is thus works well on large networks. Due to the random nature of initiation, the results can be unstable and inconsistent for certain networks.
- *Community structure detection based on the leading eigenvector* (*cluster_leading_eigen*) uses spectral clustering by computing the leading eigenvectors of a modularity matrix to find communities. The algorithm may be computationally expensive for large networks.

- *Community structure detection using the Leiden algorithm* (*cluster_leiden*) is a faster version of the Louvain algorithm and yields more accurate solutions. It iteratively refines partitions to ensure well-connected communities by optimizing either the modularity or the Constant Potts model, which does not suffer from the resolution limit. The algorithm may have difficulties finding small communities.
- *Community structure detection by multi-level optimization of modularity* (*cluster_louvain*) optimizes modularity by merging hierarchically small communities into larger ones in a greedy manner. The algorithm scales well to large networks. It may produce different results on each run and may have difficulties finding small communities.
- *Community structure detection using the maximum modularity* (*cluster_optimal*) calculates the optimal community structure by maximizing the modularity over all possible partitions. The algorithm guarantees the best modularity score, but is computationally expensive for a large network.
- *Community structure detection based on statistical mechanics* (*cluster_spinglass*) models community detection as an energy minimization problem via a spin-glass model and simulated annealing. The algorithm is computationally expensive for large networks and may be sensitive to parameter settings.
- *Community structure detection based on short random walks* (*cluster_walktrap*) simulates random walks on the network to detect community structure, assuming vertices within the same community are more likely to be visited in one walk. The algorithm may be slow for large networks.

9. (Optional) Depending on the biological question, combine network structure with prior knowledge (Fig. 5E). For example, analyze the module composition via PK resources to check for metabolite classes. Use one-tailed Fisher's Exact test to check for overrepresentation of a given class.
10. Perform analysis of network structure.

4 Conclusion

This chapter demonstrated how to perform metabolomicsdata analysis in *R* using prior knowledge in combination with enrichment analysis or network-based approaches. By integrating tools for preprocessing differential analysis, enrichment and correlation analysis, and data integration, *R* provides a comprehensive framework for deriving biological insights from metabolomics datasets and combining them with other omics such as proteomics. Parameters such as thresholds, normalization methods, and correction techniques must be carefully chosen to ensure the robustness and reliability of the results.

In order to interpret a metabolomics experiment in its entirety, it is important to understand the limitations of the prior knowledge used, the coverage of the experimental data in the prior knowledge resource, the impact of metabolite set inflation and deflation on the results and the choice of statistical method used for enrichment analysis. Researchers should not solely rely on pathway analysis and often it is more beneficial to perform literature searches for the most altered metabolites, extract metabolite receptors that could be targeted by the altered metabolites or transporters that could release the metabolites into the environment. All of this will require close collaboration between data scientists and experimentalists in the future.

References

1. Csardi G, Nepusz T (2006) The igraph software package for complex network research. Complex Syst 1695:1–9
2. Argelaguet R, Velten B, Arnol D et al (2018) Multi-omics factor analysis—a framework for unsupervised integration of multi-omics data sets. Mol Syst Biol 14:e8124. https://doi.org/10.15252/msb.20178124
3. Badia-i-Mompel P, Santiago JV, Braunger J et al (2022) decoupleR: ensemble of computational methods to infer biological activities from omics data. Bioinform Adv 2:vbac016. https://doi.org/10.1093/bioadv/vbac016
4. Schmidt C, Turei D, Prymidis D, et al (2025) Integrated metabolomics data analysis to generate mechanistic hypotheses with MetaProViz. bioRxiv. https://doi.org/10.1101/2025.08.18.670781
5. Amara A, Frainay C, Jourdan F et al (2022) Networks and graphs discovery in metabolomics data analysis and interpretation. Front Mol Biosci 8:841373. https://doi.org/10.3389/fmolb.2022.841373
6. Gegner HM, Naake T, Aljakouch K et al (2024) A single-sample workflow for joint metabolomic and proteomic analysis of clinical. Clin Proteom 21: 49. https://doi.org/10.1186/s12014-024-09501-9
7. Morgan M, Obenchain V, Hester J, Pagès H (2024) Summarized experiment: a container (S4 class) for matrix-like assays
8. Wei R, Wang J, Su M, et al (2018) Missing value imputation approach for mass spectrometry-based metabolomics data. Sci Rep 8:663. https://doi.org/10.1038/s41598-017-19120-0
9. Cox J, Mann M (2008) MaxQuant enables high peptide identification rates, individualized p.p.b.-range mass accuracies and proteome-wide protein quantification. Nat Biotechnol 26:1367–72. https://doi.org/10.1038/nbt.1511
10. Naake T, Huber W (2022) MatrixQCvis: shiny-based interactive data quality exploration for omics data. Bioinformatics 38:1181–1182. https://doi.org/10.1093/bioinformatics/btab748
11. Bijlsma S, Bobeldijk I, Verheij ER et al (2005) Large-scale human metabolomics studies: a strategy for data (pre-) processing and validation. Anal Chem 78:567–574. https://doi.org/10.1021/ac051495j
12. Lazar C, Gatto L, Ferro M et al (2016) Accounting for the multiple natures of missing values in label-free quantitative proteomics data sets to compare imputation strategies. J Proteome Res 15:1116–1125. https://doi.org/10.1021/acs.jproteome.5b00981
13. Ritchie ME, Phipson B, Wu D et al (2015) Limma powers differential expression analyses for RNA-sequencing and microarray studies. Nucleic Acids Res 43:e47. https://doi.org/10.1093/nar/gkv007
14. Huang Z, Wang C (2022) A review on differential abundance analysis methods for mass spectrometry-based metabolomic data. Metabolites 12:305. https://doi.org/10.3390/metabo12040305
15. Padhee S, Ahmed S (2024) Application of statistics in biomedical research. J Integr Med Res 2:66–71. https://doi.org/10.4103/jimr.jimr_1_24

16. Wilcoxon F (1945) Individual comparisons by ranking methods. Biom Bull 1:80. https://doi.org/10.2307/3001968
17. Welch BL (1951) On the comparison of several mean values: an alternative approach. Biometrika 38:330. https://doi.org/10.2307/2332579
18. García-Campos MA, Espinal-Enríquez J, Hernández-Lemus E (2015) Pathway analysis: state of the art. Front Physiol 6:383. https://doi.org/10.3389/fphys.2015.00383
19. Dugourd A, Saez-Rodriguez J (2019) Footprint-based functional analysis of multiomic data. Curr Opin Syst Biol 15:82–90. https://doi.org/10.1016/j.coisb.2019.04.002
20. Dugourd A, Kuppe C, Sciacovelli M et al (2021) Causal integration of multi-omics data with prior knowledge to generate mechanistic hypotheses. Mol Syst Biol 17:e9730. https://doi.org/10.15252/msb.20209730
21. Lee KS, Su X, Huan T (2025) Metabolites are not genes – avoiding the misuse of pathway analysis in metabolomics. Nat Metab 7:858–861. https://doi.org/10.1038/s42255-025-01283-0
22. Hu M, Alkhairy S, Lee I et al (2025) Evaluation of large language models for discovery of gene set function. Nat Methods 22:82–91. https://doi.org/10.1038/s41592-024-02525-x
23. Farr E, Dimitrov D, Schmidt C, et al (2024) MetalinksDB: A flexible and contextualizable resource of metabolite-protein interactions. Brief Bioinform 25:bbae347. https://doi.org/10.1093/bib/bbae347
24. Koistinen V, Kärkkäinen O, Keski-Rahkonen P et al (2023) Towards a Rosetta Stone for metabolomics: recommendations to overcome inconsistent metabolite nomenclature. Nat Metab 5:351–354. https://doi.org/10.1038/s42255-023-00757-3
25. Wishart DS, Tzur D, Knox C et al (2007) HMDB: the human metabolome database. Nucleic Acids Res 35:D521–D526. https://doi.org/10.1093/nar/gkl923
26. Kanehisa M, Furumichi M, Tanabe M et al (2017) KEGG: new perspectives on genomes, pathways, diseases and drugs. Nucleic Acids Res 45:D353–D361. https://doi.org/10.1093/nar/gkw1092
27. Pham N, van Heck RGA, van Dam JCJ, et al (2019) Consistency, inconsistency, and ambiguity of metabolite names in biochemical databases used for genome-scale metabolic modelling. Metabolites 9:28. https://doi.org/10.3390/metabo9020028
28. Dias DA, Jones OAH, Beale DJ, et al (2016) Current and future perspectives on the structural identification of small molecules in biological systems. Metabolites 6:46. https://doi.org/10.3390/metabo6040046
29. Gillespie M, Jassal B, Stephan R et al (2022) The reactome pathway knowledgebase 2022. Nucleic Acids Res 50:D687–D692. https://doi.org/10.1093/nar/gkab1028
30. Wieder C, Frainay C, Poupin N et al (2021) Pathway analysis in metabolomics: recommendations for the use of over-representation analysis. PLoS Comput Biol 17:e1009105. https://doi.org/10.1371/journal.pcbi.1009105
31. Braisted J, Patt A, Tindall C, et al (2023) RaMP-DB 2.0: a renovated knowledgebase for deriving biological and chemical insight from metabolites, proteins, and genes. Bioinformatics 39:btac726. https://doi.org/10.1093/bioinformatics/btac726
32. Djoumbou Feunang Y, Eisner R, Knox C, et al (2016) ClassyFire: Automated chemical classification with a comprehensive, computable taxonomy. J Cheminforms 8:61. https://doi.org/10.1186/s13321-016-0174-y
33. Türei D, Korcsmáros T, Saez-Rodriguez J (2016) OmniPath: guidelines and gateway for literature-curated signaling pathway resources. Nat Methods 13:966–967. https://doi.org/10.1038/nmeth.4077
34. Türei D, Valdeolivas A, Gul L, et al (2021) Integrated intra- and intercellular signaling knowledge for multicellular omics analysis. Mol Syst Biol 17:e9923. https://doi.org/10.15252/msb.20209923
35. Sciacovelli M, Dugourd A, Jimenez LV et al (2022) Dynamic partitioning of branched-chain amino acids-derived nitrogen supports renal cancer progression. Nat Commun 13: 7830. https://doi.org/10.1038/s41467-022-35036-4
36. Gaude E, Frezza C (2016) Tissue-specific and convergent metabolic transformation of cancer correlates with metastatic potential and patient survival. Nat Commun 7: 13041. https://doi.org/10.1038/ncomms13041
37. Naake T, Fernie AR (2019) MetNet: Metabolite network prediction from high-resolution mass spectrometry data in r aiding metabolite annotation. Anal Chem 91:1768–1772. https://doi.org/10.1021/acs.analchem.8b04096

Chapter 18

Metabolomics Data Analysis with the HMDB

Yamilé López-Hernández, Prashanthi Kovur and David S. Wishart

Abstract

The Human Metabolome Database (HMDB) is a web-based resource that supports metabolite identification and interpretation for human studies. This chapter outlines five protocols: (1) identifying metabolites using mass spectrometry with retention time or collision cross-section data (MS–RT/CCS); (2) identifying metabolites using tandem mass spectrometry (MS/MS); (3) identifying metabolites using nuclear magnetic resonance (NMR); (4) comparing measured concentrations to HMDB reference values; and (5) interpreting metabolomics results via HMDB's MetaboCards and PathBank links.

Keywords Database, Metabolomics, Human, Disease, Data analysis

1 Introduction

The Human Metabolome Database (HMDB) has been a cornerstone resource for metabolomics research since its inception in 2007 [1]. The initial release in 2007 contained information on just 2180 metabolites. Over the years, the database has grown significantly in size and scope, and its latest release, HMDB 5.0 [2], now features over 234,000 metabolites with rich data on metabolite structures, nomenclature, detailed descriptions, physical properties, spectra, and biological pathways. All the information in the HMDB is publicly available at https://www.hmdb.ca. The HMDB has also grown in popularity, and it now receives >5 million web hits each year.

The HMDB was originally developed as part of the Human Metabolome Project (HMP), a multi-institutional initiative launched in 2005 that was aimed at mapping the complete set of metabolites in the human body. This project, which is still ongoing, has greatly expanded our understanding of human metabolism. It has been using cutting-edge techniques such as mass spectrometry (MS) and nuclear magnetic resonance (NMR) to generate detailed profiles and metabolite concentration ranges for thousands of metabolites found in human biofluids and tissues [3–7]. The metabolites in the HMDB are categorized into four main groups: (1) Detected and quantified,

Michael Witting (ed.), *Computational Methods and Data Analysis for Metabolomics*, Methods in Molecular Biology, vol. 3063,
https://doi.org/10.1007/978-1-0716-5452-1_18,

(2) Detected but not quantified, (3) Expected, and (4) Predicted. "Detected" metabolites are those with confirmed concentrations or experimental evidence of their presence in human biofluids, cells, or tissues. "Expected" metabolites have known structures and biochemical pathways, with frequent human exposure, but have not yet been detected in the body. "Predicted" metabolites are a small set of biologically plausible compounds generated through computational biotransformations using the BioTransformer program [8–10].

In a previous chapter in this series, published in 2020 [11], we provided a general introduction to the HMDB and described the wealth of information it offers about human metabolism. For this current chapter, our focus will shift to practical applications—specifically, how to use the HMDB effectively for metabolite identification in metabolomics research.

The latest release of HMDB (version 5.0) introduces several exciting improvements, such as enhanced metabolite descriptions, an expanded spectral library, and advanced visualization tools [2]. Whether researchers are studying human serum, urine, cerebrospinal fluid, or other biological samples, the ability to confidently identify metabolites and link them to biological processes is crucial for understanding metabolic diseases, discovering biomarkers, or exploring the role of microbially or food-derived metabolites in health and disease.

One of the most notable new features of HMDB 5.0 is its expanded library of predicted and experimentally validated spectra. With over 1.4 million MS/MS spectra, as well as thousands of NMR spectra, the HMDB is better equipped than ever to help researchers identify metabolites through spectral matching. Additionally, the new search capabilities allow users to search based on a variety of parameters, including molecular mass, retention time, and other spectral properties, greatly enhancing the ability to find relevant compounds in complex experimental data. Moreover, the introduction of advanced visualization tools for viewing 3D molecular structures and visualizing interactive pathway diagrams now allows users to better understand the relationship between metabolites, proteins, tissues, and organs.

As we dive deeper into how to leverage the HMDB for metabolite identification, this guide will walk readers through the various tools available in the latest version of the database, providing step-by-step instructions on how to search, compare, and visualize data. From understanding metabolite structures and pathways to interpreting spectral data, we hope to show how the HMDB is indeed a powerful tool that can help unlock the complexities of the human metabolome. These methods provide reproducible, step-by-step strategies for integrating experimental datasets with curated HMDB content, ensuring consistent metabolite annotation and biological interpretation.

2 Materials

The HMDB is a web-based database accessible online (at https://www.hmdb.ca) from any modern computer or web-enabled device, such as a tablet, with a current web browser (e.g., Google Chrome, Microsoft Edge, Windows Internet Explorer, Apple Safari, or Mozilla Firefox) and a 25–35 cm screen with moderate resolution (such as 1080p).

3 Methods

Five methods are covered here to show how the HMDB can be used in a plasma metabolomics study of pre-diabetes: (1) How to identify key pre-diabetes metabolites using MS + RT or CCS data; (2) How to identify pre-diabetes metabolites using MS/MS data; (3) How to identify pre-diabetes metabolites using NMR data; (4) How to compare the concentrations of pre-diabetes metabolites to reference values in HMDB; and (5) How to interpret those read-outs using HMDB metabolite descriptions or HMDB pathways via PathBank [12, 13]. All activities can be done with almost any internet-compatible device, although a larger screen (25 cm or larger) is preferable.

3.1 How to Identify Pre-diabetes Metabolites Using MS + RT or CCS Data

Let us assume that plasma samples from pre-diabetic individuals have been analyzed via MS and MS/MS. For some of the most significant features, additional high-resolution MS data have been collected in the positive mode on an Orbitrap or a QTOF MS instrument. Likewise, collision cross-section (CCS) values and retention time (RT) data (reverse phase, UPLC BEH C18 column) have also been collected. For the purpose of this methodological chapter, we provide examples of experimental data (*m*/*z*, CCS, RT, MS/MS fragments, NMR peaks) that represent the type of data a researcher would typically obtain from their own experiments. These data will be used to confirm the identities of the metabolites. The names, HMDB identifiers, and corresponding experimental data for each metabolite are shown in Table 1.

1. Go to the HMDB home page (https://www.hmdb.ca) and move the cursor over the "Search" menu at the top left of the page and press the mouse button or trackpad panel. This will reveal a drop-down menu with several search options (*see* **Note 1** regarding a more detailed explanation of what is seen in this page).
2. Click on "LC-MS Search" (this is available on the site's main menu). A list of ten search or query functions will be displayed (Fig. 1). Users will be redirected to a page specifically designed

Table 1
Experimental data to use as an example for the methods described in this chapter

Metabolite name	ID	*m/z*	CCS	RT
Trimethylamine N-oxide (TMAO)	HMDB0000925	76.0757	121.7	62.5 s
Leucine	HMDB0000687	132.1019	132.0	59.5 s
Oxoglutaric acid	HMDB0000208	147.0288	132.7	61.2 s
Alpha-aminoadipic acid	HMDB0000510	162.0761	136.4	62.6 s
Palmitic acid	HMDB0000220	257.2475	172.2	0.77 min

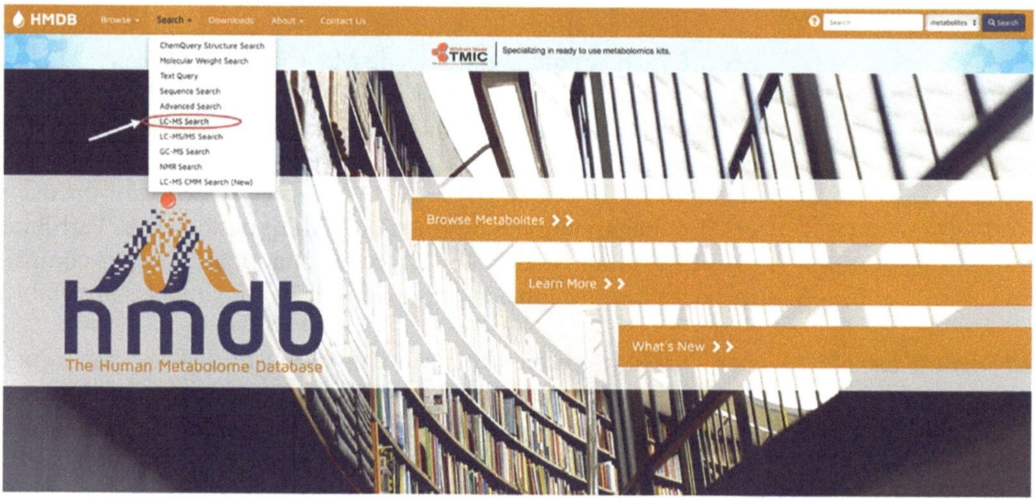

Fig. 1 Screenshot of the HMDB Version 5.0 home page

for searching using LC–MS (liquid chromatography-mass spectrometry) data.

3. In the Search box named "Query Masses (Da) and Collision Cross Section (Å^2)," enter the m/z values as well as the CCS values listed in the table above (Table 1). Ensure the "Ion Mode" is marked positive. Choose the "Adduct Type" to be "M + H." Adjust the "Molecular Weight Tolerance" settings, depending on how accurate the m/z measurements are. The tolerance is usually set as a delta m/z value (e.g., ±0.001 Da).
4. Once all the required m/z values (and optional CCS data) are entered, click on the "Search" button to start the identification process (as shown in Fig. 2).
5. Review the Search Results. The Search results will display a list of the top-scoring hits that match the input data (Fig. 3). For

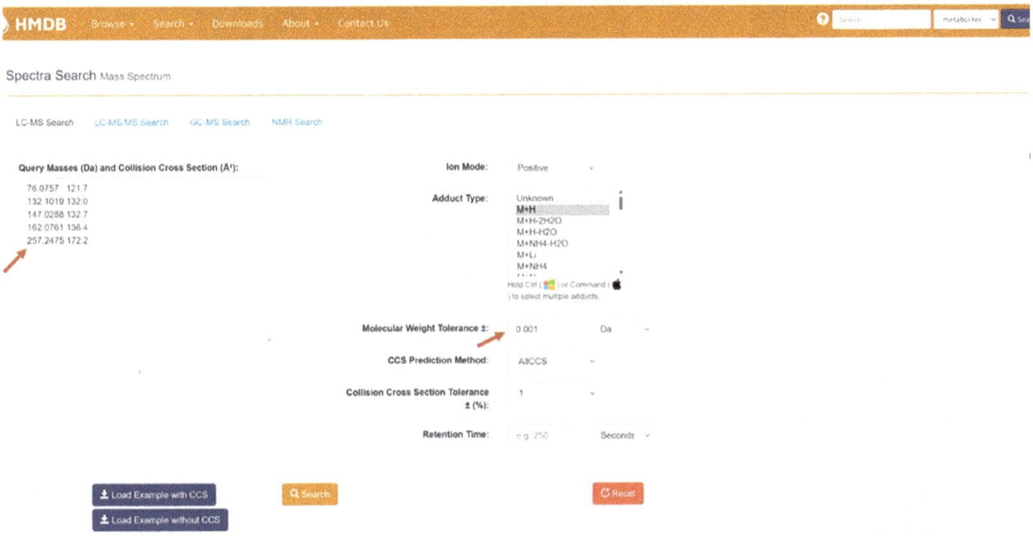

Fig. 2 A screenshot of the "LC-MS search" page for a search using different *m*/*z* values and CCSs as a text entry (two-column format)

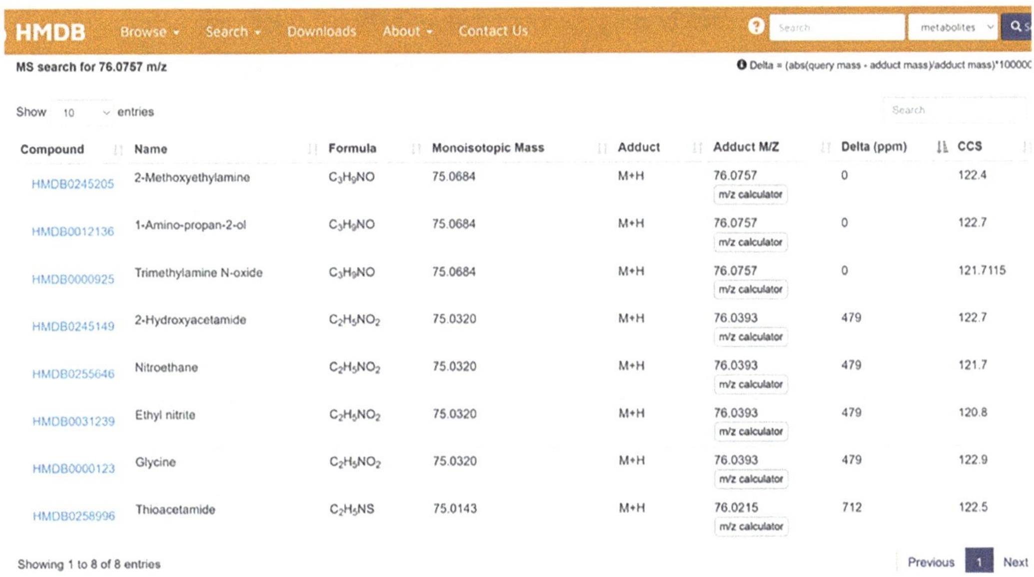

Fig. 3 Screenshot of the Search Results for all the m/z used in the example. For each *m*/*z*, users will find several possible entries that match the input data

each metabolite, users will find the following details: Compound ID, Metabolite Name, Formula, Monoisotopic Mass, Adduct, Adductm/z, Delta (ppm), and the CCS value. In this example, it is not possible to identify any metabolite with complete certainty

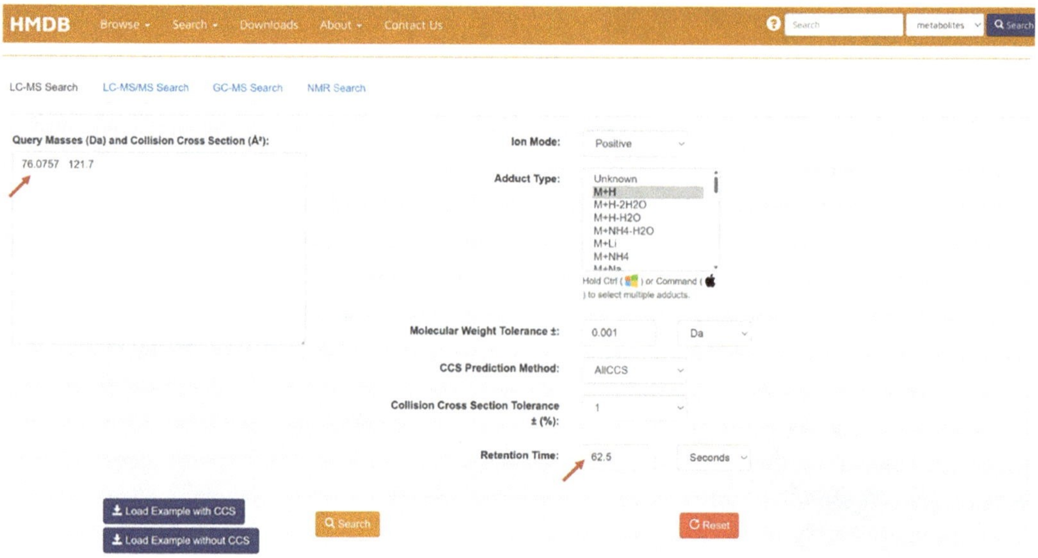

Fig. 4 Screenshot of the "LC-MS search" page for a search using *m/z* 76.0757 with CCS and RT values

using only the LC-MS search tool. At this stage, multiple metabolites match the entered data, making it challenging to definitively assign a specific identity based solely on *m/z* values.

6. Since RT data have also been acquired for each metabolite, users can enter this value in the "Retention Time" field. This step must be performed separately for each metabolite, as only one value can be entered at a time, following the Steps 3–5. For this example, we will use the first *m/z* value (76.0757), as shown in Fig. 4. Repeat the above process for the remaining masses.
7. Press "Search" and review the results. Users will notice that three metabolites have identical delta (ppm) values and match the same *m/z*. However, *trimethylamine N-oxide* is the only metabolite that matches the experimentally obtained CCS value.

3.2 How to Identify the Metabolites Using MS/MS data

Because the result of the MS-only search is ambiguous for the *m/z* value of 76.0757, let us assume that MS/MS data have been collected for that peak at a collision energy of 40 eV.

1. Go to the HMDB home page (https://www.hmdb.ca) and move the cursor over the "Search" menu at the top left of the page and press the mouse button or trackpad panel. This will reveal a drop-down menu with several search options.
2. Navigate to the "Search" section and click on "LC-MS/MS Search" (this is available on the site's main menu) (*see* **Note 2**).

Table 2
MS/MS fragments and intensities for the parent ion 75.0684

Parent ion mass (Da)	MS/MS	Intensity	CCS
75.0684	59.04	90	121.7
	76.1	40	
	58.1	100	
	60.04	80	

3. In the MS/MS Peak List (*m/z* and Intensity), paste the peak pairs as described in Table 2 (MS/MS and Intensity). Enter Parent Ion Mass (Da).
4. Define the "Parent Ion Mass Tolerance." Use 0.01 for this example.
5. Define the "Mass/Charge (*m/z*) Tolerance." Use 0.05 for this example.
6. Select Positive for "Ionization Mode."
7. Pick the bin closest to the Collision-Induced Dissociation Energy (CID Energy): "Low/Med/High" (CID energies are mapped to ~10/20/40 eV). If unsure, try more than one in separate runs. We will use High (40 eV) in this example.
8. Enter the CCS and CCS Tolerance ± (%): if CCS data has been measured, enter it (typical tolerance is 1%–2%). Try with 121.7 for the CCS and a tolerance of 1%.
9. Enter the CCS Prediction Method: choose the predictor; use only if CCS data has been supplied. Try AIICCS in this example.
10. Check the box "Include predicted spectra" to allow *in-silico* matches when the library lacks an experimental spectrum (treat these as hypotheses) (Fig. 5).
11. Click the Search button.
12. View the matching MS/MS spectra that were found. HMDB will display a mirror plot with the experimental spectrum at the top (blue) and the database reference spectrum at the bottom (red). Compare fragment peaks to confirm the metabolite identity. In this case, we have a strong match (predicted and experimental) to *trimethylamine N-oxide* (Fig. 6).
13. Repeat the same steps for the rest of the masses.

3.3 How to Identify the Metabolites Using NMR

Let us assume that the plasma samples from pre-diabetic individuals have also been analyzed via high-resolution NMR and that good-quality 1D NMR ^1H spectra have been collected. Using a

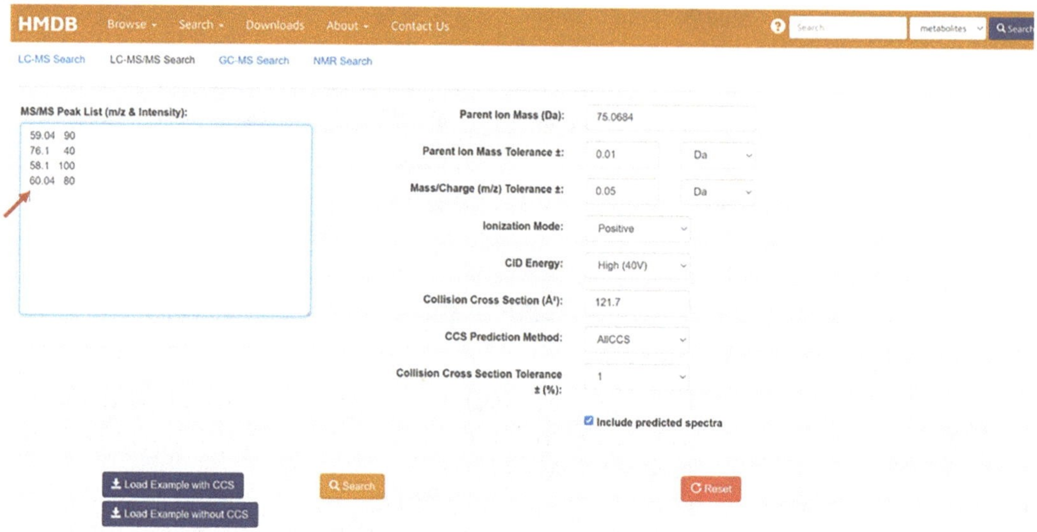

Fig. 5 Screenshot showing the "LC-MS/MS Search" page for a search using different MS/MS *m/z* values with their respective intensities and CCSs as a text entry

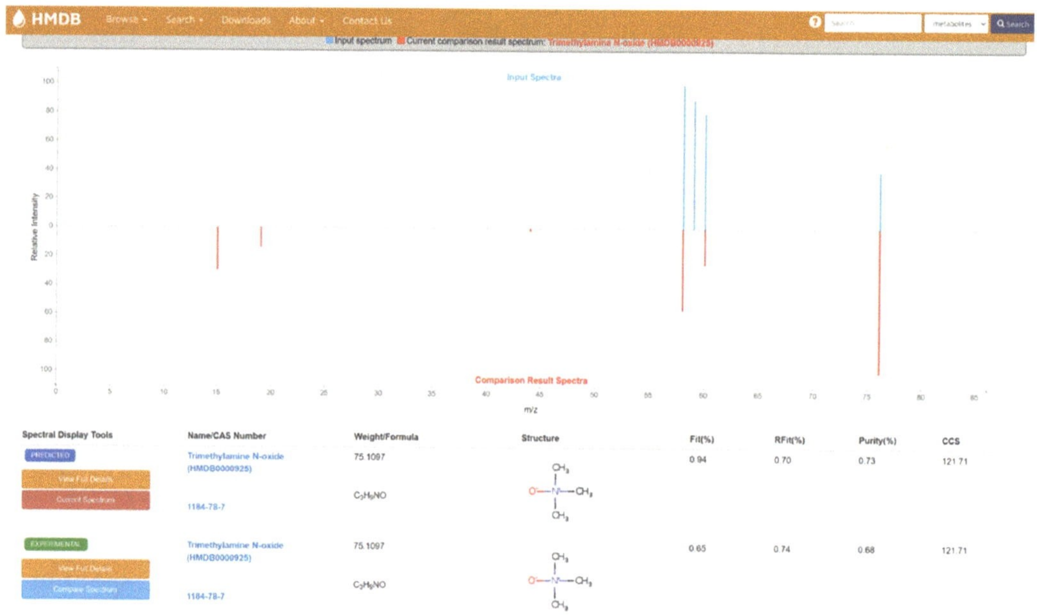

Fig. 6 Screenshot of the Search Results for the *m/z* 75.0757 identification based on the MS/MS spectra

variety of statistical methods, a peak at 3.25 ppm has been identified that is distinctly different between pre-diabetics and healthy controls.

1. Go to the HMDB home page (https://www.hmdb.ca) and use the mouse or trackpad to select "Search" on the HMDB home page menu.

2. Navigate to the "Search" section and click on "NMR Search" (this is available on the site's main menu).
3. Users will be redirected to the NMR Search section, which allows one to search using NMR spectra data.
4. Select the search type (Pure Compound or Mixture from the drop-down menu). In this case, choose Pure Compound because just one peak of interest has been found in the plasma spectrum.
5. Select the spectral library (^{1}H NMR or ^{13}C NMR from the drop-down menu). In this case, choose ^{1}H NMR.
6. Enter the chemical shift peak (or peaks) for the search. In the NMR Search page, users will see fields to input chemical peaks (in ppm) for ^{1}H or ^{13}C NMR spectra.
7. Under "Advance Search Options," adjust tolerances (optional). This chemical shift tolerance ensures that slight variations between the experimental chemical shift data and the database chemical shifts are accounted for. A good choice is 0.1 ppm.
8. Perform the search by clicking on the "Search" button. The search will take about 10 s (Fig. 7).
9. Review the search results. The search results will display a list of metabolites that match the entered NMR shifts. For each metabolite, users will see: the Metabo-Card link (*see* **Note 3**), along with the name, structure, molecular weight/chemical formula, score, match ratio, spectrum hyperlink, and a blue button to click and compare spectrum options. View the experimental or predicted 1D or 2D NMR spectra and compare them to the experimentally collected data. Look at the ^{1}H chemical shifts in the database spectra and compare them to the experimental values.

3.4 How to Compare the Concentrations of Identified Pre-diabetes Metabolites to Reference Values in the HMDB

1. Go to the HMDB home page (https://www.hmdb.ca).
2. Search for each of the identified metabolites that we have found in the previous three methods. They include *TMAO, leucine, oxoglutaric acid, alpha-aminoadipic acid*, and *palmitic acid*. Users can now proceed to search for these metabolites in the HMDB database to learn more about them and their potential role in diabetes.
3. Enter the metabolite name or HMDB ID of a compound of interest (let us choose *TMAO*) in the search box located at the top right corner of the HMDB home page. Alternatively, navigate to the "Browse" and select "Metabolites."
4. The metabolite will be listed in a multicolumn table, along with other similar-sounding metabolite names. Click on the metabolite name for *trimethylamine N-oxide* at the top of the list, which will then generate the MetaboCard for this metabolite.

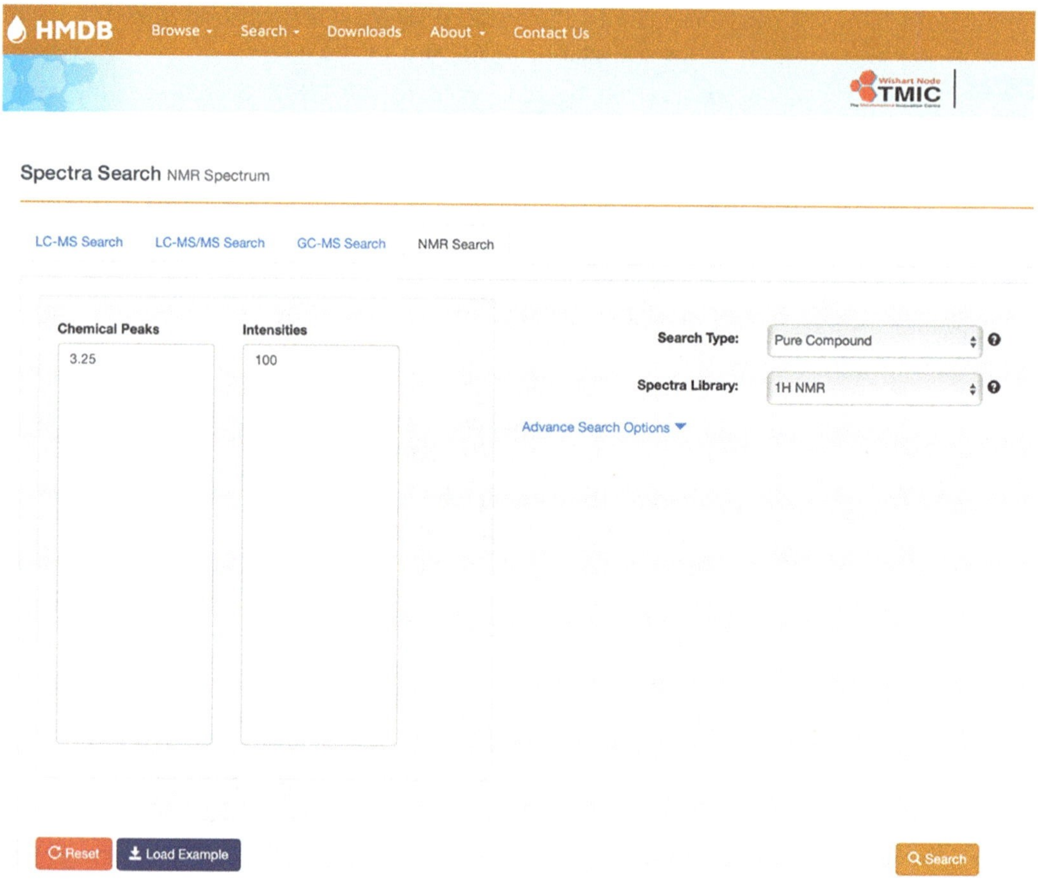

Fig. 7 Screenshot showing the "NMR Search" page for a search using a Chemical Peak with its respective intensity

Browse down the MetaboCard page for *TMAO* to locate the Concentration Data in the MetaboCard (*see* **Note 4**).

5. Look for the "Normal/Abnormal Concentration" field. This section contains reference concentration values for *TMAO* in various biofluids [e.g., plasma, serum, urine, saliva, feces, cerebrospinal fluid (CSF), etc.]. These values typically include normal and abnormal concentration ranges (e.g., concentrations in healthy individuals or in diseased states) (*see* **Note 5**) (Fig. 8).

6. Let us assume quantitative data have been acquired for *TMAO* using a targeted LC–MS assay or NMR assay. Compare the experimentally measured concentration data to the values in the TMAO MetaboCard. The median serum *TMAO* concentration of 1.54 μmol/L (IQR: 0.86–2.91 μmol/L) is listed under normal concentrations in the relevant reference data.

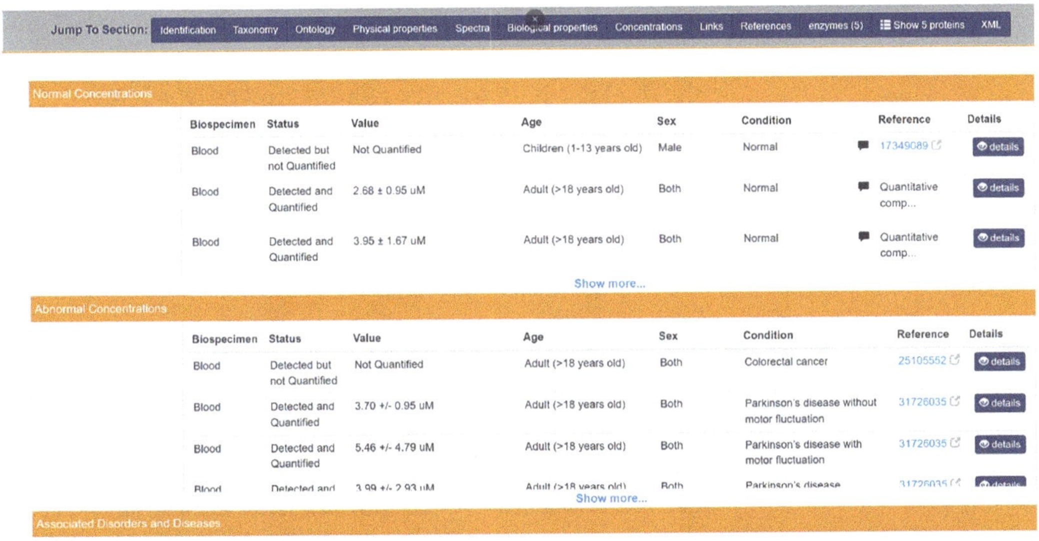

Fig. 8 Screenshot of the Normal and Abnormal Concentrations data fields from the *Trimethylamine N-oxide (TMAO)* MetaboCard

Users can scroll through the MetaboCard to find this value categorized as part of the normal range for TMAO concentrations. Look for any similarities or differences between the measured concentrations and the reference concentrations found in the MetaboCard.

7. Let us suppose the concentration determined here is above the normal value for most of the studies listed. Explore the abnormal concentrations. A median value of 4.05 μmol/L (IQR: 2.40–6.70 μmol/L) is listed as an abnormal concentration of TMAO, which is associated with an increased risk of type 2 diabetes. This could signal metabolic disturbances, potentially identifying individuals at higher risk for the condition.

8. (Optional) Download and save reference concentration data. Download relevant reference concentration data in CSV, JSON, or Excel formats if it is necessary to save or further analyze the data.

3.5 How to Interpret Those Read-outs Using HMDB Metabolite Descriptions or HMDB Pathways via PathBank

For this method, we will use one of the metabolites previously identified: *oxoglutaric acid*.

1. Go to https://www.hmdb.ca and type *oxoglutaric acid* or *HMDB0000208* into the search box located at the top right of the HMDB home page.

2. Click on the hyperlinked metabolite name for *oxoglutaric acid* at the top of the list, which will then open the MetaboCard for *oxoglutaric acid*.

3. Access the Biological Properties section for *oxoglutaric acid*.
4. This can be done by scrolling down the MetaboCard to find the Biological Properties section (indicated on the left of the Card) or using the Jump Panel (in blue) at the top of the MetaboCard to click on Biological Properties and jump straight to this data field.
5. Under Biological Properties, scroll down until the data field called Pathways is visible and click on the name. This expands a list of all curated metabolic pathways within the HMDB in which *oxoglutaric acid* participates (Fig. 9). This list contains the name of the pathway as well as thumbnail images for each pathway. Some pathways may be found in the Small Molecule Pathway Database (SMPDB)/PathBank, while other pathways can be found in the Kyoto Encyclopedia of Genes and Genomes (KEGG). Users can also access PathBank directly via its website (https://pathbank.org). PathBank contains comprehensive, curated pathway data for various organisms, including humans [12, 13] (*see* **Note 6**).
6. Review the pathway list and associations. For *oxoglutaric acid*, pathways typically include the Citric Acid Cycle (Krebs Cycle), Glutamine/Glutamate metabolism, Alanine–Aspartate metabolism, among others (users can easily navigate through the different pages). Each entry shows the pathway name, a short description, and a direct link to the detailed visualization in SMPDB/PathBank (the integrated pathway database). We will select "Citric Acid Cycle" for this example.

Fig. 9 Screenshot showing some of the pathways associated with oxoglutaric acid

7. Click on the thumbnail image of the pathway of interest (e.g., Citric Acid Cycle) to open an interactive pathway diagram. This pathway opening operation directs users to SMPDB, the human subcollection of pathways in PathBank. PathBank provides a PathWhiz-based interactive diagram:
   a) Metabolites appear as nodes (*oxoglutaric acid* highlighted).
   b) Enzymes, proteins, and reactions are represented by arrows and symbols.
   c) Users can zoom, pan, or hover over elements for more details (enzyme names, reaction IDs, PubMed references).
8. Contextualize the biological role by looking more closely at the Citric Acid Pathway. In the Citric Acid Cycle map, *oxoglutaric acid* sits at a central node between isocitrate and succinyl-CoA. By examining upstream (isocitrate dehydrogenase reaction) and downstream (α-ketoglutarate dehydrogenase reaction) steps, it is possible to place experimental findings (e.g., elevated *oxoglutarate* levels) into a systems biology context. The diagram also shows tissue localization and disease associations where relevant (Fig. 10).
9. Interpret the pathways by looking at the detailed metabolic pathway diagrams that illustrate the various biochemical reactions involving the metabolite of interest (*oxoglutaric acid*).

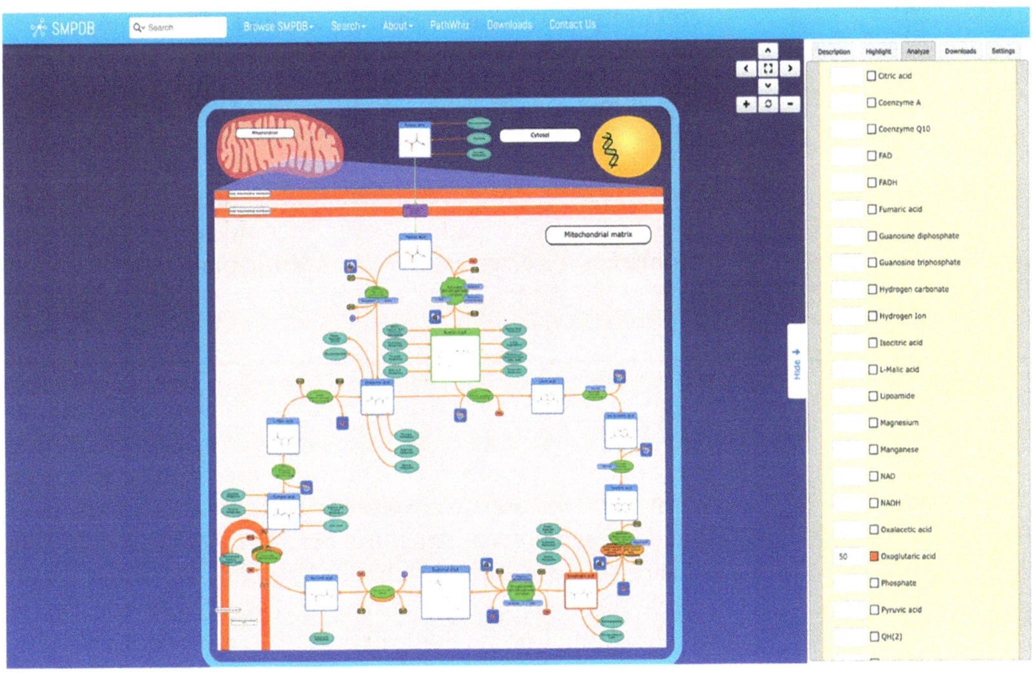

Fig. 10 Screenshot showing one of the metabolic pathways listed for oxoglutaric acid (citric acid)

10. Identify upstream or downstream metabolites. Let us focus on the Oxoglutaric Acid Pathways. In SMPDB [14] for the Citric Acid Cycle map (https://smpdb.ca/view/SMP0000057), the acetyl-CoA node has a high degree of centrality and connectivity in this pathway. In diabetes, *oxoglutaric acid* and *palmitic acid* are indirectly linked through the disruption of normal metabolic pathways, involving *Acetyl-CoA* as an intermediary, particularly within the mitochondria. Diabetes-related insulin resistance leads to increased levels of fatty acids such as *palmitic acid* in the bloodstream. This excess *palmitic acid* causes mitochondrial dysfunction and oxidative stress, altering the metabolism of tricarboxylic acid (TCA) cycle intermediates, including *oxoglutaric acid*.
11. Use the Analyze/Highlight tabs to color these nodes/edges with the experimental measurements and apply the ratio/labeling rules above to decide which direction dominates in the samples.
12. Check for pathway disruptions. If the metabolite's concentration is abnormal, it may indicate a disruption in one or more of its associated metabolic pathways. For instance, *oxoglutaric acid* plays a key role in cellular energy metabolism. An abnormal concentration of *oxoglutaric acid* could signal metabolic dysfunction, which is commonly observed in conditions such as diabetes or pre-diabetes. *TMAO* and *palmitic acid* have both been linked to disruptions in lipid metabolism, inflammation, and endothelial dysfunction. This cascade of events compromises insulin signaling, leading to increased blood glucose and the development of diabetes.
13. Visualize Pathway Interactions. Use PathBank's interactive pathway tools to zoom in on specific steps in the TCA cycle to identify *oxoglutaric acid's* role in energy metabolism. If its concentration deviates from the normal range, and it is linked to a disease in the pathway map, it could serve as a potential biomarker for diagnostic or therapeutic purposes.

4 Notes

1. The HMDB's Search menu provides a comprehensive suite of ten tools for users to explore metabolite data. These tools include ChemQuery Structure Search for substructure and similarity queries using sketches or SMILES, Molecular Weight Search for mass-window lookups, Text Query for flexible searches using Lucene Boolean operators, and Sequence Search for protein and enzymatic homology searches. The Advanced

Search offers two modes—*Metabolite Advanced Search* and *Concentration Advanced Search*—enabling multi-field queries across over 30 fields, ideal for filtering by concentration, biofluid, and units. Additionally, there are five spectral search tools for LC–MS,LC–MS/MS, GC–MS, NMR, and the new LC–MS CMM (Chromatographic Mass Matching). This workflow allows users to pivot seamlessly between exact mass, retention time, collisional cross-section, spectral data, and functional or clinical metadata, enabling a holistic exploration of metabolites.

2. In HMDB 5.0, the LC–MS/MS Search module lets users paste a centroided MS/MS peak list (*m/z*–intensity pairs), specify a precursor (parent) mass, and set instrument-appropriate tolerances for the parent and fragment *m/z*, then choose ionization mode and an approximate collision-energy bin (Low/Med/High). Optional ion-mobility CCS can be entered with a user-set ± % window, and users may enable predicted spectra to extend coverage when no experimental reference exists. The results table ranks candidates by spectral similarity jointly with CCS similarity when supplied; each row lists the compound, HMDB link, *m/z*, spectral and CCS matches, and an overall score. "Show Spectrum" control opens a JSV mirror plot with the experimentally measured spectrum on top and the library (experimental or predicted) spectrum below for fragment-by-fragment confirmation. Under the hood, HMDB 5.0 has integrated CFM-ID 4.0 [15] to generate predicted positive and negative mode QTOF MS/MS spectra at 10/20/40 eV collision energies for all the metabolites—yielding over 1,440,324 MS/MS reference spectra rendered interactively in JSV and wired directly into this search.

3. Crucially for NMR-based identification, HMDB 5.0 has deepened both experimental and predicted spectral coverage and modernized the spectral viewers. On the experimental side, the HMDB has added new experimental NMR reference spectra (e.g., 309 NMR spectra for 218 purified compounds from the Wishart laboratory). For predicted NMR spectra (needed because experimental observables cover <5% of compounds), HMDB 5.0 introduced high-accuracy predictions for 1D ^{1}H and ^{13}C NMR spectra. HMDB 5.0 contains 312,980 predicted ^{1}H/^{13}C NMR spectra rendered interactively in JSpectraViewer (JSV). These updates are visible in the GUI: MetaboCards now show 1D ^{1}H spectra and 2D ^{13}C–^{1}H HSQC overlays inside JSV, and the site's NMR pages include modern, zoomable views. In practice, users can run the NMR Search by pasting chemical shifts (ppm) for ^{1}H or ^{13}C with tolerances; results return a scored hit list with the MetaboCard link, structure, formula, match ratio, and one-click spectrum/compare options for experimental or predicted 1D/2D NMR spectra.

4. Most of the HMDB data are organized into MetaboCards, each containing over 130 fields across 14 categories, including chemical taxonomy, physical properties, spectra, biological properties, concentrations, associated disorders, and enzymes. On the Search Results page, users can filter results by metabolite status (e.g., detected, expected) and biospecimen (e.g., blood, urine, saliva) before opening the most relevant MetaboCards. HMDB 5.0 enhances metabolite identification by integrating Chemical Taxonomy and Functional Ontology. Chemical taxonomy, generated by ClassyFire, categorizes metabolites hierarchically, aiding in analysis and comparison. The expanded ChemFOnt provides rich functional metadata, enabling precise filtering (e.g., "microbiome-derived osmolytes in plasma") and cleaner pathway mapping, essential for mechanistic studies [16, 17].
5. For quantitative interpretation in HMDB 5.0, open each candidate's MetaboCard and navigate to Normal/Abnormal Concentrations to compare the measured (ideally fasting) plasma/serum values with reference entries specific to the biofluid, matrix, and assay method (including units, assay type, and cohort details). The new version update significantly enhances this process: the concentration table now includes more adult records and matrices, is better annotated with details such as age, sex, diurnal variation, disease context, assay provenance, and PubMed links, and is exportable in CSV, JSON, or Excel formats. Out-of-range findings should be treated as screening flags rather than diagnoses, and triangulated with the card's Associated Disorders and Pathways (PathBank) to assess mechanistic coherence (e.g., BCAA elevation in insulin resistance pathways).
6. In HMDB 5.0, contextualization after an ID is much richer because every MetaboCard now links to a substantially upgraded PathBank: the thumbnail pathway images open full, interactive PathWhiz [18] views where layouts, tissue/subcellular artwork, and action icons have been remediated for clarity, with alternate renderings (color, black-and-white, KEGG-like) and one-click exports to PNG/SVG as well as SBML, BioPAX, and PWML for downstream analysis.

Acknowledgments

This work was supported by the Canada Foundation for Innovation (CFI MSIF 42495), Genome Alberta, a division of Genome Canada, and the Canada Research Chairs (CRC TIER1 100628).

References

1. Wishart DS, Tzur D, Knox C et al (2007) HMDB: The Human Metabolome Database. Nucleic Acids Res 35:D521–D526. https://doi.org/10.1093/nar/gkl923
2. Wishart DS, Guo AC, Oler E et al. (2022) HMDB 5.0: The Human Metabolome Database for 2022. Nucleic Acids Res 50:D622–D631. https://doi.org/10.1093/nar/gkab1062
3. Psychogios N, Hau DD, Peng J et al. (2011) The human serum metabolome. PLoS One 6:e16957. https://doi.org/10.1371/journal
4. Mandal R, Guo AC, Chaudhary KK et al. (2012) Multi-platform characterization of the human cerebrospinal fluid metabolome: a comprehensive and quantitative update. Genome Med 4:1–11
5. Bouatra S, Aziat F, Mandal R et al (2013) The human urine metabolome. PLoS One 8:e73076. https://doi.org/10.1371/journal.pone.0073076
6. Dame ZT, Aziat F, Mandal R et al. (2015) The human saliva metabolome. Metabolomics 11:1864–1883. https://doi.org/10.1007/s11306-015-0840-5
7. Karu N, Deng L, Slae M et al. (2018) A review on human fecal metabolomics: methods, applications and the human fecal metabolome database. Anal Chim Acta 1030:1–24
8. Djoumbou-Feunang Y, Fiamoncini J, Gil-de-la-Fuente A et al (2019) BioTransformer: a comprehensive computational tool for small molecule metabolism prediction and metabolite identification. J Cheminform 11:2. https://doi.org/10.1186/s13321-018-0324-5
9. Wishart DS, Tian S, Allen D et al. (2022) BioTransformer 3.0 - a web server for accurately predicting metabolic transformation products. Nucleic Acids Res 50:W115–W123. https://doi.org/10.1093/nar/gkac313
10. Tian S, Feunang YD, Oler E et al (2025) BioTransformer 4.0 a comprehensive computational tool for small molecule metabolism prediction. bioRxiv 2025.07.28.667289. https://doi.org/10.1101/2025.07.28.667289
11. Wishart DS (2020) Metabolomic data exploration and analysis with the Human Metabolome Database. In: Li S (ed) Methods in molecular biology. Humana Press Inc., New York, NY, pp 165–184
12. Wishart DS, Li C, Marcu A et al. (2020) PathBank: a comprehensive pathway database for model organisms. Nucleic Acids Res 48:D470–D478. https://doi.org/10.1093/nar/gkz861
13. Wishart DS, Kruger R, Sivakumara A et al. (2024) PathBank 2.0-the pathway database for model organism metabolomics. Nucleic Acids Res 52:D654–D662. https://doi.org/10.1093/nar/gkad1041
14. Frolkis A, Knox C, Lim E et al (2009) SMPDB: The Small Molecule Pathway Database. Nucleic Acids Res 38:D480–D487. https://doi.org/10.1093/nar/gkp1002
15. Wang F, Liigand J, Tian S et al. (2021) CFM-ID 4.0: more accurate ESI-MS/MS spectral prediction and compound identification. Anal Chem 93:11692–11700. https://doi.org/10.1021/acs.analchem.1c01465
16. Djoumbou Feunang Y, Eisner R, Knox C et al. (2016) ClassyFire: automated chemical classification with a comprehensive, computable taxonomy. J Cheminform 8:1–20. https://doi.org/10.1186/s13321-016-0174-y
17. Wishart DS, Girod S, Peters H et al. (2023) ChemFOnt: the chemical functional ontology resource. Nucleic Acids Res 51:D1220–D1229. https://doi.org/10.1093/nar/gkac919
18. Pon A, Jewison T, Su Y et al. (2015) Pathways with PathWhiz. Nucleic Acids Res 43:W552–W559. https://doi.org/10.1093/nar/gkv399

Index

Michael Witting (ed.), *Computational Methods and Data Analysis for Metabolomics*, Methods in Molecular Biology, vol. 3063, https://doi.org/10.1007/978-1-0716-5452-1,

R

S

T

U

W

X

FSC
www.fsc.org
MIX
Papier | Fördert gute Waldnutzung
FSC® C083411

Kolibri 360 GmbH
Ferdinand-Jühlke-Straße 7
99095 Erfurt, Deutschland
produktsicherheit@kolibri360.de